EARTH FOLLIES

COMING TO FEMINIST TERMS WITH THE GLOBAL ENVIRONMENTAL CRISIS

JONI SEAGER

D0066729

ROUTLEDGE · NEW YORK

Published in 1993 by

Routledge
29 West 35 Street
New York, NY 10001

Printed in the United States of America on acid free paper.

Library of Congress Cataloging-in-Publication Data

Seager, Joni
    Earth follies : coming to feminist terms with the global
environmental crisis / Joni Seager.
      p.   cm.
    Includes bibliographical references and index.
    ISBN 0-415-90720-9
    1. Environmental policy—Political aspects. 2. Human ecology—
Political aspects. 3. Feminist theory.   I. Title.
HC79.E5S36
363.7—dc20   92-45793
                                                          CIP

EARTH FOLLIES

For Cynthia

"It is hard to be brave," said Piglet, sniffling slightly,
"when you're only a Very Small Animal."

Rabbit, who had begun to write very busily,
looked up and said:

"It is because you are a very small animal that you
will be Useful in the adventure before us."

A. A. Milne, *Winnie - The - Pooh*

Contents

Powers of Destruction
- **In War**: Vietnam; The Persian Gulf; Central America; Southern Africa;
The Drug Wars
- **In Peace**
- **Gentlemen's Agreements**

The Nuclear Club
- **Nuclear Fallout**: Fernald; Hanford; Chelyabinsk
- **Club Culture**

Out of Sight, Out of Mind
Goose Bay; The Pacific

6. Hysterical Housewives, Treehuggers, and Other Mad Women   253

LIST OF ILLUSTRATIONS

LIST OF TABLES

Acknowledgments

I started thinking about this book in 1988. As I tap out these final pages, it is now the summer of 1992. Over these four years, my reflections on the state of the earth have been enriched and sharpened by the work of dozens of women and men, some personal friends, others known to me only through their writing, who have brought voices of clarity and courage to bear on environmental debates.

Writing is perceived to be a solitary undertaking. In fact, I have always found it to be a collaborative process, and in writing this book I have drawn on the energy, goodwill, and intellectual labor of dozens of people. Cynthia Enloe tops the list of those to whom I owe acknowledgment. The mix of analytical acuity and quick wit that Cynthia brings to her own work has inspired me, as I know it does so many others. Her unstinting encouragement, ready humor, helpful editing, and friendly prodding sustained me through those many days when I was fed up with writing or glum about the state of the world's environment. Additionally, I am fortunate to have a strong network of friends whose presence in my life allows me the security and freedom to take intellectual risks. My thanks to them all, and especially to Gilda Bruckman, Judith Wachs, Mona Domosh, EJ Graff, Madeleine Drexler, Ellen Cooney, Ellen Winchester, Michael Steinitz, Laura Zimmerman, and, collectively, to the women of New Words, the feminist bookstore in Cambridge, Massachusetts. Similarly, my thanks to my family, and especially to my mother, Joan, who has always encouraged me by example and exhortation to explore new horizons and to not be afraid of new ideas.

I also especially want to acknowledge Philippa Brewster, the founder and driving force (until 1991) behind Pandora Press, the British feminist publisher. It was Philippa who first suggested the idea of this book to me, and, as my patient editor over three years, she endured my authorial delinquencies with graciousness and generosity. I regret that we are not working together on the final manuscript, but I am grateful for her early impetus and for her continuing friendship.

Over the past few years, I have been fortunate to participate in a number of collaborative environmental efforts. The people I met through each of these projects left me with a renewed sense of commitment to speaking out against environmental follies in whatever guise they appear. My thanks to Steve Chorover, Scott Paradise, and Rich Cowan for their work in convening at MIT a coalition of activists and environmentalists against the Gulf War. In the midst of writing this book, I also edited an atlas of global environmental issues, *The State of the Earth Atlas*.[1] I worked with a stellar team of researchers on this book; through their contributions to the atlas, each of them also contributed to my broader understanding of our global environmental crisis, and I extend my collective thanks to them all, with special notice to Anne Benewick, a good friend and an unsurpassed editor. My thanks to the Boston Women's Healthbook Collective, especially Norma Swenson and Judy Norsigian, and to Betsy Hartmann and Marlene Fried at Hampshire College, for taking the initiative in bringing together scholars and community activists in an ongoing effort to articulate a feminist and women-centered position on the politically charged debate over the relationships between population and environment. I helped to conceptualize and organize a feminist environmental conference held at MIT in 1992, "En/Gendering Environmental Thinking," an undertaking buoyed by a sense of common purpose and camaraderie offered by my colleagues Pat Hynes, Sandy Martin, and Ruth Perry. And I was energized and inspired by the thousands of women gathered from around the world in Miami in 1991 for the first international Global Assembly of Women for a Healthy Planet.

I have learned a great deal from colleagues and friends whose work in Geography, in Women's Studies, and in Environmental Studies has shed light on my own. I have found especially helpful the work of Marina Alberti, Sally Deutsch, Jane Dibblin, Cynthia Enloe, Yaakov Garb, Susan Hanson, Sandra Harding, Pat Hynes, Jodi Jacobson, Ynestra King, Timi Mayer, Aerial Salleh, and Ann Tickner. Many people shared their time and expertise unstintingly with me, and for assistance with particular aspects of my research I thank: Francesca Lyman, Lisa Greber, Janet Marinelli, Lisa

---

[1] New York: Simon & Schuster; London: Unwin Hyman, 1990.

Crawford, Cynthia Daniels, Priscilla Feral, Mary Fillmore, Elizabeth Odour-Noah, Marina Alberti, Peter Cebon, Judy Johnson, Sylvia Tesh, and Peggy Perri. My thanks to Montine Jordan, cartographer extraordinaire.

I am indebted to Margaret Bluman, a good friend who also helped to steer me through the perils of early publishing agreements. My thanks to my US agent, Anita McClellan, and to my editor at Routledge, Cecelia Cancellaro, for their enthusiasm and encouragement. And I owe a special thanks to Dominic Golding, Susan Cutter, and the anonymous reviewers who offered insightful critiques of my manuscript.

And finally, at further remove, my greatest debt and my greatest admiration goes to the women and men whose work has broken the ground ahead of me. I am inspired by the courage and integrity of the many people who speak out against environmental travesty, sometimes at great personal cost. I have been especially moved by the work of and the example set by Helen Caldicott, Rosalie Bertell, Rachel Carson, Magda Renner, Petra Kelly, and Wangari Maathai; may their voices never be silenced.

Introduction: What's the Problem Here?

Let's start with the bad news first. Simply put, ours is a small and much-abused planet. We have so degraded our environment, so disrupted our biosphere, so stressed physical carrying capacities that the lives of millions of people on the planet (certainly our cherished "ways of life") are at risk. The litany of contemporary environmental horrors is now familiar, even to grade-school children: ozone depletion; global warming; acid rain; chemical pollution of groundwater; the startling and escalating rate of loss of bird, animal, and plant species; tropical deforestation; increasingly massive and deadly chemical, oil, and toxic spills; and the list could go on for several pages. The daily newspapers are filled with a barrage of bad news, all of it larger than life. We hear how many acres of trees are felled in the Amazon between the time of our morning coffee and our evening meal (over 1,500 acres on an average day); we are told, with uncomfortable precision, how many of us will be likely to develop skin cancer in the coming decades, and how many of us will die from it (more than 6,500 a year in the USA alone); we are mesmerized by images of the ozone hole over the Antarctic pulsating in astral colors.

These are all Big Problems. Nature is clearly in trouble, and we with it. If we are going to solve these environmental problems, we need to bring to bear on them all of our analytical and political skills and resources, including feminist analysis.

But what can feminists contribute to our understanding of environmental problems? Is there a place for a feminist voice in the environmental analytical chorus? As a feminist and a geographer, I posed these rhetorical questions to myself a number of years ago—and, surprisingly, my first answer was "no," feminist analysis was not particularly pertinent to things environmental. The environment, after all, seems genderless, and thus not a very fruitful arena for feminist inquiry.

My training as a geographer, in both the physical and social sciences, led me to conceptualize environmental problems in their physical forms— that is, I understood environmental problems as *problems of physical systems under stress*. This interpretation of the state of our environment is reinforced by popular media reporting on environmental issues, which typically emphasizes the visual and physical manifestations of environmental problems. The media conveys very powerful and enduring images: of oil-slicked otters dying in Prince Edward Sound; of acres of scorched stumps of tropical rain forests in Brazil; of NASA computer-simulated photos of the pulsing hole in the ozone layer. These images, and the reporting that accompanies them, encourage us to think of environmental problems in their physical forms—the problem, we are told, either directly or subliminally, is too many dead trees, too big a hole in the sky, too many fossil-fuel pollutants, too much soil erosion, too much garbage, not enough water, not enough arable land, not enough fuel wood.

The tendency to conceptualize environmental problems in their physical form has a number of implications. Popular reporting on environmental affairs that focuses primarily on the scope of environmental calamity leaves causality out of the picture. The journalistic predilection for using passive language wherever possible (presumably so as to offend as few people as possible, and to limit legal liability) leaves us all ill-equipped to make sense of the environmental scene. Statements such as "we will 'lose' 50,000 animal species annually by the end of the century" or "there are more toxins in the Great Lakes than ever before" or "the ozone hole is growing" may provoke an awareness of environmental problems, but they leave the question of agency unexamined. Questions about agency— that is, the social and economic processes that create a state of scorched trees and dead otters—are placed a distant second, if they are raised at all. The physical manifestations of environmental problems are often presented as both the beginning and the end of the story.

The conventions of scientific analysis, similarly, blunt curiosity about questions of agency. Neither specialist training nor popular media coverage encourage us to ask questions about the actors, institutions, and processes behind the mutilated elephants and oil-slicked beaches. Social analysis and institutional critiques are shut out from both specialist and popular representations of the state of our earth. Feminist analysis is shut out.

If environmental problems are framed as problems of physical deterioration, this is barren territory for feminist analysis: there is no feminist analysis of the chemical process of ozone disintegration; there is no feminist analysis of soil erosion, of groundwater pollution, of the acidification process that is killing forests and lakes throughout the industrialized world. (While there *are* feminist critiques of the *construction* of scientific knowledge, the analysis of physical and chemical processes in and of themselves is not particularly suitable territory for feminist forays.)

But there is a subtext to the daily news. Reading between the lines, in the best feminist tradition, reveals the story *behind* the environmental story. The environmental crisis is not just a crisis of physical ecosystems. The *real* story of the environmental crisis is a story of power and profit and political wrangling; it is a story of the institutional arrangements and settings, the bureaucratic arrangements and the cultural conventions that *create* conditions of environmental destruction. Toxic wastes and oil spills and dying forests, which are presented in the daily news as the entire environmental story, are the symptoms—the symptoms of social arrangements, and especially of social *derangements*. The environmental crisis is not just the sum of ozone depletion, global warming, and overconsumption; it is a crisis of the dominant ideology.[1] And it is this that is grist for the feminist mill.

Scientific specialists and environmental populists leave a vacuum of curiosity, into which feminists should step. Feminists are especially acute about asking questions of agency. A feminist analysis of the environment starts with the understanding that environmental problems derive from the exercise of power and the struggle of vested interest groups played out on a physical tableau. A feminist analysis of environmental problems thus needs to be rooted in an analysis of the social, cultural, and political institutions that are responsible for environmental distress.

The first task for feminists interested in the environment is to rewrite the litany of horrors in a way that makes agency clear. Take deforestation, for example. We are not "losing" forests. In the Amazon, multinational oil companies, Japanese lumber firms, and local elites, with the complicity and cooperation of specific Brazilian government officials and international development agencies, are burning, clearing, and clear-cutting vast acreages of tropical forest at a staggering rate for mineral exploitation, timber, and cattle ranches. The rain forest in Indonesia is falling under the ax of Scott Paper Company, among others, which plans to replace 800,000 acres of tropical rain forest with mono-plantations of fast-growing eucalyptus trees for pulp paper. In Hawaii, power-plant developers are clearing the last rain forest in the USA. Deliberate deforestation is a tool of the trade in the dozens of dirty little wars raging in Central America and Asia.

But naming names is not enough, and feminists are not the only ones to point to the structures behind the environmental symptoms. Radical economic analysts have long pointed, for example, to the role of capitalist economic structures as the cause of much environmental damage (an analysis that falls short when environmental degradation is found to be as prevalent in noncapitalist structures as in capitalist). But the analysis that feminists offer—an analysis rooted in uncovering the workings of gender—is unique, and as of yet, not widely applied to environmental questions.

I started this chapter by saying that "we" have so degraded our environment, so disrupted our biosphere, so stressed physical carrying capacities that the continued existence of life on the planet is at risk. But "we"—an undifferentiated humanity—have not done so. Rather, our environmental problems are the progeny of very particular clusters of powerful institutions acting in particular ways. A handful of large institutional structures largely control the state of our environment. Large-scale environmental degradation—not litter on the streets, but the really major environmental problems that may well kill us all—is the product of three or four clusters of large institutions that include, prominently, militaries, multinationals, and governments (which are often in collusion with, or indistinguishable from, militaries and multinationals).

On the other side of the fence, there is the environmental movement. While this "movement" is neither homogeneous nor uniform, the environmental agenda is one that is increasingly coordinated by a handful of large, powerful, and very well-funded environmental groups that I call the "eco-establishment." Grassroots environmental groups, "fringe" groups (such as deep ecology), and individuals fill out the rest of the environmental slate, but their presence is overshadowed by the eco-establishment. Governments, militaries, and multinational institutions are primary agents of environmental degradation; eco-establishment institutions are setting the environmental agenda, and framing the ways in which we perceive environmental crisis.

Militaries, multinationals, governments, the eco-establishment. When I write down this list of institutions on a piece of paper, the first thing that I notice, as a feminist, is that these are all institutions of men. These institutions and groups are controlled by men (and a mere smattering of women). The culture of these institutions is shaped by power relations between men and women, and between groups of men in cooperation or in conflict. Institutional behavior is informed by presumptions of appropriate and necessary behavior for men and for women. Their actions, their interactions, and the often catastrophic results of their policies cannot be separated from the social context that frames them. And, on twentieth-century Earth, the large social frame is one of gender difference. Everywhere in the world, men and women lead different lives; everywhere in the world, men have more institutionalized power, more autonomy, more money, and more privilege than their female counterparts; even when the "pie" of social power is small, women's share is smaller still.

The institutional culture that is responsible for most of the environmental calamities of the last century is a masculinist culture. The "expert structure"—of scientists, environmentalists, and bureaucrats—that interprets and assesses the state of the earth is, for the most part, one of men. *As a feminist then, the first environmental question to ask is whether or not it "matters" that the institutions that for the most part control our collective environmental fate are constructs of male culture.*

This book is the first to apply feminist theory to a close analysis of the workings of power in those institutions that have the greatest bearing on the state of our environment—militaries, corporations, governments, and environmental organizations. The primary purpose of this book is to suggest the ways in which gender and gender relationships enter into environmental issues, and the ways in which environmental issues are shaped by gender-specific constructs. My research is also shaped by (and, I hope, contributes to) the lively feminist debates over the role of "motherist" politics in feminism, a controversy about ecofeminism and the role of spirituality politics in feminism and environmentalism, and the complicated historical relationships between "women" and "nature."

As a feminist who is extremely leery of essentialism ("men are essentially destructive . . . women are essentially nurturers"), I try to examine environmental agency without reinforcing essentialist fallacies. The fact that we all know individual women who thrive in the institutions I name, and that we all know individual men who suffer the predations of this institutional culture does not undercut the saliency of the argument that these institutions are structured around masculinist presumptions and prerogatives. The feminist environmental question that I pose is not so much about men *qua* men, nor even about women as individual agents, but rather whether it matters that the institutions that control virtually all decisions and actions that shape our environment are institutions shaped by male culture. The common-sense answer to this question is a resounding "yes"—common sense suggests that a skew of power and representation in favor of men within these institutions has to "matter"; feminist theory and women's history tell us it matters. But in what ways does it "matter"?

The feminist challenge is to identify the ways in which gender and gender relations enter into the environmental arena through the workings of institutions, and through the role of individuals working within particular institutional contexts. Men and women have different relationships to militaries, multinationals, governments, and large environmental organizations. Similarly, the implications and experience of environmental decay are often different for men and women, rich and poor, elites and disenfranchized. The task for feminists is to unravel the ways in which gender operates as a structuring condition within the institutions that hold the balance of power on environmental issues.

I firmly reject "biology is destiny" arguments in whatever guise they take. But I do, just as firmly, believe that we cannot answer hard questions about environmental agency without looking closely at the nature of power, the workings of power, *and the gender of power* in the institutional arrangements and groups on both sides of the environmental coin—those that are primarily responsible for destroying the environment, and those that have set themselves to saving the environment. It is folly to ignore the fact that virtually all of the institutions, bureaucracies, and groups fanned out across the environmental spectrum are run by men in pursuit of male-defined objectives.

◐

Women are fed up. Women carry much of the burden in this world of sustaining daily life, and they are angry—not only because they see their burdens becoming heavier as the state of the earth deteriorates, but because they see that as environmentalism is becoming a "big" game in the men's world of politics, science, and economics, women are once again excluded from the ranks of the powerful players. In 1991 thousands of women gathered in Miami for what was billed as the first Global Assembly of Women for a Healthy Planet. Listen to the voices of some of the women from that conference:

> We are here to say that this mad race toward self-destruction must stop . . . and that the overwhelming exclusi on of women from national and international decision-making, their exclusion from economic and political power, must end.
>
> Bella Abzug (USA)

Women are constantly being told that their analysis is political hysteria.

Vandana Shiva (India)

The current state of the world is the result of a system that attributes little or no value to peace. It pays no heed to the preservation of natural resources, or to the labor of the majority of its inhabitants, or to their unpaid work, not to mention their maintenance and care. This system cannot respond to values it refuses to recognize.

Marilyn Waring (New Zealand)

Only those who have fought for the right to protect their own bodies from abuse can truly understand the rape and plunder of our forests, rivers, and soils.

Margarita Arias (Costa Rica)

If women are to 'clean up the mess,' they have a right to challenge the people and institutions which create the problems.

Peggy Antrobus (Barbados)[2]

For people who do not encounter feminist analysis in their daily lives, these voices may sound strident. The response of many environmentalists when women start talking feminism is to draw the ideological wagons into a defensive circle, with environmental reason inside, threatened but valiant, against the crazy assault of women's "paranoia" and "hysteria" without.[3] Men are unaccustomed to hearing women express their anger, and some men feel personally attacked by a feminist analysis that lays bare the gender skew of power and responsibility. This has become especially clear to me over the past two or three years as I have given talks across the country on various aspects of the themes I explore in this book. It is apparent that a feminist analysis of environmental issues makes some people—and especially many men—anxious.

The purpose of noticing that environmentally instrumental institutions are run by men is not simply to ask, "Well, what would be different if women were in charge?" For one thing, the state of the world being as it is, we can't answer that question yet: we do not have many examples of women-led, non-masculinist institutions (or, even, of male-lead, non-masculinist institutions). This question, which is often raised as a challenge whenever feminists talk about gender and power, sidesteps the point that most feminists are making: "the problem" with masculinist institutions is not primarily that men are in charge, but that structures can be so rooted in masculinist presumptions that even were women in charge of *these* structures, they would retain the core characteristics that many feminists and progressive men find troubling.

In a new book on gender in international relations, a field particularly relevant to environmental relations, Ann Tickner elaborates this point:

> *Rather than discussing strategies for bringing more women into the international relations discipline as it is conventionally defined, I shall seek answers to my questions by bringing to light what I believe to be the masculinist underpinnings of the field* [emphasis added]. I shall examine what the discipline might look like if the central realities of women's day-to-day lives were included in its subject matter. . . . Making women's experiences visible allows us to see how gender relations have contributed to the way in which the field of international relations is conventionally constructed and to reexamine the traditional boundaries of the field. It is doubtful whether we can achieve a more peaceful and just world . . . while these gender hierarchies remain in place.[4]

Over the past two decades, feminists have brought similar gender-based analysis to illuminate the workings of social institutions as diverse as the law and the family, and to virtually every academic field from the sciences through the arts. My contribution here is to bring this kind of feminist analysis to bear on our understanding of the global environmental crisis.

In point of fact, most men, like most women, lead humble lives; most men probably don't feel that they are in charge of anything, or that they have more power or privilege than their female counterparts. At first glance, this is largely true, and, as I and other analysts point out, most feminists are not necessarily interested in "picking on" men (or women) as individual actors. More to the point, when male power is aggrandized by institutional power, it generates an impression of a more distorted gender dichotomy than may actually exist. The magnifying lens of power creates what one observer calls "hegemonic masculinity," a type of culturally dominant masculinity that, while it does not correspond to the actual personality of the majority of men, sustains patriarchal authority and legitimizes a patriarchal social and political order.[5]

At the same time, it is not coincidental that masculinist structures are run by men; there is a synchronicity between "hegemonic masculinity" and ordinary manhood; there is a continuity between the exercise of institutional power by "extraordinary" men and the privileges, however minor, that most men, even the powerless, share in . . . and that most women, even the powerful, do not. Gender hierarchies privilege all men: even if individual men do not feel specifically enriched or empowered by them, gender hierarchies that universally install men at the top universally privilege men's knowledge and men's experiences—experiences that are then assumed to be normal, or ubiquitous, or, simply, the most important. It is still the case that the world is a mirror, in the words of Virginia Woolf, that reflects men (even "little" men) at twice their natural size.

Where then, some critics ask, is the room for the "good men"? Men have been prominent in the struggles for peace, for social justice, and against systems that are environmentally destructive—does a *feminist* environmentalism exclude them? Not at all. Ordinary men, the many "good men" among us, can distance themselves from the hegemonic masculinity that they feel does not represent them. Men need not defend, nor be defensive about, the institutions that are wreaking environmental havoc. Patriarchy is not only oppressive to women. Men, like their feminist sisters, can insist on a more clear-headed analysis of how power works in institutions, and men can also be clear about the extent to which they are or are not complicit in perpetuating that power or those institutions. Paraphrasing the words of another feminist scholar, I believe that because gender hierarchies have contributed to the perpetuation of environmental catastrophe, *all those concerned with the environment—men and women alike*—should be concerned with understanding and overcoming their effects.[6] The fact that men sometimes find this particularly hard to do underscores the strength of the symbiotic bonds between institutionalized masculinist prerogative and ordinary manhood.

Because patriarchy is not something in which membership is optional, for women or for men, the only way that most women and men can "get by" in this world is by making alliances and compromises with male power structures; indeed, all of our major social institutions are supported by the labor and compliance of women as well as men. In writing a book about the gendered nature of institutional power, and the environmental consequences of the gendered wielding of that power, I do not expunge the reality that there are men and women on "both sides" of the environmental coin. This is *not* a book that reduces environmental understanding to simplistic categories of "wonderful women" and "evil men." It is a book that asks us all to take a hard look at how environmental realities are shaped by institutional realities that are, in turn, shaped by distinctive gender assumptions and dispositions.

◖

Feminism and environmentalism are among the most powerful social movements of the late twentieth century. The vision of promise—the carrot on the stick—of both movements is the possibility that personal interactions and institutional arrangements can be transformed into non-exploitative, nonhierarchical, cooperative relationships. Both are progressive movements, both offer a challenge to mainstream "business as usual" standards, both assert the need for reordering public and private priorities, and the constituencies of both overlap. Intuitively, it would seem that the feminist and the environmental movements should be closely allied. And yet they are not. If anything, as we go deeper into the 1990s, the gap between feminism and environmentalism appears to be widening.

The environmental agenda in Europe and North America is increasingly orchestrated by a handful of large environmental organizations, "the eco-establishment"—groups such as the Worldwide Fund for Nature, the Sierra Club, Friends of the Earth, and even more "radical" groups such as Greenpeace, all of which now control multimillion dollar budgets, all of which support expensive lobbying offices and hundreds of paid staff members. One reason for the widening gap between feminism and environmentalism is that the institutional culture of this eco-establishment is, by and large, hostile to women. Resistance to feminism seems to be as firmly entrenched in the environmental establishment as it is in society at large, perhaps only taking more surprising and more subtle forms. The leaders of the eco-establishment, most of whom are men, appear unwilling to entertain a critique of their institutional culture, and in fact they are increasingly looking to the conventional male worlds of business and science as exemplars of organizational behavior. As environmental organizations take on the coloring of big business, and the leaders seek to be "taken seriously" by the movers and shakers in business and government—in the terms defined by business and government leaders—they compromise their credibility as "outsiders." More to the point, the replication of conventional organizational culture within the environmental movement alienates many women working for environmental change, and at the same time it excludes feminist analysis.

On the other hand, feminist theorists have been slow to address environmental concerns. There is only a nascent feminist analysis of ecological issues, perhaps because feminist scholars are concentrated in fields such as history, literature, and art, while environmental issues are assumed (wrongly, I believe) to belong primarily to the realm of the physical sciences. Over the last three decades of the recent wave of the women's movement, feminists have developed finely honed analyses of social domination and the workings of power in personal and institutional life. We have yet to apply these analyses to environmental issues.

One of the few environmental issues that feminists *have* addressed is the thorny issue of the presumed bond between women and nature. Feminist historians and particularly feminist historians of science have identified that a central dualism—the men-culture and woman-nature dichotomy—is pivotal to the development of Western civilization, and of

Western patriarchy with it.[7] A number of feminist writers posit that the *domination* of women and that of nature by men are linked, and are linked to this dualism, and that a deconstruction of one leads to an illumination of the other.

Feminist response to the presumed woman-nature bond has taken a number of tacks. In the first instance, and most noticeably in the 1970s and early 1980s, many feminists seemed to agree that the way to advance the cause of feminism was to deny the potency of difference between women and men. Early feminist agitation—especially in Western industrialized countries—was aimed at equalizing the relations between men and women. This might be characterized as a "rationalist" or anthropological feminist position. In terms of the Nature question, writers such as Simone de Beauvoir and Sherry Ortner argued that the woman-nature connection should be seen as a male cultural artifact, the product of a particular historical period, with little contemporary relevance or value other than as a tool of the patriarchy in justifying the ongoing oppression of women (and of nature).[8] Ortner, for instance, argued that women should reject their presumed link with Nature, and should seek to be integrated into the ("men's") world of Culture, and that feminists should explore theoretical work that exposes the presumed woman-nature bond as a bankrupt male artifice.

In the late 1970s, other feminists, influenced by prominent theorists such as Susan Griffin and Mary Daly, called for a feminist revaluation and reclamation of the woman-nature connection.[9] They argued that while the woman-nature bond had been defiled and denigrated by patriarchal culture, the bond of women with nature in fact represents a significant and empowering bridge for women—a bridge to their past, and a bridge to the natural cycles that can seem to have such significance in women's lives. For these feminists, a celebration and affirmation of women's distinctive culture offers an avenue out of and away from the dominant male culture. "Ecofeminism," virtually the only ecological ideology to self-consciously bridge feminism and environmentalism, derives from this second feminist analysis.[10]

Both strands of feminist analysis are problematic. The first feminist position draws on historical research and contemporary explorations of cross-cultural relativism. It posits that the woman-nature bond is an Anglo-European male cultural construction of a particular historical period, the efficacy of which women should resist and deny. From this position, some feminists pose arguments against women's separatism—arguments that, in some instances, privilege heterosexual and "mainstream" women's organizing, and which can merge with a conservative agenda, or that at best lead to a liberal, "reformist" feminist stance. The second feminist position, by claiming a distinctive women's culture, lends support to women's (and lesbian) separatism, but it relies on ahistorical, universalizing, and essentialist arguments about the inherent bond of women with nature.

A third feminist approach to the woman-nature debate is just now emerging, one that draws on historical research charting the contours of gender difference across time, and on psychological work that maps out the nature of gender difference in particular cultural contexts of work and play.[11] This research suggests that there *are* differences between men and women in moral behavior and character, the reality of which shouldn't be denied or avoided, but that these are culturally constructed and variable across cultures and time. The promise of this research is that the elucidation of these differences may provide the sense of a common bond among women, and, further, it asserts that women *do* have a distinctive culture from that of men, and that theirs may suggest alternative "ways of being" in the world. When applied to the woman-nature debate and in assessing the relevance of feminism to environmentalism, this new feminist insight suggests a middle path:

> that while recognizing that the nature-culture dualism is a product of culture, we can nonetheless consciously choose *not* to sever the woman-nature connection by joining male culture. Rather, we can use it as a vantage point for creating a different kind of culture and politics that would integrate intuitive and rational forms of knowledge . . . and enable us to transform the nature-culture relationship.[12]

It is not clear how this "middle path" can be "operationalized" in terms of changing the "realpolitik" of environmentalism, but it does suggest the saliency of critiquing institutional arrangements from a distinctively woman-centered stance.

◐

My research adds another dimension to making feminist sense of our environmental crisis. I pursue a more materially grounded, structural analysis of our environmental problems; my work here is nested within the feminist genre of deconstructing the workings of institutionalized power. Rather than asking where "we" went wrong, I ask "who is 'we'?" Rather than thinking about women's relationship to nature, I think about men's and women's relationships to the institutions that structure daily life. I feel strongly that to turn the tide of environmentally destructive behavior, we must understand how and why certain institutional cultures create environmentally untenable ways of being.

The cast of institutional characters that I introduced earlier—the militaries, corporations, bureaucrats, and environmental organizations—hold the balance of power in determining our collective environmental fate. We need to transform our "culture of pollution" by transforming the core institutions that shape that culture. In some cases, institutions can be "reformed." Reformism should not necessarily be slighted as a liberal band-aid—feminist transformation within institutions, for example, has always necessitated a substantial reworking of both the presumptions and the operations of institutional culture, whether implementing affirmative action in hiring, ensuring sexual harassment protection procedures in the workplace, or introducing women's sports into the Olympic games. In other cases, though, feminist transformation has required the wholesale dismantling of "men's club" institutions. Global demilitarization, for one, needs to be high on a feminist environmental agenda. But the dismantling of militarism is not going to be achieved only by beating swords into ploughshares. Dismantling militaries necessitates dismantling the bonds of masculinity that prop up and sustain military powers.

I started this chapter with the bad news about the state of our earth. The good news is that we have the analytical skills to expose the structures of environmental destruction. If we are willing to take seriously the implications of our understanding, we can change course. Feminism, and feminist transformation of environmentally instrumental institutions, is not a magic balm—it will not solve all environmental problems, and it will not save the Earth. But it is perilously evident that "salvation" will not come through the masculinist structures that have brought us to the brink of environmental collapse. The African American poet Audre Lorde, speaking of women's multiple oppressions (of homophobia, sexism, and racism), reminds us that "the master's tools will never dismantle the master's house."[13] It is a warning to which environmentalists, and all concerned global citizens, should pay heed.

Chapter 1

Up in Arms Against
the Environment: The Military

Vietnam, Fernald, Ohio, and the Marshall Islands in the mid-Pacific, places that are thousands of miles apart in culture and geography, seem to have little in common. But in fact they share a lot—they are links in the chain of militarized environmental destruction that stretches around the world.

Militaries are major environmental abusers. All militaries, everywhere, wreak environmental havoc—sometimes unintentionally (though seldom unknowingly), more often as predetermined strategy. If every military-blighted site around the world were marked on a map with red tack-pins, the earth would look as though it had measles.

Militaries are privileged environmental vandals. Their daily operations are typically beyond the reach of civil law, and they are protected from public and governmental scrutiny, even in "democracies." When military bureaucrats are challenged or asked to explain themselves, they hide behind the "national security" cloak of secrecy and silence—and it is military men themselves who get to define what "national security" is. In countries that are in the grip of martial law, militaries have an even more free and unhindered reign: with wide-ranging human rights abuses the norm under militarized regimes, environmental transgressions are often the least of the horrors for which critics try to hold militaries accountable—and thus even the fact that militaries *are* agents of major environmental degradation is often overlooked.

Militaries are powerful environmental ravagers. The reach of militarized environmental destruction is global. The most powerful military contrivance, "nuclear capability," pushes environmental capability to the limits; past the limits already for some of the radioactivated, blighted wastelands created around the world by military testing, dumping, and adventurism.

Although mainstream environmental groups only commit about one percent of their resources to weaponry and other military issues,[1] environmental activists around the world *are* paying more attention to the role of militaries in environmental destruction. But throughout environmentalist discussions, from earnest ecology tabloids to the academic literature, there is neither curiosity nor discussion about the fact that militaries happen to be almost entirely male. There is no speculation about whether it matters that the global devastation caused by militaries is the product of a cult of masculinity.

But it does matter. Natural environments are increasingly the direct target of military aggression. In militarizing the environment, militaries feed on and fuel the masculinist "prerogative" of men conquering nature. Even when the environment isn't a target of war, it is threatened by "normal" military activities conducted in the pursuit of protecting "national security." All of the people who are defining national security in every country of the world are men, and part of their agenda in defining national security is to protect their male *and* military privilege. Militaries cloak their environmentally dangerous activities by hiding behind the "national security" defense, and they discourage environmentally responsible consciousness. In the military ethos, environmental consciousness is placed a distant second to more "serious" concerns about national security— and thus, men who would place environmental priorities above military ones are often cast as unpatriotic, even effeminate. This gender dynamic is hardened by the fact that, worldwide, most of the environmental grassroots activists are women, and most of the military bureaucrats they have to confront are men.

Militarized environmental destruction is more global, more ubiquitous, and more protected than the actions of even the most flagrantly irresponsible multinational corporations or governments. Whether at peace or at war, militaries are the biggest threat to the environmental welfare of the planet.

POWERS OF DESTRUCTION

In War

War always damages the natural and built environment—to some extent, this is an inevitable side-effect when nations bomb, invade, and

generally attempt to destroy one another.[2] In the midst of an armed conflict, though, it is the direct human toll of war that captures our attention.
Environmental damage typically doesn't attract much public concern:
there was scant attention, for example, given to the environmental impacts
of the oil that gushed for weeks into the Persian Gulf as the Iranians and
Iraqis bombed each other's oil fields in the late 1980s, or when the armies
of Britain and Argentina blew up much of the coastlands of the Malvinas
in their struggle over territorial rights, or, more recently, when the US-led
coalition reduced much of Iraq's social and environmental infrastructure
to rubble. In contemplating the effects of war, it *is* necessary to focus on
the direct human suffering, but it is delusionary to suppose that the human
toll of war is a tragedy separable from the environmental toll—if there is
anything we have learned in recent years, it is that when you damage the
environment, you damage people. Environmental destruction intensifies
and extends human suffering, even long after peace accords have stopped
the immediate bloodletting. The severity and the persistence of health
damage and social disruption from environmental damage is a fact not
lost on military strategists—it is a fact that they have turned recently to
their advantage.

Since the early 1950s, war has taken a deliberately sinister environmental turn. Wreaking environmental destruction has become part of planned

Figure 1.1.   The military leaves its mark on the land. Minefields still dot the
countryside in the Falkland Islands, reminders of the war ten years earlier.
(Credit: Nathaniel Nash, NYT Pictures.)

military strategy in "modern" warfare. The British army was, by all ac-counts, the first to use chemical herbicides and defoliants for military purposes in its anti-insurgency Malaysian campaigns of the 1950s—the first use of poisons aimed not directly against combatants, but against the environment itself. A few years later, environmental chemical warfare came into its own in the US-Vietnam War.

Vietnam   Over the course of the Vietnam War, the US military ravaged and purposefully poisoned an entire country; it is not clear whether the land, air, water, and people of Vietnam will ever fully recover.

The American military dumped approximately 25 *million* gallons of defoliants and environmental toxins on Vietnam over the course of the war. Almost half of the forests in the southern part of Vietnam were soaked, many repeatedly, with a chemical soup of herbicides and assorted toxins, especially defoliants and napalm—more than 4 million gallons of "Agent Orange" alone and 8 million gallons of other herbi-cides. During the war, almost 5 million acres of inland tropical forests were heavily damaged, many irreparably so. Since the mid-1940s, Vietnam's tropical forest acreage has diminished by half. As the forests shrink, so does the tropical community of animals and birds they support. Between 1963 and 1968, the US Air Force dropped almost 400,000 tons of napalm on Vietnam. To destroy the Viet Cong's transport system, the Americans bombed and napalmed elephants. The US military dropped more than 25 million bombs on Vietnam; each bomb blew away a crater of precious topsoil; an estimated 25 million acres of farmland was destroyed. Over half of the coastal mangrove swamps in the south were entirely destroyed by bombs, napalm, defoliants, and bulldozers. The mangrove swamps, which when healthy are major ecological stabiliz-ers for the entire Southeast Asian ecosystem, will take more than a century to recover—if ever they do.[3]

In this wasteland, a war-numbed, politically isolated nation has tried to rebuild—and with little foreign aid or development assistance. The ecological catastrophe wrought by the war has snowballed, inevitably compounded by people trying to eke out a living in a blighted, poisoned land. Without the formerly rich forest cover, soil quality has deteriorated. Without the forest, which acts as a water-collection sponge, unimpeded runoff water carries away valuable topsoil, silting the rivers, and causing major flooding; in southern Vietnam, hazardous floods now occur with three times the frequency as before the war. Farmers, trying to grow food on marginalized land, in an ironic twist have turned to heavy pesticide use to try to boost production; in consequence, chemical runoff from agricultural pesticides is now poisoning much of Vietnam's water supply.

Faced with degraded agricultural conditions, farmers have pushed into the remaining forests; forests are stripped for fuel and to clear land for growing food. When food is scarce—as it now is, with an estimated 8 to 9 million starving, largely because the land can no longer sustain even subsistence agriculture—women eat less and last.[4] Famine, like the war that spawned it, is not gender-neutral.

The much publicized illnesses that American veterans now suffer as a result of exposure to Agent Orange pale in comparison with the legacy the Vietnamese live with: the Americans went home, but the Vietnamese are still living in their chemical-doused country. The Vietnamese government estimates that there have been 3,500 deaths as a direct result of American chemical spraying, and, overall, 2 million victims suffering from exposure to chemical weapons.

The health of Vietnamese women has suffered, more so than men, from the poisoning of Vietnam. Dioxin, a primary contaminant of Agent Orange, persists in the food chain for decades. In human tissue, dioxin has a halflife of 12 years, which means that dangerous levels of dioxin are passed from generation to generation; lactating mothers are at greatest risk. Dioxin is highly carcinogenic, even in minute quantities, causing genetic mutations and any number of cancers. It also is a major teratogenic (birth-deforming) chemical. Vietnamese women today have the highest rate of spontaneous abortions in the world; birth defects occur at alarming rates; 70 to 80 percent of women in Vietnam suffer from vaginal infection; cervical cancer rates are among the highest in the world; in Vietnam, fetal death rates in pregnancy were 40 times higher in the early 1980s than they were in the 1950s.

When a delegation of American veterans recently returned to Vietnam, the two nurses in the delegation were greeted by Dr. Nguyen Thi Ngoc Phuong, the Director of Tu Du Obstetrical Hospital in Ho Chi Minh City. The most disturbing stop on their tour of the hospital was a room in the clinic where all four walls were lined with shelves of hundreds of bottles of fluid containing deformed fetuses—a grim representation of the 40,000 women Dr. Phuong is working with who suffer reproductive disorders (see Figure 1.2).[5] Dr. Phuong and her research team have been collecting the fetuses as part of their effort to assemble the evidence of the lingering health effects of Agent Orange. As she says, "It's no use just having these fetus jars sitting on the shelf. It's no use people intuitively knowing that Vietnam has suffered and is suffering a high incidence of serious health disorder because of Agent Orange. Somebody has to prove that this is the case."[6]

Modern militaries rely on a threatening enemy to make soldiering attractive to men who are nervous about their masculinity.[7] In modern warfare,

Figure 1.2. The legacy of Agent Orange: Deformed Fetuses, TuDu Obstetrical Hospital, Ho Chi Minh City, Vietnam. (Photograph by Marcus Halevi.)

the environment has become a militarized target, and "ecocide" provides another arena for the play of militarized manhood. US herbicide teams in Vietnam rallied with the humorous macho motto "Only We Can Prevent Forests," a play on the popular conservation slogan "Only You Can Prevent Forest Fires." The fact that nature is widely conceptualized as female adds the "allure" of a sexualized assault to military attacks against the environment. The military pendulum swings between protecting and assaulting feminized lands.[8] Men pitting themselves against the environment is not new; what military bureaucracies are doing is shaping this nature-conquering impulse into a war strategy.

Militaries use sanitized language to describe their activities as a means of distancing themselves from the consequences of their own actions. "Techno-strategic" language, as Carol Cohn terms it,[9] plays an important role in warfare—it allows individual soldiers to perform sometimes horrible tasks with equanimity. Thus, human death is talked about as "collateral damage" . . . plans to incinerate cities are described as "countervalue attacks" . . . fusion bombs are called "clean weapons." Similarly, militaries use familiar and friendly terms to describe the environmental damage they inflict. The patterns that bombs make on the landscape are called "footprints"; saturation bombing is referred to as "carpet bombing"—as

Helen Caldicott remarks, "Carpets are nice and soft and homely and domestic . . . so carpet bombing must be OK."[10] Media commentators on military activities are often complicit in perpetuating deceptive descriptions of military damage. Helen Caldicott describes the commentary on the start of the Persian Gulf War: "By God did they have a ball! You'd think it was a Superbowl match! Who were the commentators? White male Americans . . . and were they excited! 'Did you see that one? Look, Baghdad's lit up like a Christmas tree!' Christmas tree???"[11]

The Persian Gulf   The most recent Gulf War, pitting a US-led multinational coalition of armed forces against the army of Iraq is almost a casebook study of the environmental costs of war.[12] In the three months immediately after the war, a UN observer team described conditions in Iraq as "apocalyptic," while a spokesperson for the Kuwaiti government spoke of an ecological catastrophe unparalleled in global history. The war left large parts of Iraq, Saudi Arabia, and Kuwait in tatters, with "collateral damage" felt throughout the entire Middle East region.

Iraq's urban infrastructure was reduced to rubble by the most intensive air bombardment in the history of warfare. In every major settlement in the country, coalition bombing destroyed water supply systems, electrical systems, fuel supplies, food stocks, sewage systems, transportation systems and public-health delivery systems—in short, everything that sustains organized urban life—and a year later many of these services are still not restored. A UN task force that visited Baghdad immediately after the war concluded that the bulk of the damage to these civilian support structures was neither coincidental nor accidental, but rather the consequence of a successful and intentional campaign to destroy Iraq's war machine by attacking its urban and industrial base. Tens of thousands of people were left homeless from bomb damage to private homes. Epidemics and grave food shortages swept through Iraq in the aftermath of the war, killing thousands more. A public health study six months after the war concluded that infant mortality rates in Iraq tripled as a result of the war, internal strife, and the international blockade.

Iraq's industrial, chemical and nuclear plants were early targets in the war. Within the first two weeks, the US military reported that over 500 "sorties" had been flown against 31 Iraqi chemical and nuclear plants. (This bombing is in direct contravention of a United Nations resolution that specifically prohibits attacks on nuclear facilities). While Pentagon spokesmen assured the outside world that no chemical or radiation contamination resulted from these attacks, most observers consider this implausible.[13] Since most of these plants were located along the Tigris River, there is, in fact, a strong likelihood that contamination was extensive and was flushed through the river system into the Persian Gulf.

The Gulf itself suffered extreme environmental damage. Several oil spills, some caused by the Iraqi army, others by the coalition military, dumped millions of gallons of oil into the Gulf. Grassbeds were coated with oil, beaches were fouled, wetlands were destroyed, and mousse-like slicks several miles wide threatened coastal ecology from Kuwait to Oman. Fragile coral-reef systems were damaged, nesting grounds for endangered turtles and dugongs were fouled, and the fishing industry in much of the Gulf was all but closed down within a month of the start of the war. Early in the war, wildlife specialists in the region predicted a massive die-off of marine animals, including birds, dugongs, dolphins, and turtles. Birds were particularly hard-hit by the oil pollution; Saudi officials estimate that at least 14,000 birds were killed along the Saudi shores alone. It is still not clear what will be the long-term damage to the marine ecosystem. What is clear is that the US-led military force, which had warned throughout the months leading up to the conflict that oil *would* be used as a weapon of war, had made no plans to contain or lessen the damage of this certain environmental disaster.

The oil fires in Kuwait caused a massive "pollution incident" (in the words of experts) on a scale not previously seen; the US EPA estimated that roughly 10 times as much air pollution was emitted over Kuwait as by all US industrial and power-generating plants combined. Spewing smoke, soot and particulate matter over hundreds of miles, the oil fires caused temperature inversions over the region, localized holes in the ozone layer, regional acid rain, and damage to agricultural lands and water supplies in several countries downwind of Kuwait. Residents of Kuwait City suffered respiratory diseases, some of which may be long-lasting. After the fires themselves were extinguished, damaged oil wells continued to gush oil into the Kuwaiti desert, pooling in massive "oil lakes." Thousands of birds, unable to distinguish between the sheen of water and the reflection of the oil, have died in the 'lakes,' and the desert ecology in large parts of Kuwait has suffered what will be virtually irreparable damage.

The ecosystems of Kuwait and Iraq suffered the most severe battle damage. But the environments of Saudi Arabia, and to a lesser extent the Gulf rim states of Oman, the United Arab Emirates, and Bahrain, were all significantly degraded by the sheer weight and nature of the military presence in the region. The thinly populated desert of north and northeast Saudi Arabia experienced an unprecedented population boom: within a period of five months, hundreds of thousands of soldiers, accompanied by thousands of tanks, fighting vehicles, and other heavy equipment, set up base in fragile arid lands. There are spiraling impacts from this "invasion": there is the (unconfirmed) possibility that underground aquifers in Saudi Arabia were contaminated by the hasty disposal of wastes and sewage from a population of a half-million; it remains unclear how the military handled or disposed of hazardous wastes; heavy-equipment maneuvers over the fragile desert disrupted the entire ecosystem; both sides of the Iraq/Saudi border were laced with landmines, most of which remain in place; unexploded ordnance from "live-fire" exercises litters the desert; camel herds were frequently caught in the crossfire of war exercises, and then in the war itself. Elsewhere in the region, amphibious-landing exercises on the coasts of the Gulf states, especially in Oman, Kuwait, and Saudi Arabia, wreaked havoc with coastal ecology: wildlife habitats in coastal regions were disrupted, ordnance and mines litter the beaches and coastal waters, abandoned vehicles and fortifications constructed from military "junk" blockade access to the sea, and coastal archeological sites have been destroyed.

The burden of war damage is never borne evenly. In the aftermath of the war, women in both Kuwait and Iraq—even urban women unaccustomed to fashioning family provisions from raw materials—became hewers of wood and carriers of water. A brief UN report from Iraq, several months after the war, observed that women and children were spending large parts of their day searching out food, fuel, and water, often carrying these supplies for miles (see Figure 1.3). Indigenous peoples, too, pay a particular price for militarized environmental damage. The lands of the Bedouins have been devastated: the desert ecosystem on which they rely has been mined, bombed, and debased by the military presence in Saudi Arabia and the war in Kuwait and Iraq. Military activity, pollution, and oil spills dislocated the Marsh Arabs, a little-known population that lives in the wetlands and marshes of the Persian Gulf.

Beyond these grim indicators, there is much about the environmental impact of the war in the Gulf that remains unknown—in part because the US government has been actively suppressing postwar scientific assessments of environmental damage.[14] The oil spills and the oil fires, in particular, occurred on such an unprecedented scale that the usual predictive models are proving inadequate. We will witness the larger story of this war unfold over the next few years. However, "we"—the global community—will not suffer the consequences equally, if at all. The peoples of the Gulf region are global guinea pigs in this environmental catastrophe, a fact that may explain the relative complacency in Western coalition countries about the war "victory" and the startling lack of curiosity among Western media and government representatives about the larger environmental state of affairs in the Gulf.

Figure 1.3. Women bear a disproportionate burden of war-related environmental damage. Women and children gather water and fuel near a refugee camp in Iraq, March 1991. (Credit: Angel Franco, NYT Pictures.)

Central America   The damage inflicted on Vietnam's environment is one of the most extreme examples of ecological destruction—truly "omnicide," as Rosalie Bertell calls it—seen in the world to date. It is too early to assess the long-term consequences of the environmental damage of the Gulf War. But another disaster looms, of proportions that may easily overshadow both of these disasters: Central America is on the brink of catastrophic ecological collapse, a consequence of the fact that this is one of the most heavily militarized regions in the world.[15]

Environmental problems in Central America are deeply rooted in a long history of natural resource plundering by foreign corporations operating in conjunction with local elites—and a concomitant history of the use of military force to sustain those elites and to protect foreign economic interests. The intertwined grips of military rule and foreign exploitation set the stage for the current environmental crisis.

In El Salvador, Guatemala, Honduras, Panama, and until recently in Nicaragua, large landholders own the vast share of agricultural land. In El Salvador, 2 percent of landowners own 60 percent of the arable land;

in Guatemala, 2 percent of landowners control 80 percent of the country's farmland; in Brazil, the top 2 percent of landowners own 60 percent of the arable land.[16] For this landowning elite, farmland is a source of income, not of subsistence. They devote their land largely to growing export cash-crops such as cotton, coffee, bananas, timber, and beef. More than 85 percent of farmland in El Salvador, for example, is currently used for export crops.[17] Local and foreign militaries prop up this system, and often military men are themselves among the largest of landowners. But the tropical soils and climate of Central America cannot support large-scale monoculture, and to sustain the levels of production needed for the export cash-crop market, landowners have to take extreme measures. Heavy pesticide use is their first resort. By the mid-1960s, 40 percent of all US pesticide exports went to Central America, and the region currently has the highest per capita use of agricultural pesticides in the world. Not surprisingly, it also has the world's highest rate of pesticide poisonings. Landowners and their military protectors, putting profits first, have been reckless in their exploitation of the environment. The consequences are becoming more clear every year: soil erosion, pesticide poisonings, water pollution, and wildlife extinctions are rampant, and environmentally dead zones are rippling throughout Central America.

Against this backdrop, the environment is now under even more direct and literal assault. Nicaragua bore the most recent brunt of deliberate war-related environmental damage. When the revolutionary Sandinista government came to power in Nicaragua in the early 1980s, they found an environmental crisis of massive proportions. Among other dubious distinctions, Nicaragua held the record as world leader in deaths from pesticide poisonings. The new government committed itself to cleaning up the environment, reforesting regions that had been clear-cut, restoring old windmill energy facilities that had been idled by the Samoza regime, and imposing pollution controls. In response, the "contras" targeted environmentalists and environmental projects for attack. Between 1982 and 1989, the contras killed 30 environmentalists and kidnapped 70 more, forcing the closure of national parks and halting several water pollution control projects. They burned newly-reforested acres, and blew up dams.

The war precipitated an economic crisis, and in the late 1980s the Sandinista government, desperate for foreign exchange, resumed the export of wildlife. Central America is one of the world's major suppliers for the luxury wildlife trade in reptiles and birds—exporting annually millions of parrots, songbirds, snakes and lizards, live or killed for their skins and feathers. Several species of snakes, turtles, and lizards are teetering on the brink of extinction, and governments in the region have few resources to protect the remaining animals. Similarly, the timber trade out of Central

America has resumed, as governments, strapped by military expenditures, sell lucrative logging concessions to international timber companies. The Chamorro government, newly in power in Nicaragua, is considering deals with Taiwanese timber firms that would give forestry concessions for more than one-sixth of all the forests left in the country.[18] Lumber is now one of the top five exports in Honduras, and the country's forests are being commercially logged at a rate of almost 100,000 acres a year. At the current rate of deforestation, by the year 2025, there will be no tropical forests left anywhere in Central America, with the exception of a few small stands in Guatemala.

The story is much the same in El Salvador, Guatemala, Honduras, even in relatively unmilitarized Costa Rica. The El Salvadoran landscape bears the scars of napalm and white phosphorus bombs. Years of carpet bombing, aerial strafing, napalm and gasoline bombs, phosphorous rockets and defoliants have decimated the forests, which are reduced by 80 percent, according to a Salvadoran environmental group; the World Wildlife Fund has categorized the wildlife population in El Salvador as "seriously depleted."[19] In Guatemala, the army is imposing a "scorched earth" policy as part of its new "low-intensity conflict" strategy against the guerilla movement in that country. Throughout the region, military roads, bases, and airstrips slash through remote wilderness, and heavily armed soldiers on boring patrols shoot up wildlife for "fun." Throughout the region, human and economic resources are diverted from social programs, including environmental protection and pollution control, into military needs—or into avoidance of being drawn into direct conflict.

The 1989 US invasion of Panama, although a smaller and shorter war excursion than most in the region, appears to be having widespread environmental consequences. During the invasion, US planes bombed Panama City, destroying several neighborhoods, and forcing more than 30,000 people to flee the capital. These newly homeless people are rapidly moving into the forests surrounding the city. The new Minister for Natural Resources reports that, "Since January [1990], more than 3000 acres of forest from the protected park areas have been destroyed."[20] Additionally, displaced people are setting up makeshift homes in the Panama Canal watershed area. Under this new pressure, the rate of siltation in the canal, already plagued by environmental problems caused by years of deforestation along its banks, is accelerating, posing a long-term problem for the future of navigation through the canal.

The significance of the Central American environmental tragedy is not just local or even regional. The World Health Organization calls the rainforests of Central America the "lungs of the world." Destruction of the rainforests, loss of wildlife, and environmental degradation threatens global health at the same time as it creates widening zones of unsustainability, deepening poverty and social inequities in a region plagued by a history of resource-based social injustice.

Southern Africa   An even more direct link between militarization and resource depletion appears to be operating in southern Africa. The antigovernment guerilla forces in Angola, "UNITA" and the Mozambican rebel group, "RENAMO," are implicated in an illicit elephant-poaching ring and a smuggling trade in ivory and tropical hardwoods.[21] Until the recent cease-fire, the South African military routinely aided UNITA forces by delivering arms and weaponry to them in return for ivory and timber. An estimated 100,000 elephants have been killed by the right-wing UNITA rebels (who are also backed by the US government) to finance their twelve-year war against the Angolan government. UNITA officials initially denied reports of their complicity in the wildlife slaughter, but in a 1988 interview, UNITA leader Jonas Savimbi admitted that his troops paid the South African Defense Forces for military assistance with ivory and teak. Even then, however, UNITA leaders made the ludicrous claim that any ivory they traded was procured from "peasants, delivering the tusks of elephants who had died of natural causes."[22] Official secrecy about the ivory trade was broken by a dissident South African paratroop commander who told a South African newspaper in 1989, "Elephants were mown down indiscriminately by the tearing rattle of automatic fire from AK-47 rifles and machine guns. They shot everything—bulls, cows, calves—showing no mercy in a campaign of extermination never seen before in Africa."[23]

A second smuggling route operated through Mozambique, where RENAMO guerilla forces, assisted by South Africa, have killed tens of thousands of elephants in recent years to finance their insurrection.[24] Elephant poaching throughout Africa has reduced herds by about 70 percent, and many conservationists believe that elephant populations may already be below replacement levels. In Angola, which once boasted one of the largest elephant populations in Africa, herds have been reduced so dramatically that the prospects for elephant survival are grim. In Mozambique, the elephant population has declined 70 percent in ten years; fewer than 15,000 elephants remain.

Ivory is not the only ecological contraband that South Africa has helped market to pay for its agents' arms. Horns from the endangered rhinoceros are equally valuable, and commonly found in shipments of ivory. African rhinoceros populations have fallen from 60,000 in 1970 to about 3,500 today. Specialists in the wildlife trade believe that South Africa is the largest exporter of rhinoceros horns in the world.

Similarly, the military trade in tropical timber in southern Africa may be pushing deforestation over the threshold of no return. The tropical forests of southern Africa are in no better shape than the elephants—within twenty years, there will be almost no tropical forests remaining in most of western and southern Africa.

UNITA and RENAMO, both outnumbered, fought scorched-earth wars against their governments. Both rebel groups attacked rural communities, schools, health-care centers, and agricultural cooperatives, and destroyed agricultural fields to cut off food supplies to the cities. Predictably, warfare in the hinterlands of both countries has driven millions of people off the land; refugees, at least as many as 4 million, have fled into camps on urban fringes. Concentrated there, the refugees live in utter destitution, left no choice but to pick the earth bare for fuel and shelter. Around Maputo, Mozambique's capital, the deforested "fuelwood ring" is 55 kilometers wide, and in southern Malawi, where hundreds of thousands Mozambicans have sought shelter, the land is stripped of trees.[25]

The Drug Wars   Another, more bizarre kind of war, with a very distinctive pattern of environmental damage, is now under way in Central and South America. With the end of the "Cold War" in the early 1990s, NATO governments, and especially the American government, need to find new military objectives to justify their bloated military budgets and armed strength—the greatest threat to a military is to be without an enemy. The "drug war" offers a convenient venue for increased military activity, and its militarization will escalate as more conventional military threats diminish; as one US Commander quipped, "The Latin American drug war is the only war we've got."[26]

When the American government declared a "war on drugs" in the late 1980s, Guatemalan forests were the first casualty. In 1987, officials in the Guatemalan government and the US Drug Enforcement Agency (DEA) entered into an agreement to defoliate vast areas of Guatemala's north and northwest—a region that contains a wildlife refuge and the largest area of unplundered rainforest remaining in Central America. Starting in April 1987, DEA and Guatemalan military planes dumped a barrage of chemicals that included paraquat, glyphosate, and chemical components of Agent Orange. By that summer, one-third of Guatemala had been sprayed, hundreds of cattle died from drinking contaminated water, dozens of Guatemalans had died, and uncounted more were sick.[27]

The DEA asserted that the spraying was intended to destroy poppy and marijuana fields. Critics note that the sprayed areas happened to be war zones, and argued that the US government was in fact using chemical warfare to assist the Guatemalan military's counterinsurgency operation under the guise of drug eradication. Drug-crop spraying does, in fact, usually herald increased militarization. In 1989, for example, low-flying US government herbicide-spraying planes in Guatemala were fired on on a number of occasions; in response, the US State Department now sends armed helicopters to accompany the spraying planes. In Guatemala and elsewhere, increasingly massive military intervention will likely be necessary to protect drug eradication programs; military activities associated with the "drug war" and conventional military activities feed off each other, causing spiraling militarization.

The aerial spraying in Guatemala has escalated since 1987, and the American agencies in charge of the operation asked for a 1991 budget of $1.2 million, almost double the expenditures on the program in 1989.[28] The American government exerts heavy pressure on the Guatemalan regime to exact cooperation with the drug-spraying program: in Guatemala, and elsewhere, cooperation is rewarded with increased American military and development aid. A government that does not cooperate with aerial herbicide-spraying faces drastic cuts in international development funds. Desperate for foreign exchange, the Guatemalan government now reportedly allows the US "virtual free rein" in conducting aerial spraying operations.[29] The American DEA relies on Guatemala's military intelligence division, the G–2, for support with their drug-eradication program. DEA officials remain unconcerned about the consequences of strengthening the power base of a military unit that is notoriously connected with the death squads operating throughout the country. In the words of one official, "As long as they keep doing good work, you don't ask."[30]

Chemical defoliation is the primary weapon in the militarized "war on drugs." Herbicides and other defoliants are powerful, and usually carcinogenic, chemicals. Aerial spraying is the least discriminating way to apply herbicides: everything in the path of the spraying planes—people, food crops, livestock, domestic animals, drinking water supplies—gets doused. Many of the chemicals now being used in the drug war were intended for small-scale, controlled agricultural applications, and designed to eradicate specific weed plants. The main herbicide currently being used in Guatemala, for example, is produced by Monsanto Chemical Company for domestic use in the US under the name "Roundup." The containers of Roundup marketed in the US carry extensive warnings against, among other things, application to water and wetlands, application to "desirable" plants, and contact with domestic animals and children; Roundup, the manufacturers warn, is intended for "spot applications." While it is clear that the large-scale and indiscriminate application of these chemicals is environmentally unsound and dangerous to human health, aerial spraying in the name of the drug war is spreading throughout South and Central America.

The American government has set its sights on Peru, Bolivia, and Colombia for the next round of militarized attacks on drug fields. In 1990, these three governments, anxious not to lose foreign aid, signed an agreement with the US committing themselves to "a full-fledged military campaign" against the cocaine industry.[31] Within a few months of the agreement, US support to the Bolivian armed forces escalated from $6 million to $36 million; US military aid to Peru has increased from $0.4 million in 1988 to $35 million in 1992.[32]

Latin American human rights observers are disturbed by the rapid escalation of military support to regimes already notorious for human rights abuses. In Colombia, for example, the Colombian Defense Ministry has made combating "insurgents" its first priority, and in 1990 Colombian military officers stated flatly to a Congressional subcommittee that millions of dollars in US "counter-narcotics" aid was, in fact, helping to fight political insurgency.[33] Many outside observers agree that the Peruvian military uses the drug war as a pretense to fight its interminable war on the poor.[34] By strengthening the hands of militaries in Central and South America, the US drug policy could destabilize the civilian governments newly in office in many of these countries. At a minimum, the reinforcement of military power will hamper the effort to rein in the numerous and well-documented abuses of military might that have plagued this region for the past two decades. In 1992, a former top drug enforcement official in Peru was quoted as saying, "Interdiction . . . could push Peru down the road of Lebanon, Burma, Afghanistan, and Laos," referring to countries where the drug trade strengthened warlord armies at the expense of government authority.[35]

Prior to this most recent agreement, which gives US forces direct leverage within the governments and militaries of Andean nations, the Reagan administration proposed spraying the highlands of Peru in 1988 with a powerful herbicide, marketed in the US under the tradename "Spike." Spike kills every broadleaf plant it contacts, can sterilize an area for as long as five years, and has never been adequately tested for human health effects.[36] The US government claimed that Spike would only be applied in isolated areas where no food crops are grown—a ludicrous claim at best, as a UN agronomist working in the region pointed out: "That's pure fantasy. The vast majority of peasants cultivate less than a half hectare of coca to supplement their income, and it's mixed with other food crops."[37] In a surprise twist, executives at Eli Lilly, the manufacturer of the herbicide, refused to sell the poison to the US government, saying that what the government proposed was an inappropriate and probably dangerous use of their spray. The Reagan administration, stung by Eli Lilly's refusal, retaliated in a press conference, calling into question the executives' patriotism and manhood: Reagan aides accused Eli-Lilly of "going AWOL in the war on drugs" and called the executives "hysterical," an intentionally feminized slur.[38] Eli Lilly stood firm, and the government simply took their business elsewhere.

Drug spraying is part of a global "geopolitical" strategy; it is a strategy that causes dozens of direct deaths, and untold hundreds of indirect poisonings. When American drug enforcement agencies sprayed poppy fields in Burma in 1987, for example, 10 people died from poisoning within the week.[39] All of the victims were elderly women. In Burma, as elsewhere, it is women's work to tend the fields. In agricultural economies around the world, it is women's work to fashion food and clothing for themselves and their families from local resources. Since chemical defoliation is invariably concentrated in rural and agricultural areas, it is thus women who typically come in closest contact with poisoned fields and sprayed produce; it is women who till the fields, turning up clouds of pesticides; women who wash in, launder in, and draw water from contaminated sources; and women who handle and clean poisoned food. Many of the toxins used by militaries are carcinogenic, and in addition, these chemicals often are also teratogenic—causing birth defects and damage to women's reproductive systems.

When an agricultural environment is poisoned by military activities, whether in the pursuit of drugs or combat, whether in Central America or Southeast Asia, women suffer first and most. When an environment is so poisoned or degraded by military activities that it is unable to sustain everyday life, women's labor becomes a substitute for environmental sustainability: it is women who have to walk farther for clean water and women who are responsible for feeding their families food that won't poison them. And when, finally, families are forced to flee a blighted environment, joining the growing ranks of environmental refugees around the world, it is women who bear the greatest burden of relocating their families and resuming "normal" life elsewhere.

In Peace: Maintaining a Military Posture

Militaries are at war only some of the time. The rest of the time, they are "maintaining a military posture"—mostly, this means developing and buying and testing newer, bigger and "better" weapons to justify their existence. This apparently more passive side of military activity is, if anything, even more dangerous than war for the health of the environment and for the health of anyone who happens to live downwind or downstream of a military base.

Bases and other military installations were the sleeping environmental hotspots of the late 1980s, and will be the source of some of the biggest environmental catastrophes around the world in the 1990s. In the US, military sites already head the list of the most toxic, most dangerous, and least regulated sites yet found.[40] The US Pentagon produces more toxic waste than the five largest American chemical companies combined. And yet many people appear to be taken by surprise when atrocious military environmental violations start to leak out (sometimes literally so). Military bases are usually considered friendly neighbors to nearby towns: they provide employment (often the only local employment) for local civilians; people in the military are considered to be "first class citizens"; and militaries polish a public-relations image that they "serve to protect."

Typically, it is women who first notice degradation of their local environment. Women who spend more time in the home and local neighborhood, who spend more time tending to children, and who do most of the household chores, are the first to notice unusual patterns of illness in the community, the first to notice persistent chemical odors or residues. In 1985, residents of a Jacksonville, Florida, suburb were forced to board up their homes and abandon their neighborhood, which, unknown to them, was built on top of a former Navy waste disposal site. Residents describe their growing awareness of something going wrong in their neighborhood:

> My wife and I, she noticed it first. She said, "Did you notice a different taste in the water?" and I said, "I can smell it, but I really can't taste it. . . . I could smell it, especially when I'd take a shower, it'd smell up the whole house."
>
> Carolyn remarked: "You'd wash your clothes the very best you knew how and they still got dingier every wash. And I tried everything, you know, bleach and all the home remedies and it got worse and worse and worse."[41]

Around the world, this scenario is played out over and over. Women start to notice things going wrong in their home and in the neighborhood. Many of these women, reluctantly at first, become activists. They form neighborhood monitoring committees; they map out patterns of illness; they contact local and national media; and, eventually, in their crusade to find out what's going wrong in their neighborhood and to find someone responsible for cleaning it up, they confront the military.

Confronting the military is something few people, especially women, are prepared for. Lois Gibbs, the whistleblower in the Love Canal (New York) toxic waste site, reminds us that "Many, if not most, women leaders in the [American] hazardous waste movement are low and moderate income people, have formal educations that ended with a high school diploma, do not have any formal organizing training, have never before been involved in any other social justice issue, and come from and live in a very 'traditional' kind of lifestyle."[42] Yvonne Woodman, an activist in the Jacksonville campaign, describes herself in this way: "I was raised to respect the American government and I love America. I'm proud to be an American citizen. My husband was in the Navy; my son is now in the Navy. I've never been a radical, I've never gone out and laid in front of a bulldozer. I've never done any of that stuff."[43] Woodman is now one of 150 residents suing the Navy. Meantime, crushed, leaking cans and drums of paint thinners, solvents, cleaners, and unknown chemicals, many with military markings on them, continue to surface in neighborhood backyards in Jacksonville; part of a fuselage and the wing of an aircraft worked their way to the surface. The Navy denied all knowledge of dumping, refused to meet with Jacksonville residents, and has retreated into a "no comment" posture.

Militaries seldom have to respond to questions about their actions, and they don't like being challenged. Military men especially don't like being confronted when the challengers are women—worse, "mere housewives." When forced into a confrontation, sexism is their first resort. Women community leaders repeatedly report that they have to endure arrogant, patronizing, and sarcastic military officials who try first and foremost to denigrate the authority and knowledge of the women challenging them. The fact that the community activists are women and therefore presumed to know little about "military matters" is the first challenge raised.

If sexist intimidation doesn't quell community activism, militaries typically then threaten to close the bases that are under public scrutiny. In many communities, this poses a serious economic threat. Not coincidentally, this is also the tactic most likely to divide the men and women of the community. Typically, more men than women are employed on military bases, or are employed at higher-paying jobs. Men are more likely than women to have emotional or career attachments to militaries. Thus, the threat of closure often pits men and their jobs on one side versus women and public health on the other. Militaries consciously use this threat to isolate women activists from family and community support.

But military efforts to stall—or forestall—environmental inquiries at the local community level are becoming less successful as the full scope of military culpability becomes evident. Military sites in the US are among the most polluted ever found (see Table 1.1). One of the most toxic spots on earth, for example, is the American Army's Rocky Mountain Arsenal in Denver, Colorado. Environmental cleanup of American military bases and installations is estimated to cost, by the mid-1990s, upwards of $400 billion—and that is just the cost for cleaning up military sites *in the US*;

it takes no account of the military chain of pollution, toxic waste, and hazardous materials that stretches around the world. In reality, no matter how much money may be thrown at the problem, "cleanup" may be illusory. A 1991 Congressional report concluded that "despite the spending of billions of dollars, it may be impossible with current technology to remove contaminants from ground water and deeply buried soils near many of the [military] installations."[44]

Increasingly, the presence of a military facility is the most reliable single predictor of environmental trauma. In the US, the military currently controls 17 nuclear production facilities, and thousands of other nonnuclear military installations—most of which are sources of environmental hazards. The US military is responsible for one-third of all the hazardous waste produced in the US. In their daily operations, militaries use a wide range of chemicals, solvents, propellants, fuel, cleaning materials, oils, paint thinners. They routinely generate toxic waste that includes cyanides, acids, heavy metals, PCBs, phenols, paints, and contaminated sludges. In addition to these materials, many of which are also common in "civilian life," militaries generate a wide range of hazardous materials that are particular to military missions: bomb materials, radioactive materials, metal solvents, and ammunition components, many of which are secret.

Even less is known about conditions at US installations overseas, though contamination is likely to be extensive. On Guam, the US Air Force and Navy dumped large quantities of a solvent, TCE, and untreated antifreeze solutions onto the ground and into storm drains, contaminating the aquifer that supplies drinking water for three-quarters of the island's population. Landowners in Iceland are seeking compensation for damage caused by toxic wastes dumped at a US radar site more than 20 years ago. At Subic Bay in the Philippines, large quantities of toxic wastes have been released into the bay. Other militaries share the US' record for environmental calumny. The Soviet military has grossly contaminated the air, land, and water of much of Hungary, the former East Germany, Poland and Czechoslovakia. The French military has left behind a legacy of poison and pollution in its former African colonies.

Chemical weapons are among the most secret military compounds—and, second only to nuclear weapons, among the most hazardous. Militaries excel at generating chemical weapons, but they do not know how to safely store or dispose of them. The American military currently has stockpiles of millions of pounds of mustard gas, nerve gas, and other equally deadly and obsolete chemical weapons, many dating from World War II, stored at eight sites across the US and at several sites overseas—and they resumed production of new chemical weapons in 1987, after a 26-year hiatus. They admit that some of these storage sites are leaking. In

Table 1.1. Cleaning Up After the Military[a]

Site, Location	Type of Contamination	Est. Total (in millions $)
Hanford Nuclear Reservation Richland, Wash.	Plutonium and other radioactive nuclides, toxic chemicals, heavy metals, leaking radioactive-waste tanks, groundwater and soil contamination, seepage into the Columbia River.	30,000–50,000
McClellan Air Force Base Sacramento, Calif.	Solvents, metal-plating wastes, degreasers, paints, lubicants, acids, PCB's in ground water.	170.5
Hunters Point Naval Air Station San Francisco	Chemical spills in soil, heavy metals, solvents.	114.0
Lawrence Livermore National Laboratory Livermore, Calif.	Chemical and radioactive contamination of buildings and soil.	1,000 plus
Castle Air Force Base Merced, Calif.	Solvents, fuels, oils, pesticides, cyanide, cadmium in soil, landfills and disposal pits.	90.0
Edwards Air Force Base Kern County, Calif.	Oil, solvents, petroleum byproducts in abandoned sites and drum storage area.	53.4
Nevada Test Side Near Las Vegas	Radioactive and ground water contamination.	1,000 plus
Idaho National Engineering Laboratory Near Idaho Falls	Radioactives wastes, contamination of Snake River aquifer, chemical-waste lagoons.	5,000 plus
Tooele Army Depot Tooele County, Utah	Heavy metals, lubricants, paint primers, PCB's, plating and explosives wastes in ground water and ponds.	64.4
Rocky Mountain Arsenal Denver	Pesticides, nerve gas, toxic solvents, and fuel oil in shallow, leaking pits.	2,037.1
Rocky Flats Plant Golden, Colo.	Plutonium, americium, chemicals, other radioactive wastes in ground water, lagoons and dump sites.	1,000 plus
Los Alamos National Laboratory Los Alamos, N.M.	Millions of gallons of radioactive and toxic chemical wastes poured into ravines and canyons across hundreds of sites.	1,000 plus
Tinker Air Force Base Oklahoma City	Trichloroethylene and chromium in underground water.	69.7

Site	Contamination	
Twin Cities Army Ammunition Plant New Brighton, Minn.	Chemical byproducts and solvents from ammunition manufacturing.	59.9
Lake City Army Ammunition Plant Independence, Mo.	Toxic metals and chemicals in ground water.	55.1
Louisiana Army Ammunition Plant Doyline	Hazardous wastes, ground water contamination.	66.9
Oak Ridge National Laboratory Oak Ridge, Tenn.	Mercury, radioactive sediments in streams, lakes and ground water.	4,000–8,000
Griffiss Air Force Base Rome, N.Y.	Heavy metals, greases, solvents, caustic cleaners, dyes in tank farm and ground water and disposal sites.	100.0
Letterkenny Army Depot Franklin County, Pa.	Oil, pesticides, solvents, metalplating wastes, phenolics, painting wastes in soil and water.	56.2
Naval Weapons Station Colts Neck, N.J.	Heavy metals, lubricants, oil, corrosive acids in pits and disposal sites.	33.8
Aberdeen Proving Ground Aberdeen, Md.	Arsenic, napalm, nitrates and chemical warfare agents contaminating soil and ground water.	579.4
Camp Lejeune Military Reservation Jacksonville, N.C.	Lithium batteries, paints, thinners, pesticides, PCB's in soil and potentially draining into New River.	59.0
Cherry Point Marine Air Corps Station Cherry Point, N.C.	Untreated wastes soak creek sediments with heavy metals, industrial wastes and electroplating wastes.	51.6
Savannah River Site Aiken, S.C.	Radioactive waste burial grounds, toxic chemical pollution, contamination of ground water.	5,000 plus
Mound Laboratory Miamisburg, Ohio	Plutonium in soil and toxic chemical wastes.	500 plus
Feed Materials Production Center Fernald, Ohio	Uranium and chemicals in ponds and soil.	1,000–3,000

[a] Removing the contamination at US military bases and Energy Department sites could cost as much as $400 billion over the next 30 years. This list includes sites that will be among the most expensive to clean up.

Source: Keith Schneider, *New York Times*, August 5, 1991.

1988, Army officials told a US Congressional committee that the 25-year-old stockpile of chemical weapons includes "more than 1,000 leaking containers."[45] The US is the only nation in the world to store chemical weapons on the territory of a foreign country—it stores chemical weapons in Germany, much to the consternation of the German government, and there is little information about the condition of this foreign stockpile.

When pressed, military spokesmen admit that they have no way of disposing of these chemicals safely: incineration and chemical neutralization, two favored proposals, generate mountains of hazardous waste. Ocean dumping, their preferred disposal method, was banned by Congress in 1972. In the meantime, these chemicals sit on American bases, leaking into the air, water, and land, and creating zones of unknown toxicity and persistence. France, Iraq, and the former Soviet Union also have stockpiles of chemical weapons; as many as twenty other countries *may* have them. These militaries are no better than the US military at destroying or storing chemical weapons safely—and some are worse. The French military, for example, is suspected of continued ocean dumping of chemical weapons.

The production of toxic wastes is at the top of the list of military environmental violations—but, it is a long list. • Modern armed forces are land-grabbers—they appropriate increasingly large expanses of land and airspace for their exclusive use, and their appetite for land increasingly collides with other needs, such as wilderness protection, agriculture, recreation, and housing. Modern militaries' appetite for land is growing: a World War II US Army battalion needed 4,000 acres to practice maneuvers; today's battalion needs more than 80,000 acres.[46] In 1981, it was estimated that militaries controlled approximately 1 percent of total territory in the top 13 industrial nations—this may sound small, but it represents an area roughly the size of Turkey.[47] In Hawaii, one of the most militarized of American states, estimates suggest that the military owns or controls from 10 to 25 percent of the land on Oahu, one of the main islands, and over 10 percent statewide.[48] In West Germany, when it was still a separate nation, the military forces controlled about 5.6 percent of total land area; in what used to be East Germany, the Soviet military alone controlled 4 percent of all land. In their wake, militaries leave tracts of land strewn with unexploded bombs and munitions, and create widening, uninhabitable "dead zones" throughout the world, from the American Midwest to the Scottish Outer Hebrides. • Militaries consume large amounts of resources. The Pentagon, for example, is the largest domestic consumer of oil—it uses enough energy in 12 months to run the entire US urban mass transit system for almost 14 years. • Additionally, militaries contribute a growing share of "conventional" pollutants. According to a 1983 estimate, emissions from the operations of armed forces, worldwide, alone account for 6 to 10 percent of global air pollution.[49] Militaries are the largest consumers of certain CFC substances, which are responsible for ozone depletion. • Military technicians kill thousands of animals annually in weapons, chemical, and materials tests; the US military alone uses an estimated 540,000 dogs, cats, primates, and other animals yearly in torturous tests—such as a recent one in which measuring devices were implanted into the spines of primates who were then restrained in vibrating machines for four days.[50]

Gentlemen's Agreements

What gives militaries such license? How are they able to appropriate resources, produce hazardous wastes, stockpile poisons, destroy whole ecosystems, flaunt safe environmental practices, and to do so with increasing temerity, even as knowledge of their culpability spreads?

Some observers point to the distorted sense of self-importance that imbues most militaries. Representatives from one American environmental organization that is currently suing the commander of an Air Force base comment that, "infected with the importance of their purpose, the military's fundamental mentality seems to be that they are simply not interested in mundane matters like environmental conservation."[51] Environmental safety is low on the list of military priorities. As one observer notes, "There is a bunker mentality . . . every penny that goes to safety programs is a penny taken from manufacturing nuclear warheads."[52] In another case, the commander of a military installation in Virginia, in a community confrontation over leaking PCBs from his base, explained his nonchalance about the environmental contamination by saying, "We're in the business of protecting your country, not protecting the environment."[53] This attitude results in one of the great ironies of our age: that in the name of protecting us, militaries are poisoning us.

This sense of a higher purpose, combined with male arrogance in the face of community pressure from activists who are predominantly women, allows the military to consider itself above the law. But this deceit is not only of their own making: they are encouraged in this by their counterparts in the national security bureaucracies. In virtually every country, military facilities are exempt from environmental regulations and monitoring requirements. Government commissions, fact-finding task forces, and public inquiries have their hands tied by military stonewalling and claims of secrecy in the national-security interest. Even when an investigation does turn up damning evidence of environmental wrongdoing, the military is left to be its own watchdog—because "national security" means that "outside" groups are denied access to military sites. Military environmental immunity makes a mockery of the distinctions between "democratic" and "authoritarian" regimes. The ecological results of US military actions are indistinguishable from the unregulated environmental atrocities perpetrated in many parts of Eastern Europe by the Soviet military.

The "national security" knot is a perfect defense: it allows the military to deflect all questions, including questions about what they are doing in the first place that is supposedly in the interest of national security. "National security" is a vague and constantly shifting concept—it has no real or absolute meaning; it is whatever the military defines it to be (with the agreement of other men in the national security loop.) While concepts of national security shift over time, one consistent hallmark of "national security" is that it is a realm of men—it is men who define it, and men who defend it. A recent survey of American government offices in which national security policies are formulated found that women occupy only 44 of some 1,015 policy-making positions. Nor are women represented in the international arms-control bureaucracy: for example, of the 13 top-level positions at the newly created US Institute of Peace, none are held by women.[54] Even at a time when women appear to be gaining access to conventional arenas of political power, defense, intelligence, and arms control are among the most jealously guarded of male inner sanctums.[55]

Meanwhile, the men in government (and those in government are almost all men) who should be *monitoring* the military are typically in government positions of some significance. They have worked hard to arrive in positions of trust and to be privy to the most serious levels of government business.[56] They enjoy the perquisites of power, and enjoy being part of a privileged club, where men talk to one another about "serious" matters—like national security. Many of these men don't want to fall outside the national-security "loop"; most of all, they don't ever want to be accused of not taking seriously national security or, worse, of being unpatriotic. Hedrick Smith, in his much-vaunted analysis of the workings of the US government, *The Power Game*, describes the seductive appeal, and significance, of being "inside the loop":

> Access in the power game is not merely physical; it is mental too. It is not only entry to the inner sanctum; it is being in the power loop—being chosen to receive the most sensitive information, as fresh grist for the policy struggle. Being "cut out" on information, or being "blindsided" as the power lingo has it, can be crippling. . . . In the national security power fraternity, the put-down comes in the form of one official asking another: "Are you in the loop?"[57]

Later in his book, Smith quotes a US Senator describing his initial introduction to the cosy relationships between the military and the government:

> It's slightly incestuous. I'm three months in office and I get invited to the Pentagon. An Army car picks me up. I arrive, and I'm taken to a nice office and order my breakfast. Well-dressed stewards, a four-star general on my right, the secretary of the Army to my left, one- and two-star generals around the table. I told General Wickham, the Army chief of staff, "If my battalion commander in Korea could see me now, he'd never believe it." Wickham laughed. He told me, "There is a kind of awe, and I hope you'll get over it. But there is a close relationship between Congress and the military."[58]

The workings of the sense of shared purpose that bonds men in government and men in militaries are seldom evident to outside observers. But in 1987, public hearings on the American "Iran-Contra" affair gave a glimpse of this dynamic in action:

> Public men use verbal rituals to blunt the edges of their mutual antagonism. A congressman would, for instance, preface a devastating attack on Admiral Poindexter's rationale for destroying a document by reassuring the admiral—and his male colleagues—that he believed the admiral was 'honorable' and a 'gentleman.' Another congressman would insist that, despite his differences with Reagan officials Robert McFarlane and Oliver North, he considered them to be 'patriots.' Would these same male members of Congress, selected for this special committee partly because they had experience in dealing with military officers and foreign-policy administrators, have used the word 'honorable' if the witness had been a woman? Would 'patriot' have been the term of respect if these men had been commending a woman? There appeared to be a platform of implicit trust holding up these investigations of foreign policy. It was a platform that was supported by pillars of masculinity. . . .[59]

The intangible bonds of fraternity between men in government and men in the military are hardened by more concrete partnerships among military services, defense contractors, and members of Congress from states where military spending is heavy and visible—a network of overlapping financial, industrial, and policy agendas that President Eisenhower loosely identified as the "military-industrial complex." Revolving-door job connections and interlocking networks among defense contractors, military officials, and ex-government officials ensure a mutual self-interest in keeping military programs going, and in keeping critical questions to a minimum.[60]

The military has managed to set up a seductive dichotomy, with men, the military and national security on one side versus women, the environment and trivial concerns on the other. By setting up an opposition between "serious" national security concerns that are the province of men, and "trivial" environmental concerns that are the province of women, military officials manipulate ideologies of gender to protect their turf. Given this dichotomy, it is clear why most of the men in the "men's club" of government ally themselves with the military.

Male bonding in the cause of national security is a powerful force, one that translates into specific policy initiatives. For example, it means that militaries are routinely exempted from the restrictions of domestic environmental laws and international environmental treaties, usually on the rationale that environmental regulations might interfere with "military readiness." In the US, the exclusion of militaries from environmental accords started in the 1950s, when government agencies were first coming to terms with the potential health hazards from the country's nuclear programs. The Atomic Energy Act of 1954 subjected civilian nuclear facilities to licensing procedures that included public information disclosure, but government-owned nuclear weapons facilities were exempted from compliance with the Act. In 1959, the chairman of the Atomic Energy Commission appeared before a congressional committee to request military exemption from any national health or safety standards the commission might set. He argued that, "as atomic weapons, nuclear propelled vessels and other military reactors in the custody of military departments become more numerous, the requirements of military readiness in training and maneuverability may exert more and more influence toward less restrictive safety procedures."[61]

Militaries are exempt from all kinds of environmental agreements, many that appear to have little to do with "military readiness" or "maneuverability." For example, in 1988, a 29-nation treaty banning plastics dumping at sea specifically excluded military vessels, despite the fact that American Navy ships alone dump more than 5 tons of plastic waste overboard *daily*.[62] In 1986, the US Congress passed a radical "Right to Know Act" that required industries and companies to report their toxic releases; the Department of Defense exempted itself from reporting its toxic chemical releases, and in fact it maintains that it does not know how much hazardous emissions it produces or what happens to them.[63] In both national and international negotiations, militaries are routinely treated as "special case" agencies that need to be exempted from environmental controls.

In the name of national security, militaries are largely left to be their own environmental watchdogs. The result of *this* particular gentlemen's agreement is a callous disregard for public health and safety impacts of military activities. The military record on environmental accountability is characterized in its entirety by suppression of health information, subterfuge, falsification of documents, duplicitous practices, cover-ups, flagrant disregard of public health, and harassment of activists.[64]

In addition to the rewards of male bonding, there are more tangible rewards for cooperating with the military. As militaries control increasing shares of national budgets, their economic clout increases, as does their influence across all levels of government. One of the concomitant perquisites of this influence and clout is the militaries' privilege of being left alone to do whatever they want to do, unchallenged, regardless of the environmental consequences. Worldwide, the military share of national budgets is increasing at an astonishing rate. Ruth Sivard, a tireless researcher who monitors international military spending, estimates that in constant dollars, world military expenditures in 1987 were about 2.5 times the level of 1960.[65] Escalating military spending and increasing military control of national budgets undermines the ability of citizens, elected government officials, and bureaucrats alike to monitor and control military activities.

In terms of military appropriation of budget priorities, there are more direct environmental trade-offs. Since military and social programs have to compete for shares of limited national revenues, every increase in military budgets means a decrease in spending in the public sector. This is not a startlingly radical observation—in the 1950s, Dwight Eisenhower made the point directly: "Every gun that is made, every warship fired, signifies in the final sense a theft from those who hunger and are not fed, those who are cold and not clothed." Comparisons make clear the environmental cost of military spending:[66] for the price of one British Aerospace Hawk aircraft, 1.5 million people in the Third World could have clean water for life; the budget for the American "stealth bomber" program represents two-thirds of the costs of meeting US clean water goals by the year 2000; the money spent on one nuclear weapons test could provide installation of 80,000 hand pumps to give Third World villages access to safe water; the money spent on operating a B–2 bomber for one hour could provide maternal health care in 10 African villages to reduce infant mortality by half; the money needed to supply contraceptive materials to women around the world already motivated to use family planning is the equivalent of 10 hours of global military spending; the West German military procurement budget for 1985 would pay for the clean up of the West German sector of the North Sea; the annual cost of a proposed anti-desertification program for Ethiopia is the equivalent of two months of Ethiopian military spending at 1989 levels; three days of global military spending would fund the Tropical Forest Action Plan over 5 years; in ten days of the Persian Gulf War, the US military spent the equivalent of the entire annual domestic budget for energy development and conservation.

In India, a country with over one-third of its population living below the poverty line, the government spends 14 percent of its revenue on defense; in Saudi Arabia, military spending, as a percent of GNP, jumped from 5 percent in 1960 to 22 percent by the late 1980s; in the Sudan, in the same period, military spending increased by a factor of four; the Canadian government, usually considered to be a bit player in global militarism, spends twelve times as much on the military as it does on the environment.[67] The price of military expansion, everywhere in the world, is environmental neglect, increasing social inequality, and deterioration in the daily quality of life for hundreds of thousands of people. The effects of reductions in social welfare programs, and deterioration in the environmental quality of life, ripple through society unevenly: people on the economic margins, the poor, and the disenfranchised bear the brunt. Since, worldwide, *women* comprise the largest population living in poverty, when spending on social programs is reduced as a "trade-off" for military spending, it is women who suffer first and most deeply. Ironically, as militaries gain economic clout and their environmental record deteriorates, environmental monitoring programs themselves, usually chronically underfunded in the first place, are often early victims of budget cuts on the "social" spending side of the national ledger.

In thinking about the relationships between masculinist prerogatives and military imperatives, there is a further, more complex set of questions that needs to be broached: Is it possible that the current global system of sovereign states (and the wars and national security mechanisms put into place to protect those private territories) is *itself* a product of *male* consciousness? Are sovereignty, nationalism, territoriality and wars particularly *male* constructs? These are provocative questions, but ones increasingly salient for environmentalists as we are slowly coming, in the 1990s, to realize the costs and frustrations of trying to solve common global environmental problems through an international system geared exclusively to sovereign solutions. (The links between masculinity and sovereignty are explored more fully in chapter 3.)

The evidence to support the assertion that there is a "positive" link between masculinity and wars ranges from biological-determinist arguments ("men are more aggressive and fascinated with death") to analyses of the many arenas in which social conditioning and unequal power relationships between men and women are expressed; militarism and warfare are themselves such arenas, but, more potently, serve to prop up male domination throughout civic society.[68] While determinist arguments are largely discredited, the social-structure arguments are not easily dismissed, especially those focused on the relationships between the violence of war and the socialized role of violence in men's lives as a tool of domination, and the links between a social system of patriarchy and militarism and warfare as enforcement systems for patriarchy. Further, there is considerable evidence of a "gender gap" on militarism—that men and women hold distinct views of and have different relationships to war, militaries, weapons, and the arms race.

Feminist arguments that posit a close connection between masculinity and war do not indict men as individual actors—we all know men who have devoted their lives to antimilitarism and peace—nor do they blame individual men for the arms race. Rather, "making a feminist analysis of the arms race . . . means looking at the causes of war in the structures of a male-dominated society, and looking beyond those structures to an underlying male cosmology."[69]

So, back to the original question. Sovereignty, nationalism, territoriality, wars: are these particularly *male* constructs? If one answers "yes," then it is arguable that the most potent, the most universal, and perhaps the most environmentally unsound of 'gentlemen's agreements,' the one in which men in governments and men in militaries have the most entrenched interests, may well be the very global system itself—a system that is defined by chopping up the earth into privatized sovereign states, which are then militarized and "protected."

THE NUCLEAR CLUB

Fallout

The more dangerous, complex, and secret that military activities are, the more grim and dangerous is the environmental fallout. The combination of toxic materials and Byzantine bureaucracies involved in *nuclear* weapons production, storage, and transportation represents a grave threat to neighborhood, national, and ultimately, planetary environmental viability.

Fernald None know the consequences of military nuclear malfeasance better than the people of Fernald, Ohio. Fernald, a small rural community northwest of Cincinnati, is the starting gate for the US nuclear arms race. It is the site of the Department of Energy's "Feed Material Production Center," where uranium, the basic ingredient necessary to produce plutonium for warheads, is chemically processed, smelted, and machined as weapons reactor fuel. Since it opened in 1952, this facility has helped produce the plutonium for the equivalent of 26,000 nuclear warheads. In the course of producing plutonium, the factory has also produced vast quantities of toxic and radioactive wastes—which have been released into the air and water around Fernald, producing a potentially massive public health crisis. Recent admissions that the dumping occurred with the full knowledge of officials at all levels of the nuclear bureaucracy have, in addition, produced a massive confidence crisis.

The scale of environmental violation at Fernald is staggering. Since 1952, the facility has released about 265 tons of uranium into the environment, and another 337 tons are unaccounted for; radioactive and toxic wastes stored in pits, tanks, and corroding drums have leached into the ground, tainting water supplies; drinking wells throughout the Fernald region were contaminated at several hundred times natural levels, something that the Department of Energy knew for several years without informing residents; over a billion pounds of radioactive wastes are currently stored at the site in silos and shallow pits that are now leaking; every year of its operation, until the practice was uncovered in 1988, the plant illegally dumped 109 million gallons of highly radioactive wastes into storm sewers.[70]

In late 1988, the federal government acknowledged that officials knew about these environmental problems for decades and did nothing to change disposal practices at the plant (of course there really are NO "safe" disposal options for nuclear waste in any event), nor to alert the local community to the potential hazards. It was a reluctant acknowledgment, forced to the surface by years of agitation on the part of a few activists.

One of the first people to probe the Fernald situation was a 71-year old Quaker named Polly Brokaw, a long-time antinuclear activist.[71] In the late 1970s, she saw a map showing nuclear weapons facilities in Ohio; Fernald was on the map. Wondering what it was and why she had never heard of it before, Polly drove out to the plant and found a creek, Paddy's Run, flowing through the site. With a teakettle she collected some water from the creek and took it to a lab at the University of Cincinnati. As she says, "It tickled the geiger counter." Follow-up samples of the soil and water around the plant showed unmistakable evidence of elevated radiation levels. In 1983, Brokaw helped to spark an antinuclear demonstration outside the gates of the plant. This was the first public squall around Fernald, and it angered the leaders of the Fernald Atomic Trades and Labor Council, who asked for a postponement of the demonstration. Workers at the Fernald plant went on strike in 1985 and again in 1986 over safety and wage issues, but their participation in community efforts to expose the hazardous practices at the facility has been uneven. In fact, as often as not, workers at the plant resisted activist efforts—a classic example of the environmental divide between organized labor and community grassroots organizations. Fernald workers, worried about keeping their jobs, labeled the protestors as un-American, and the rift between the workers and the activists has continued since.

Lisa Crawford is the spokeswoman for Fernald Residents for Environmental Safety and Health (FRESH), a group founded in 1984. FRESH currently has a core working group of about 30 people, two-thirds of whom are women, and a larger support network of 200 to 300. FRESH is very much a community-based group, and has gone to lengths to remain on good terms with the Fernald workers—for example, FRESH deliberately does not take an anti-nuclear stand, nor does it sponsor public demonstrations or protests.[72] But even with this middle-of-the-road posture, FRESH activists are accused of being troublemakers who are blamed for, among other things, causing a decline in property values—an accusation that diverts responsibility for the crisis from the Department of Energy to the activists. Lisa, who had never been politically active until she found out that her family well was contaminated with uranium, defends her activism: "People say we're un-American for attacking the government. In fact, we're hardworking, tax-paying citizens who have been used and lied to. We're just fighting for our rights."[73]

Often, unfortunately, in order to reach the officials and bureaucracies they need to fight for their rights, environmental activists have to first come to terms with the resistance and obstructionist tactics of the predominantly male weapons-facility work force—their husbands, brothers, fathers, and sons. At its peak, the Fernald plant had almost 3,000 employees, mostly men, making it the largest employer in the region. Although the employees themselves were at greatest risk from the environmental and health violations at Fernald, they could neither afford to complain, nor share any suspicions they may have had. Working for the government atomic program, "If you talked about your job, you lost your job," one former employee recounts.[74]

As suspicions of environmental calamity at Fernald began to grow in the mid 1980s, and as the activists gathered "hard evidence" of environmental violations, Department of Energy officials pulled the cloak of secrecy tighter around Fernald. Secrecy compounded the incompetent and illegal actions of the plant's managers. Government officials simply ignored the early warnings of dangerous pollution from surveyors and consultants who correctly foresaw air and water pollution problems. Record keeping was deliberately lax at the plant. When two tanks on the site containing radioactive liquid started to leak, instead of fixing the leaks, officials ordered soil banks to be packed around the base of the tanks. Large uranium dust emissions were the norm throughout the operating years of the plant: a July 1956 "Stack Loss Report," for example, recorded a release of 2,747 pounds of uranium dust; 300 pounds of uranium dust were released in one day in December of 1984. When some of the FRESH women activists asked about dust releases, plant officials patronizingly told them not to worry, because "it is heavy dust, and won't travel beyond the fence line of the plant." Echoing this ludicrous claim, one official who was the site manager for 23 years maintains that "no releases beyond the [plant] limits occurred while I was manager." FRESH has now won a court settlement in which the Department of Energy admitted cover-ups and malfeasance, but, in the words of Lisa Crawford, ". . . they basically said in the court deposition, "yes, we did it, we knew about it, but there's nothing you can do about it because we're working on national security."[75]

Fernald is far from unique. The problems at Fernald—both the environmental problems and the fractures in community response—resonate in small towns and communities throughout the US, wherever there is a military nuclear facility. The staggering scope of military-inflicted damage to the American environment was confirmed in 1988, when the Department of Energy released its first systematic assessment of its nuclear weapons program. Some of the more egregious revelations include the following: at the Pinellas Plant in Largo, Florida, toxic substances were routinely discharged into the county sewer system; the Idaho National Engineering Laboratory dumped, for years, radioactive and toxic wastes into unlined lagoons and the waste leached into the aquifer that supplies most of the drinking and irrigation water for eastern Idaho; at the Portsmouth Uranium Enrichment Complex in Ohio, 36 pounds of known carcinogenic chemicals are released daily into the atmosphere through the plant's cooling towers; at the Savannah River Plant, substantial leaching of radioactive materials buried in unsecured pits is contaminating nearby aquifers and groundwater.[76]

In Hanford, Washington, the environmental "fallout" of the military nuclear weapons program precipitated a public health catastrophe on a scale not usually seen during "peacetime."

Hanford   During the Second World War, a semidesert area in south-central Washington state was chosen as the site for one of three atomic cities to be built by the US Army to support the atomic bomb project. The Hanford reservation, as the installation is known, supplied the plutonium for the 1945 atomic bomb test at Alamogordo, New Mexico, and for the bomb that a month later destroyed Nagasaki. The secrecy surrounding the production of the first atomic bombs and the early years of the nuclear weapons buildup also concealed the fact that radioactive by-products of Hanford plutonium contaminated a large part of the Pacific Northwest— and worse, that officials at the plant were aware of the contamination, and worse still, that some of it was deliberate.

Over the years, in their zeal to produce plutonium, Hanford's managers routinely allowed clouds of radioactive iodine, ruthenium, cesium, and other elements to be released from the processing plants' emission stacks. Liquid wastes were dumped directly into the Columbia River. Reports that were declassified in 1986 establish that pasturelands, croplands, and forests hundreds of miles away were repeatedly contaminated; radioactive iodine is now present in the aquifers (the main water source for the region) under the Hanford reservation, and in farmers' wells across the Columbia River. Further, the reports give clear evidence that health specialists at Hanford recognized at the time the health risks of releasing so much radiation.

The worst release of radioactive substances was during a deliberate experiment in 1949. The Atomic Energy Commission (in charge of the plant at the time) and the Army assumed that the Soviets were rushing to produce atomic bombs; further, they believed that the Soviets were using technology that would leave a measurable radioactive trace. They decided to test their assumptions in eastern Washington by creating radioactivity levels in the Hanford area similar to those that presumably existed near Soviet plants. The experiment, called "green run," released into the atmosphere some 5,500 curies of iodine 131 and a still-classified inventory of other fission products, secretly measured by the AEC, in a 200-by-40 mile plume. There was no public health warning. (By contrast, when the 1979 Three Mile Island nuclear accident released 15 to 24 curies of radioactive iodine into the countryside, people around Harrisburg were evacuated and milk was impounded.) The government, with cold calculation, used the population around Hanford as guinea pigs, releasing radioactivity into the food, water, milk, and air, without consent or warning.

The nine reactors at Hanford are all now closed—in response to the public expose of safety hazards. But 60 million gallons of high-level radioactive waste remain stored "temporarily" at the site, and waste disposal ditches and lagoons remain full.

Since the beginning of its nuclear program in the 1940s, the government has steadfastly denied that people living near its weapons plants are in any health danger. But, as Rosalie Bertell points out, it is the US military itself that controls most of the radiation-related health research. And with the revelations about government cover-ups of the Hanford contamination, confidence in government assurances is running low in eastern Washington. At least 13,000 people in Hanford received radiation exposure comparable to that suffered by Chernobyl disaster victims; independent scientists say that the doses of radioactive iodine received by hundreds of the children who lived near the plant were high enough to cause thyroid abnormalities in 80 to 90 percent of the cases; the 1953 class of a high school downwind from Hanford now has a 52 percent incidence of thyroid dysfunction.[77] Betty Perkes, a 54-year old mother of five who lives on a farm downwind from Hanford believes that the loss of an infant and the thyroid diseases that she, her husband, and three daughters suffer from were caused by the plant. Laverne Kautz, a wheat farmer north of Hanford, is angry about government complacency: "They say there were no observable health effects from radioactive releases. How do they know? In the past 15 years, I have had 10 cousins, 5 aunts, nine dear friends, and my mother suffer from cancer. Of these 25 people, 15 have died."[78] A recent epidemiological study of the Hanford region established that women run twice the risk of men of developing thyroid cancer.[79]

The women of Hanford have suffered additional health consequences. When Hanford opened, the town was a closed base. The only people who lived there were the plant workers and their families—a community isolated by both geography and secrecy. This isolation took a special toll on the wives of Hanford workers.[80] They were not allowed to know the details of their husbands' work, they were far from friends and families, and they were generally excluded from the "man's world" mission of Hanford. If they worried about their husbands' safety in what they knew only as top-secret atomic work, their concerns were dismissed—or, worse, they were accused of trying to undermine a project vital to national security. One consequence of this schism between pressure-driven men and their isolated wives seems to have been an increase in domestic violence. Although domestic violence from the early years of Hanford is not documented, the conventional wisdom among the women of Hanford at the time was that "you never put a person working with atomic weapons on the spot"—the words of a woman who was describing the increasingly irrational and volatile tantrums of her husband. Ironically, the *closing* of Hanford has also brought an upsurge in domestic violence. Men, under pressure as they see their jobs disappear, and some of them resentful of the environmental activism of the women in their community, have taken out their frustrations in the home. In 1987 and 1988, as the last reactors at Hanford were being shut down because of safety violations, calls to "Safe Place," the local battered women's shelter, jumped by 75 percent.[81]

◐

Military mismanagement of hazardous substances, especially nuclear materials, threatens the health of widespread populations—but it threatens workers' health first. Dr. Alice Stewart, a British epidemiologist, was the first to correlate the high rates of deaths and disease she found among Hanford workers with the low-level radiation exposure that was part of their everyday work lives.[82] The US government has kept detailed medical records for its 600,000 American nuclear weapons employees since 1942. But, these records were kept secret, and not even available to the workers themselves. When Dr. Stewart presented her findings of elevated cancer levels in nuclear weapons workers to the US government in the late 1970s, she challenged the policy of medical secrecy, saying it was scientifically and morally indefensible. In 1989, the Department of Energy agreed to open its medical files.

Recent press reports have focused on the plight of the "downwinders," people who lived downwind of the above-ground atomic tests in Nevada and New Mexico who were not warned of the dangers from fallout and contamination. But in late 1989, researchers compiled a preliminary estimate of all Americans who have been directly exposed to radiation since the early 1940s by the programs to develop, build, and test atomic weapons—"downwinders" are only a minority of those bearing the costs of the fallout.[83] The list in Table 1.2 is at best preliminary and speculative, but it makes an important point about the spiraling health and environmental costs of a "national defense" based on nuclear weaponry.

Chelyabinsk   Chelyabinsk, a dusty industrial province in the Ural Mountains, is home to a million inhabitants; it is also home to the (former) Soviet Union's primary facility for making nuclear weapons—the Kyshtym complex, constructed as a Soviet response to the American Manhattan project.

From the late 1940s until the middle 1950s, radioactive wastes were dumped directly into the nearby Techa River and into small lakes near the site. Tens of thousands of people living downstream received average doses of radiation four times greater than those received by the victims of Chernobyl; one of the lakes is still so contaminated with waste that officials say that the only "cleanup" solution is to fill the lake completely with concrete.[85] Local residents were not warned about the releases until several years after the dumping began, and even then information was deliberately falsified. The cancer rate in Chelyabinsk is now the highest for the entire Soviet Union; health officials estimate that people living along the Techa River are twice as likley to contract leukemia as people in the surrounding areas. During the last decade, diseases of the circulatory

Table 1.2. Americans Exposed to Radiation in the US Atomic Weapons
Program: 1940s—late 1980s[84]

Nuclear Workers	
Workers in atomic weapons plants and laboratories, 1942 to present	600,000
Military Personnel	
Personnel involved in atmospheric tests and cleanup in the South Pacific	168,500
Personnel involved in atmospheric tests in New Mexico in 1945 and Nevada 1951–1962	62,000
American Civilians	
Workers at the Nevada Test Site	1,200
People living downwind of Nevada Test Site	25,000
Uranium miners for companies under contract to the Atomic Energy Commission	500
People living near the Hanford Reservation, late 1940s through the early 1960s	120,000
People living near the Idaho National Engineering Lab	85,000
People living near Rocky Flats in Colorado, exposed by fires in 1957 and 1969 and other mishaps	100,000–500,000
People living around the Feed Materials Center in Fernald	14,000
Americans in the South Pacific	
Exposed to fallout from atmospheric testing	2,150

Source: Keith Schneider, "Opening the record on nuclear risks," *New York Times,* December 3, 1989.

system have increased by 31 percent among the local population, bronchial asthma increased by 43 percent, congenital deformities by 23 percent.[86]

Several accidents plagued the Kyshtym operations. A massive waste-dump explosion in 1957 released 20 million curies of radiation into the immediate region, and sent a deadly radioactive plume far into Siberia that exposed at least 270,000 people. The Soviet government and military leadership denied reports of the explosion, and took little remedial action. The American CIA suppressed its own intelligence reports about the accident—to publicize such a horrific event might have alarmed Western citizens and politicians at the very time that Washington was mounting intensive public-relations campaigns on behalf of its own atomic programs. In 1967, a drought reduced water levels in one of the lakes used for radioactive dumping, leaving behind an intensely radioactive surface film; the following summer, high winds swept the deadly silt across the landscape, irradiating a 25,000 square-kilometer region containing 436,000 people. The total amount of radioactivity dispersed was about 5 million curies, equivalent to the radioactivity released when Hiroshima was bombed.

Soviet military and government officials consistently denied all reports of accidents, and actively suppressed information about the health effects of the environmental contamination from the nuclear weapons site. The local hospital was built with only a 50-bed capacity. One of the doctors explained the rationale for the inadequate health facilities:

> When this facility was built, the authorities were reluctant to set up a large facility because the local people might draw conclusions about what kind of complex it really was. And, of course, this was a military secret. When local people were evacuated from their homes and taken to hospitals in the 1950s, they weren't told the real reasons why. And the health workers in this establishment were obliged to sign forms before starting work here that prohibited them from revealing why these people were in the hospital, or even from saying or writing anywhere that they had radiation sickness.[87]

When an American reporter pressed this doctor about the medical malfeasance she and her colleagues had perpetrated, she replied, "Yes, just like Hanford."

Club Culture

After-the-fact investigations of military environmental neglect and damage, whether in the former Soviet Union, Britain, or the USA, reveal that, in virtually every case, at least some base commanders, plant managers, and government officials had full knowledge of ongoing environmental violations, and did little to prevent or stop unsafe practices. While this fact may raise questions about the moral character of the individuals involved, more to the point, it raises serious questions about the institutional and structural contexts that shape military culture—a culture that appears to have a certain predictable universality: militaries from cultures and states that have little else in common share a distinctive environmental sensibility—namely, one of disregard.

Militarized secrecy is one key to understanding the social architecture of military culture. The veil of secrecy surrounding military programs protects militaries from public accountability, it imbues military actions with a sense of higher purpose, and it allows military men to feel invulnerable. The secrecy surrounding *nuclear* programs is much more dense than in other military programs; it derives from the early wartime ethos of nuclear weapons production. A spokeswoman for the US Department of Energy, commenting on the environmental disasters at nuclear weapons facilities, explained, "We did inherit a lot of stuff. In 1943 the guys were doing the best they knew how, and the people who came after them didn't question what they'd done."[88] Robert Alvarez of the Environmental Policy Institute also identifies the early years of the nuclear program as crucial in setting the tone and direction for today's military nuclear programs: "From the beginning the program has been run with isolation, secrecy, self-management and compartmentalization."[89]

Isolation and secrecy in the nuclear weapons program is not just a legacy of the 1950s. Managerial attitudes about environmental regulation of nuclear facilities appear to be almost as disdainful in the 1990s as in the midst of World War II; one Environmental Protection Agency official recently characterized the Department of Energy's approach toward environmental laws as, "Look, Buster, don't bug me with your crap about permits. I'm building atomic weapons."[90] Surveillance and harassment of whistle-blowers has been documented at several Department of Energy nuclear sites, including the Oak Ridge National Laboratory and Hanford.[91] In the early 1990s, an independent government task force found evidence that managers at Hanford and several other weapons facilities in the US were using wiretaps and surveillance to intimidate and monitor workers within the plants who had voiced health and environmental concerns.[92] One of the targets of surveillance, a Hanford engineer who was dismissed because of his exposé about faulty monitoring equipment in the plant, reflected: "In a lot of respects the activities I have personally observed are similar to what we've heard about the KGB and the Gestapo. If you want to describe it as a police state, I think that's fair."[93] Another Hanford employee who voiced concerns about safety problems said she was threatened with dismissal, ordered to see a psychiatrist, harassed, and trailed by security agents.

The balance of power in the world of nuclear weapons and nuclear states pivots on a pyramid of secrecy, exclusivity, and fraternity.

Secrecy is the gatekeeper of power. All elites use secrecy to privatize access to knowledge. Men in power—male elites—have often used secrecy to exclude women, explicitly, from their preserve. Secrecy allows men to mystify the work that they do; it allows them to perpetuate the notion that what they are doing is "too difficult" or "too complex" for women to understand—whether it be mathematics or welding.[94]

There are only five states in the world with declared nuclear weapons, another four or so that are essentially nuclear states; this is an exclusive group, often referred to as the "nuclear club." And a club it is, an exclusive knot of superstates. "Nuclear capability" is much coveted and jealously guarded, not only as a technics, but as a symbol of the full flowering of military might and state maturity—the signifier of a manhood of sorts.

Within nuclear states, the even more exclusive club is the fraternity of weaponeers—the men who dream up, design, and engineer the superweapons of the nuclear age. In *Fathering the Unthinkable*,[95] Brian Easlea, a former nuclear physicist, argues persuasively that the context of the early atomic bomb projects was not only driven by masculinist paradigms, but was framed implicitly and explicitly by a "cult of masculinity." The development of nuclear weaponry brought together, for the first time, two of the most entrenched men's clubs—science and the military. (Science has long been an almost entirely male preserve and male scientists have long used secrecy, mystery, and obfuscation to protect science as a male enterprise.[96]) This heady combination resulted in an impenetrable fusion of secretive masculine power. Physicists managed to harden the gender barriers of their discipline by allying themselves with the military; previously caricatured as wooly-headed, mild-mannered "sissies," physicists improved their public image (and their ability to garner government grants) inestimably by producing monster weapons in the 1940s and 1950s.[97] By combining forces, male scientists and military men forged an impenetrable alliance: military men got "the ultimate weapon," and physicists got to "prove their manhood." The hyper-machismo and hyper-secrecy that characterize nuclear weapons programs today—behind which egregious environmental violations are hidden—is a product of the fusing of masculinized science with militarized masculinity, both wrapped in the protective cloak of secrecy.

The world of the early nuclear weaponeers was a paramount men's club. Historical and contemporary accounts of the communities of scientists and military men convened in Chicago, Berkeley, Los Alamos, Alamogordo, and Hanford in the 1940s and 1950s paint a "boy's world" picture.[98] With very few exceptions, women were bit players on the edges of the nuclear scene—they were mostly present as scientists' wives, dutifully packing up one temporary home after another, or facilitating the requisite social gatherings for their socially awkward husbands and colleagues. Wives were supposed to know nothing and ask nothing about their husbands' work. Richard Rhodes, in his classic 800-page account of the making of the atomic bomb, describes very few instances in which women had any entree to their husbands' world: "[Arthur] Compton, who described himself as 'one of those who must talk over important problems with his wife,' arranged uniquely to have Betty Compton cleared. None of the other wives was supposed to know about her husband's work. Laura Fermi found out, like many others, only at the end of the war."[99]

Descriptions of life in the closed, secretive communities of the bomb-makers echo with the same despair that seemed to envelop the materials-supply community at Hanford. The intensity of living within a cramped and confined community with men working on weapons of mass destruction seemed to take a special toll on the marginalized women:

> Whatever his burden of morale and work in those years, Oppenheimer also carried his full share of private pain. He was kept under constant surveillance, his movements monitored, and his rooms and tele-phones bugged; strangers observed his most intimate hours. His home life cannot have been happy. Kitty Oppenheimer responded to the stress of living at isolated Los Alamos by drinking heavily.[100]

Laura Fermi recounted: "We were a high-strung bunch of men, women and children. High-strung because altitude affected us, high-strung be-cause we were too many of a kind, too close to one another, too unavoid-able . . . high-strung because we felt powerless under strange circum-stances."[101]

The women of these communities, silenced and marginalized as they were, occasionally found creative means of self-assertion. An amusing story about Edward Teller's wife suggests an uncanny continuity in forms of women's resistance:[102]

> Mici Teller waged rebellion saving the backyard trees to preserve a playground for her son. "I told the soldier in his big plow to leave me, please, the trees here . . . so Paul can have the shade, but he said 'I got orders to level off everything so we can plant it'. . . The soldier left, but was back the next day and insisted he had more orders to 'finish this neck of the woods.' So I called all the ladies to the danger and we put chairs under the trees and sat on them. So what could he do? He shook his head and went away and has not come again."[103]

For the early nuclear bomb projects, women served primarily as mothers and wives, and they bore the brunt of the stress, isolation, and the tension of their husbands. But mostly, women were irrelevant to the "serious" work that the men did—work that was self-consciously designed to change the course of the world, to gain mastery over great powers, and to build monster weapons. Women still are largely outsiders in the man's world of weaponeers: while there are a few more women scientists, technicians, and mathematicians at Los Alamos, Sandia, or Lawrence Livermore (the three primary sites in the US for nuclear weapons research and design) these days, nuclear weaponeering is still a male enterprise.[104]

More than just a cult of masculinity though, the historical record suggests that the early nuclear weapons program was fueled by images of overt competitive male sexuality.[105] Easlea suggests that

> Physicists were not only anxious to stamp and be the first to stamp their masculine power of intellect upon the world through their historic building of the atomic bomb but that, driving them relentlessly forward, was a subterranean desire to demonstrate once and for all the unique creativity of the male vis-à-vis the female. (97)

Carol Cohn adds to Easlea's analysis in her study of the contemporary defense community—a community she describes as burdened by "the ubiquitous weight of [male] gender, both in social relations and language," that thrives on sexually explicit imagery and sanitized abstraction.[106] Cohn talks about the surprisingly transparent phallic imagery that pervades military discussions of military strength—a language of soft laydowns, deep penetration, hard missiles. The history of the development of the atomic bomb is rife with "male birth" imagery: the Los Alamos bomb was referred to as "Oppenheimer's baby," "Teller's baby" at the Lawrence Livermore labs; those who wanted to disparage Teller's contribution claimed that he was not the bomb's father, but the mother—i.e., that Teller merely "carried" someone else's idea.[107] In the early tests, before they were certain that the bombs would work, the scientists expressed their concern by saying that they hoped the baby was a boy, not a girl—that is, a dud; when the bomb worked, the telegrams sent around the world crowed that the "baby is a boy." Cohn reports that mothering (denigrated) vs fathering (valorized) imagery, male (positive) vs female (negative) sex typing, is still entrenched in the nuclear mentality.

New light is shed on the culture of nuclear destruction if it is understood as a private men's club, within which masculinity is both an explicit sexualized expression and an implicit taken-for-granted context. The dominant representation of the early years of the atomic project is overwhelmingly one of nuclear scientists giving birth to male progeny with the ultimate power of violent domination over female Nature. The horror of work that consists of creating and "improving" tools of mass destruction is blunted by a culture that inculcates and values the male-socialized traits of separating morality and emotionalism from working and thinking. Disregard for the consequences, including the health and environmental costs, of the nuclear arms race is normalized within an insular, privileged fraternity. In thinking about the development of nuclear weaponry, we

WHITE MEN IN TIES DISCUSSING MISSILE SIZE
Figure 1.4.   Surprisingly transparent phallic imagery pervades military rhetoric.
(© Ken Brown, courtesy of artist.)

might reflect on the wry observation made by poet W. H. Auden about the
militarized space program:

> It's natural the Boys should whoop it up for
> so huge a phallic triumph, an adventure
> it would not have occurred to women
> to think worthwhile. . . .[108]

OUT OF SIGHT, OUT OF MIND

The inertia, resistance, and obfuscation encountered in trying to force
domestic militaries to be environmentally accountable is compounded
when dealing with a foreign military presence. It is all but impossible for
governments or citizens to force accountability from a foreign-owned and
operated military facility—worldwide, it is standard operating procedure
for the military facilities of a foreign state to be explicitly beyond the reach

of local law or sovereign rule. Around the globe, there are about 3,000 military sites controlled by one country but situated in another. There are British and American bases in former West Germany, Cyprus, the Falklands and Belize. The Indian government has soldiers on station in Sri Lanka. Cuban bases exist in Angola, the successor states of the Soviet Union maintain bases in Vietnam and Eastern Europe. The French military maintains facilities in Chad, Gabon, Senegal, the Ivory Coast, and Djibouti, among other places.[109] None of these militaries is any more environmentally responsible than the American military, and some are worse. Not all militaries produce or store nuclear weapons and waste, and not all 3,000 foreign installations produce toxic waste. But a large proportion of them do, and there is every reason to believe that most of these sites represent lurking environmental disasters.

If anything, the environmental record of militaries operating on foreign soil is *worse* than home-based military operations. Military policy and attitudes on foreign bases are often shaped by overt racism: it is often assumed that the local hosts are less important, less worthy of respect, even less "human" than the foreign power—a dynamic that is most noticeably at work when there is an ethnic or racial difference between the foreign military and the local citizenry. Weapons testing, chemical dumping, war simulation exercises—activities that are so intrusive or offensive that they would never be tolerated in Devon or Brittany or Massachusetts—are simply exported, often in a close replication of colonial "trading" patterns. Environmental concerns are not taken seriously when the environment of concern belongs to someone else—especially some "lesser" someone else. Racism and sexism intertwine to defeat activists trying to monitor environmental atrocities by foreign forces: as in the US and Britain, most grassroots environmental activists everywhere around the world are women. American military commanders are patronizingly dismissive of *American* "housewives" trying to impose ethical and environmental responsibility on the military—this pales beside the scorn and contempt that greets Filipina or Micronesian or Innu women activists who dare to challenge European or American military men.

Militaries operate under principles of what can best be called the "fallacy of remoteness"—they deflect criticism of their most hazardous testing and dumping activities by siting them in "remote" areas. "Remoteness," though, is a slippery geographical concept: it is a highly subjective designation, one typically defined from a Eurocentric vantage point. A place that is "remote" from Britain or the US—and thus judged to be "safe" for inherently unsafe activities—is not remote for the people who live there. The criterion of "remoteness" has guided British nuclear testing in the Pacific and waste dumping on Australian Aboriginal lands, French waste dumping in Chad, Japanese proposals to dump nuclear waste in the underwater Pacific trench, Soviet testing in Kazakhstan, and French and American nuclear testing and disposal in the Pacific Islands.

Goose Bay   It is often difficult for a community to assess, in advance, what the environmental effects of military activities may be—partly because militaries usually refuse to disclose details of their activities. People in the small settlement of Goose Bay, Canada, recently wrestled with this Catch-22 dilemma. In the 1980s, the Canadian government and NATO entered into a ten-year memorandum of agreement to expand NATO activities in northern Labrador, as a prelude to permanent NATO expansion of the existing Air Force base in Goose Bay. NATO military planners intended to use northern Labrador (which they consider to be "empty land") for low-level flying exercises. Their plans would commandeer for military use as much as 100,000 square kilometers of Labrador for low-level flying, bomb runs, fuel-dumping exercises, mock air-combat exercises, and weapons training. The intent was to bring NATO men through Goose Bay in short training rotations—up to 3,300 "transient personnel" every two weeks.

Goose Bay is a small community of whites and native Innu. Almost everyone is dependent on the land to make a living. Fishing, trapping, and hunting are the major commercial *and* subsistence activities. A small group of residents, and a particularly active mixed Innu and white women's group, mounted a challenge to the NATO plans, which they pursued through the Canadian courts. Faced with the prospect of up to 17,000 low-level "training sorties" a year (almost 50 a day) out of Goose Bay, and 3,300 new men through the community every two weeks, the protestors were concerned that Goose Bay would suffer extreme environmental and social trauma. There is already evidence that low-level flying has disrupted the migration, reproduction, and feeding patterns of the wildlife on which everyone in Goose Bay depends. Even before the proposed expansion, jet bombers fly lower in Labrador than is allowed almost anywhere else in the world. No flights are permitted under 250 feet in the former West Germany, for example, yet in Labrador they are permitted as low as 100 feet. In the words of one observer, "the Canadian Department of National Defense and NATO behave as if [the region] is uninhabited; the 15,000 Innu who live there know otherwise."[110]

The Canadian government, acting on behalf of NATO, commissioned a $6 million "environmental impact assessment" of the military expansion plans.[111] The impact assessment report, over a thousand pages long, is little more than a whitewash: it identifies dozens of "project elements" that may have environmental effects, and then, for each one, concludes

Figure 1.5. Militaries like to use "remote" places for their most hazardous activities. But remote for whom? (© Pacific Conference of Churches/Lotu Pasifika Productions, Fiji. By permission.)

that environmental damage can be "mitigated," "avoided," or "reduced." For example, while recognizing that "Spills of fuel have been a concern at the base for a number of years and we recognize that the present fuel storage and distribution system does not meet current industry or government standards," the environmental experts conclude that "contingency plans" to contain fuel and hazardous materials spills will reduce the impact of accidents to a "minor" concern. The report assures the residents of Goose Bay that the military will "avoid" or "mitigate" the impacts of fuel dumping, accidental hazardous materials spills, extreme noise pollution from low-level flying, routine use of weapons ranges, construction of remote facilities, and aircraft crashes, but gives almost no specifics of how these problems will be contained. Militaries have bankrupted their credibility as responsible environmental managers, and the fears of many of the residents of Goose Bay were not calmed by the government report.

Opponents to the NATO plan compiled their own environmental impact report in reply to the official assessment.[112] The major deficiencies of the official report, as identified by the opposition report, are those that seem to characterize most military environmental assessments: complacency about the effects of military activities on the physical environment, and ignorance about the social (especially gender and racial) dynamics of the military presence. Most noticeably, NATO plans in Goose Bay take little account of the effects on the women of Goose Bay—and the *different* effects on white and on native women—of having 3,000 transient, foreign men on tours of duty through the community every two weeks.

The struggle in Goose Bay is an internationalized version of the now-familiar domestic drama of community opposition to military activities. The women activists of Goose Bay, like their counterparts in Hanford or Fernald, have been accused of destroying the economic prospects of the town, and they are the butt of sexist jokes and harassment. Having failed to stop the opposition to the base with these divide-and-conquer tactics, NATO is now looking to take their business elsewhere. NATO is considering a site in Konya, Turkey, as an alternative low-flying training center. As one NATO spokesman said, "there are no court battles in Turkey to stop low-level flying."[113]

Because their reach is global, militaries can implicitly (or explicitly) pit poor countries against rich, environmental concerns against economic need, women in Canada against women in Turkey. Poor countries (or poor regions within countries) are often placed in a position of bidding *for* environmentally unsound military activities that people in rich countries have the "luxury" of rejecting. Environmentalists often cannot afford to make alliances across countries, cultures, and languages. International antimilitary environmental alliances are tremendously expensive to sustain: the women in Goose Bay are unable to work with their counterparts in Turkey. While environmentalists can usually coordinate opposition to military activities only on a local or regional level, international military alliances such as NATO can move their environmentally destructive activities around and across countries as though they were moving players across a global board game. Militaries have the international advantage and they control the moves in the global environmental shell game.

The Pacific   Most of the nuclear weapons systems stationed in Europe and the US have been tested on indigenous peoples' lands in the Pacific, without their consent and often without warning.

France currently operates 15 military bases and installations in the Pacific. In 1966, with the end of its colonial control in Algeria, where it previously tested nuclear weapons, France began testing nuclear weapons at Moruroa Atoll in French Polynesia. After 41 above-ground nuclear blasts, bowing to protests over open-air testing, France took its testing program underground in 1974. Rocked by dozens of nuclear explosions, the basalt beneath Moruroa has crumbled, and the whole atoll has subsided nearly 5 feet. In the course of the testing program, France reportedly turned the north end of the atoll into a radioactive waste dump, and spread 10 to 20 kilograms of plutonium-impregnated tar across one section of Moruroa as part of a security exercise for nuclear accidents; a cyclone in March 1981 swept the entire mess into the sea.[114] Between 1966 and 1988, the French military exploded 108 underground nuclear tests in the Moruroa Atoll, and three at Fangataufa Atoll. They continue an aggressive program of nuclear testing, averaging five to ten tests a year.[115] Significantly, the French government has been reluctant to cooperate in supplying statistics for a South Pacific Cancer Register, but the limited information so far released by the French indicates that stomach and lung cancers are now increasing in French Polynesia, and Polynesian women show a particularly high rate of breast cancer.[116]

Britain, which still has 13 installations in the region, tested nuclear weapons at Christmas Island in the Pacific before moving on to Aboriginal lands in Australia in the 1950s. With the help of the Australian government, the British military appropriated 800,000 square kilometers of land, mostly Aboriginal territory, in central Australia—"the vast empty spaces in the centre of Australia" in the words of one military official—rounding up and forcibly removing entire Aboriginal communities with no compensation.[117] Hundreds, possibly thousands, of Aborigines are estimated to have been affected by the fallout from nuclear open-air testing, but British and Australian government cover-ups have so far frustrated efforts to discover the

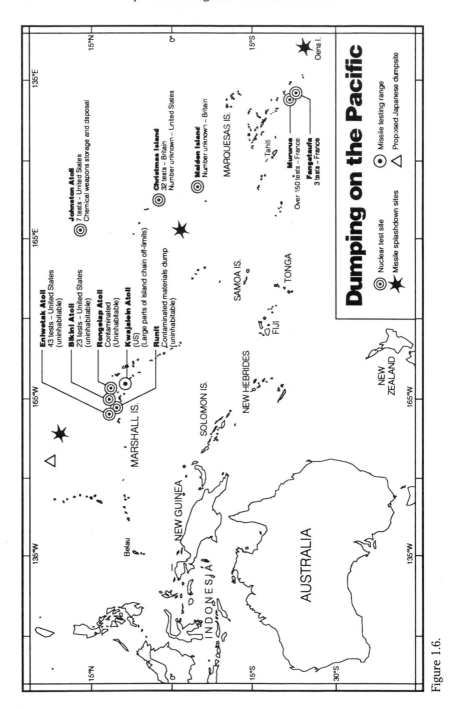

Figure 1.6.

full extent of the tragedy.[118] Most of the test sites remain highly radioactive and off-limits; the British and Australian governments remain reluctant to commit themselves to cleaning up the sites.[119]

Since the Second World War, the Pacific has become the dumping ground of first (and last) resort for the US military, which owns 167 military bases and installations throughout the region. In 1983, the US appropriated one atoll, Kwajalein, the world's largest coral atoll, as a missile testing range; MX missiles are fired at the atoll from the Vandenberg Air Force Base in California, 4,200 miles away, to test the accuracy of the "delivery system." As one Pentagon official remarked in 1985, "If we didn't have Kwajalein, we wouldn't be able to test such long-range stuff over open, largely uninhabited areas of the earth's surface. So it's important to have a place like this in the middle of the Pacific Ocean."[120] Many of the residents of Kwajalein have been displaced and resettled to barracks on a densely populated island called Ebeye; the US military has restricted access to the lagoon, which the Kwajaleinese used to use for fishing, and to their 93 islands, which they used for agriculture. A US doctor assigned to Kwajalein reported to a US congressional committee that, "The Army's position was summed up for me one day when a high-level command officer at Kwajalein remarked that the sole purpose of the Army at Kwajalein is to test missile systems. They have no concern for the Marshallese and it is not of any importance to their being at Kwajalein Missile Range."[121]

The South Pacific is "important" to the US military for other reasons too. In 1990, mounting protests in Germany forced the US military to remove their stockpile of chemical weapons, mostly nerve gas, stored on German territory. The military transferred the deadly stockpile of 100,000 volatile and leaking containers to Johnson Atoll in the Pacific, where it plans to destroy them by incineration—despite the protests of the South Pacific Forum, an association of 15 Pacific states.

On July 1, 1946, the US began its "peacetime" nuclear weapons testing program in the Pacific by dropping a nuclear bomb on Bikini Atoll, part of the chain of Marshall Islands. It was the first of more than 200 American nuclear explosions in the Pacific between 1946 and 1962, 66 of which took place in the Marshalls.

Enewetak Atoll, west of Bikini, has suffered 43 nuclear bomb blasts. In 1982, in an unsuccessful effort to make the Enewetak islands habitable again, the US Army scraped the radioactive topsoil off the southern islands, dumping it into a bomb crater on nearby Runit Island. A 370-foot-wide, 25-foot-high, and 18-inch-thick concrete dome covers the dump. Conservative estimates are that Runit Island will have to remain "off-limits" for human life for the next 25,000 years; the "protective" concrete dome has a life expectancy of only 300 to 1,000 years, and is already cracked.[122]

Over the more than four decades of nuclear testing, some Pacific islands have disappeared entirely ("vaporized" in military language), and many more have been rendered uninhabitable—and they will remain radioactive, poisoned, and uninhabitable for tens of thousands of years. The Bikini Islands, for example, were bombed in 1954 with a US 17-megaton hydrogen-bomb blast, code-named "Bravo," about 1,000 times the force of the bombs used at Hiroshima and Nagasaki; the islands have now been declared uninhabitable for 30,000 years. The inhabitants of Bikini, formerly self-sufficient islanders who were forcibly removed from their islands for what they were told was a temporary dislocation, have led a rootless existence for the past 45 years, shunted from island to island, and have ended up subsisting largely on surplus US Army food, living in grim barracks-like shelters on Ebeye.

The inhabitants of Rongelap, about 100 miles away from Bikini, suffered direct fallout from the 1954 "Bravo" tests, and there is mounting suspicion that this was intentional—part of a military plan to have a pool of human experimental subjects to track long-term radiation effects. This suspicion was reinforced by the recent release of a 1958 US government health report on the Marshall Islands. In this report, Dr. Robert Conard, head of the study, noted that "greater knowledge of [radiation] effects on human beings is badly needed." Then, he described the opportunity presented by the Rongelapese: "Even though the radioactive contamination of Rongelap Island is considered perfectly safe for human habitation, the levels of [radioactive] activity are higher than those found in other inhabited locations in the world. The habitation of these people on the island will afford most valuable ecological data on human beings."[123] In the mid-1950s, immediately after the "Bravo" bomb blast, 33 percent of pregnancies among women who had been exposed to the fallout resulted in fetal death; fetal death rates still remain high. In 1966, 52 percent of the people on Rongelap Island who had been under ten at the time of the test were found to have thyroid cancer or abnormalities; twenty-two years after the test, 69 percent of the same group had developed thyroid tumors.[124] The Rongelapese live today with a health legacy of respiratory diseases, thyroid tumors, eye cataracts, miscarriages, stillbirths, grossly deformed fetuses that live only a few hours after birth, and genetic disorders.

Darlene Keju-Johnson, a Marshall Islander, describes the terror of life in the Pacific at the hands of the US military[125]:

One important date that I never forget was in the year 1946. In that year the navy official from the US government came to Bikini Island. He came and told the chief Juda—and I quote—"We are testing these bombs for the good of mankind, and to end all world wars."

When the navy official came it was too late. There were already thousands of soldiers and scientists on the atoll and hundreds of airplanes and ships in the lagoons. They were ready to conduct the tests. The Bikinians had no choice but to leave their islands, and they have never returned. The navy official did not tell the chief that the Bikinians would not see their home again. Today Bikini is off limits for 30,000 years. In other words Bikini will not be safe for these Bikinian people ever again.

The Bikinians were promised that the United States only wanted their islands for a short time. The chief thought maybe a short time is next week, maybe next month. So they moved to Rongerik.

Rongerik is a sandbar island. There are no resources on it. It was too poor to feed the people. We live on our oceans—it's like our supermarket—and from our land we get breadfruit and other foods. But on Rongerik there was nothing. The United States put the Bikinians on the island and left them there. After a year they sent a military medical official to see how they were. When he got there he found out that they were starving. Imagine; move someone else from their own home, by your power. Dump them on a little sand. And don't even bother to go back and see how they are doing for a year.

The people of Bikini have been moved, or relocated, three times. The people of Enewetak Atoll were also relocated. You cannot imagine the psychological problems that people have to go through because of relocation.

In 1954 the United States exploded a hydrogen bomb, code named BRAVO, on Bikini. The Marshallese were never told about this bomb. We were never even warned that this blast was about to happen on our islands. Instead we experienced white fall-out. The people were frightened by the fall-out. The southern area of our islands turned yellow. And the children played in it. But when the fall-out went on their skins, it burnt them. People were vomiting.

The people of Rongerik and Utirik were not picked up until three days after the explosion. It was horrible. Some American soldiers came and said, "Get ready. Jump in the ocean and get on the boat because we are leaving. Don't bring any belongings. Just go in the water." . . . They were taken to Kwajalein. It took one night to get there. They didn't even give the people a change of clothing, so it meant they had to sleep in their contaminated clothing all the way. You imagine. They are burnt, they are vomiting. When they got to Kwajalein they were given soap and were told to wash in the lagoon. The soap and salt water was supposed to wash off the radiation. They were not told what had happened, what was wrong with them. Their hair was falling out, finger nails were falling off . . . but they were never told why. . . .[An American serviceman] has told us that the United States knew that the wind was blowing towards islands where people lived, but that they went ahead and tested anyway. It was not a mistake. It is interesting that the United States government moved the Marshallese in the 1940s when the small bombs were tested, and then when the biggest bomb ever was tested the Marshallese were not even warned.

Since the testing there has been a tremendous increase in health problems. The biggest problems we have now, especially among women and children, is cancers. We have cancers in the breast. We have tumor cancers. The women have tumors in their private places. Children are being deformed. I saw a child from Rongelap. It is an infant. Its feet are like clubs. And another child whose hands are like nothing at all. It is mentally retarded. Some of the children suffer growth retardation.

Now we have this problem of what we call "jelly-fish babies." These babies are born like jelly-fish. They have no eyes. They have no heads. They have no arms. They have no legs. They do not shape like human beings at all. But they are being born on the labor table. The most colorful, ugly things you have ever seen. Some of them have hairs on them. And they breathe. This "ugly thing" only lives for a few hours. When they die they are buried right away. A lot of times they don't allow the mother to see this kind of baby because she'll go crazy. It is too inhumane

Many women today are frightened of having these "jelly-fish" babies. I have had two tumors taken out of me recently and I fear that if I have children they will be jelly-fish babies also. These babies are being born not only on the radioactive islands but throughout the 35 atolls and five islands of the Marshalls.

It is not just the people who have been affected but also our environment. For example, we have breadfruit—it is like potato. Instead of green and healthy-looking, it looks muted. It is deformed. Just like our infants. A lot of these foods are no longer edible and we don't know why. . . .

And as if it wasn't enough for us Marshallese to live under this reality the United States decided to use Kwajalein Atoll for missile testing. Once again our people were relocated by force. Two-thirds of the Kwajalein lagoon, which is the largest in the world, was taken for missile testing and the people shoved off from the many islands on to tiny Ebeye. Ebeye is only 66 acres. Today there are more than 8500 people living on that island.

There are many problems on Ebeye. . . . The people on Ebeye have to survive on canned foods, rice and bread. We can't eat our traditional food. We have a problem with malnutrition. Children are not healthy because their diet is very poor.

A lot of Marshallese are fed up with the DOE and the United States government. . . . We are only small—a very few thousand people out there on tiny islands, but we are doing our part to stop this nuclear madness.

According to a 1980 report by the UN International Commission on Radiological Protection, the combined effects of the American, British, and French nuclear tests will kill more than 15,000 people in the Southern Hemisphere.[126] Andrei Sakharov, the creator of the Soviet hydrogen bomb, has made an even more startling estimate that for every megaton of nuclear weaponry exploded in the atmosphere, with fallout spread worldwide by rain and wind, 10,000 people will die prematurely of cancers they would not otherwise have developed.[127]

The military presence in the Pacific is not only directly causing death, disease, and destruction—it is disrupting traditional lifeways, destabilizing social relations, and is most notably changing gender relations in island societies. A report commissioned by the US Air Force in 1974 in anticipation of opening a major base in the Marianas Islands, just now released, explores the scope of the problems generated by a large military presence superimposed on small, tightly knit indigenous societies:

> The homogeneous closely-knit community has contributed to years of peaceful existence on Tinian [in the Marianas], one devoid of major crimes. . . . with the development of military facilities the breakdown of family ties, personal conflicts, and social problems because of urbanization, competition, and cultural transition will increase. . . . The impact on public safety will be felt because of the anticipated increase in bars, drinking, fights, and the possible introduction of new elements such as drugs, prostitution, and other major criminal activities.[128]

Prostitution, public safety issues, drinking, and the breakdown of family ties all affect women in very specific and distinctive ways, and it is women who bear the brunt of militarized social dislocation. Recognition of the links between a foreign military presence, exploitative prostitution, sexist socioeconomic segregation, the deterioration of public safety, and environmental degradation has mobilized women's antimilitary movements throughout the Pacific.

Women in the Pacific are mindful of Rongelapese "jelly-fish babies," Filipina prostitutes, and the slum-like conditions created on Ebeye, and it is women who are the principal movers behind the decade-old movement for a nuclear-free Pacific. The women of Belau (also called "Palau" in some news reports) have waged the most prolonged protest against the militarization of the Pacific, but they have paid a huge price for their activism. The nation of Belau, with a population of 15,000, is a small chain of islands 500 miles east of the Philippines in the South Pacific. In 1979, Belauans wrote and voted overwhelmingly for a constitution that, among other things, barred nuclear weapons and materials from its territory. This outraged the American government, which views Belau as a "strategic" link in its Pacific "defense arc"; the US has since forced nine plebiscites on Belau in an effort to overturn the anti-nuclear provisions of the constitution. Each plebiscite has failed. The US government is now proposing, in lieu of the constitution, a "Compact of Free Association" between the US and Belau. The Compact would give Belau 50 years of American aid, in exchange for which the American military would have rights to one-third of Belau for military bases, including a 30,000-acre jungle warfare training base, two military airfields, ship anchorages, and a 2,000-acre storage base for munitions.

Women elders are in the forefront of the Belauan antinuclear independence movement. Belau is a matriarchal society in which women exercise considerable influence, and as Gabriela Ngirmang, one of the elders, modestly explained, "Many people were afraid to speak out against the American plan, especially people who work for the government. They were afraid they'd lose their jobs. So we were just a few people who were able to stand up and speak out."[129] As each plebiscite has failed, pressure on the Belauan antinuclear activists has increased. The controversy has now felled two Belauan governments, and violence, once virtually unknown on Belau, has become commonplace. All of the 22 women elders who filed suit against the last plebiscite have been physically threatened; Gabriela Ngirmang's house was fire-bombed; the father of another activist was murdered. Lorenza Pedro, a teacher from Belau, summarized the hopes of the Belauans: "I question why the US uses these tiniest islands in the Pacific to defend this great big country. Are the American people so important but not Belauans? . . . The US is a great big country and we Belauans are just a tiny island which we are grateful for and we wish that the US will leave us in peace."[130]

While the Marshallese continue to wrestle with the consequences of nuclear testing, a new proposal is on the table that will make the islands a dumping ground for American garbage—literally.[131] An American waste disposal company, Admiralty Pacific, proposes to ship household waste from the west coast of the US to the Pacific islands—an estimated 34 billion pounds of waste in the first five years of the program alone. Although "household waste" sounds benign, even ordinary garbage contains detergents, bleaches, acids, and other chemicals of high toxicity; Greenpeace estimates that approximately 125 million pounds of acutely hazardous wastes can be expected in the first five years' shipments to the islands. This plan is backed by the Marshallese president, Amata Kabua, largely because it would bring in needed foreign exchange. The other perceived "benefit" of the plan, though, is that the garbage would create landfill—which is desperately needed in the low-lying Marshalls, likely to be swamped with the sea level rises anticipated with global warming. In a terrible irony, one environmental tragedy is thus providing the leverage for another.

The Japanese government has also approached the Marshall Islands' president with plans to dump its nuclear waste in either the Bikini or

Enewetak lagoons.[132] By the beginning of 1990, Japan had accumulated a million drums filled with nuclear reactor waste—and nowhere to dump them at home. The rationale for using the Marshalls as a dumping ground is that "the poisons are already there. The uninhabited islands are forever contaminated. Why not turn a liability into an economic asset?"[133]

◖

Antimilitarism in the past four decades has been fueled by fears of the ultimate threat posed by militaries as harbingers of nuclear war. Although the threat of global nuclear war appears to have diminished in the early 1990s, this does not mean that the environmental threats of militarism have receded. The militarized threat to the planet—the power of militaries to render our physical support systems inoperable, the power to transform living environments into poisoned, uninhabitable, unsustainable places— is a present and growing danger. For clusters of cancer victims in the American Midwest, deformed children in the Pacific Islands, slowly dying veterans in Southeast Asia, and poisoned peasant women in Central America, it is a present reality.

Considered singly, military environmental transgressions and the human tragedies they spawn around the world may appear to be random, unpredictable, even accidental; they may appear to be isolated incidents, or the fault of irresponsible individual base commanders and plant managers; they may appear to be inevitable, essential, unavoidable. But they are none of these things. Militaries are keenly aware of the environmental and health consequences of their actions—in many instances, the consequences are provoked specifically as part of a grand military plan. The complacency and disinterest with which militaries regard their environmental transgressions hang on a scaffolding of military privilege—a privilege that, in turn, is deeply rooted in a structure of militarized masculinity.

Chapter 2
Business as Usual

The Train of Progress: A Fable

There is a modern environmental fable about in the land. Simply put, it is this: that environmental degradation is a silent and *inevitable* companion of industrialization—a caboose hitched on the train of progress that no one can unshackle. If we want refrigerators, modern medicines, cars, computer chips, and polyester (and who does not?), then the price we pay is in toxic by-products, industrial waste, chemical effluent, environmental degradation, and occasional spectacular disasters. It appears to be a pretty straightforward pairing, a coupling that is somehow simultaneously the responsibility of each and every one of us, and yet the responsibility of none. This train that we're all on, with its unwanted and unshakable caboose, is fueled only by the press of consumer demand, we are told. The men who run the huge multinational corporations that run the world's largest three "smokestack industries" (oil, cars, and chemicals) only steer the train, they say; they didn't lay the rails, and they certainly didn't hitch up the toxic caboose. No one knows who did.

In the real world, this fable plays itself out in deadly ways. There are places in the world today where people routinely wake up gasping for air, where the paint on cars and houses peels away from chemical corrosion, where cows drop dead from drinking the water in drainage ditches around factories, and where this is considered to be normal. Or, at least, is taken to be the ordinary price one pays for needing a job in and trying to live in an industrial or industrializing, chemically dependent, technologically sophisticated culture. This is the price, we are told, of living in an age of modern miracles.

People in West Virginia live downwind from modern miracles. The Kanawha Valley in West Virginia is home to a dozen major chemical factories owned by the giants of the industry—DuPont, Monsanto, Union Carbide, among others. It is dubbed "Chemical Valley" by some locals, "Cancer Valley" by others.[1] Business is always good at the auto-body shops in Kanawha Valley—car surfaces, pitted by chemicals, need constant upkeep. One auto-body shop owner says that the local chemical companies employ insurance adjusters to quietly settle claims from residents with damaged autos, at about a rate of 100 cars a month. A couple of years ago, he says, a single emission from one plant led to claims by more than 300 people. As for the effects of the ubiquitous chemical drizzle on people, he says, "That's a good question. Maybe it makes you die young."[2] This casual observation in fact appears to be the case. State health department records, for example, show that between 1968 and 1977, the incidence of respiratory cancer among Kanawha Valley residents was more than 21 percent above the national average;[3] a 1981 survey in a neighborhood not far from the plants uncovered a cancer rate twice the national average.[4]

Kanawha Valley is not unique. The "chemical corridor" of Louisiana, between New Orleans and Baton Rouge, is home to 25 percent of the US chemical industry; a Greenpeace survey team found "the highest rates [in Louisiana] of mortality from all diseases, cancer deaths, and infant deaths" along this 150-mile stretch.[5] Industrial "death valleys" dot the global landscape—places like Bhandup, outside Bombay, and Cubatao, near Sao Paulo. These places, where the chemical factories and manufacturing plants of dozens of industries cluster together, have rates of infant mortality, birth defects, childhood leukemia, and adult cancers that skyrocket off the charts. Other places, like Seveso, Italy, and Times Beach, Missouri, are so poisoned by the detritus of industry that no one can live there anymore, and may not be able to for hundreds of years. The Vistula River in Poland is so polluted that it is virtually a dead river. Breathing the air in Krakow is a health hazard. Twenty million Europeans get their drinking water from the Rhine River, along which about one-fifth of the world's total chemical production is located. The Rhine's industrial and chemical effluent belches into the North Sea, one of the most severely polluted bodies of water in the world. A no-man's land of environmental blight surrounds the refineries of Sudbury, Ontario, Teesdale, UK, and the chemical belt of New Jersey. A map of the major anthropogenic environmental disasters of the last century is also a map of industrialization, and as industrialization reaches into new centers in Brazil, South Korea, and India, environmental degradation follows in its wake as inevitably as a caboose follows a train.

People around the world are sickening and weakening and dying from the air they breathe, the water they drink, the food they eat. Almost 6 million new cancer cases each year are recorded by the World Health Organization. More than four and a half million people die of cancer in the world each year, a Hiroshima every month. No one really knows how many of these cancers and illnesses are directly caused by environmental derangement, but the limited analysis available to us seems to be the visible tip of an invisible iceberg.[6] In a 1985 congressional survey, American chemical companies admitted that they annually leak or vent 196 "extremely hazardous" compounds into the air. The US Environmental Protection Agency (EPA) cautiously estimates that as few as 15 to 45 of the hundreds of released air toxics directly cause up to 1,700 cases of cancer each year.[7] American industry alone generates annually 280 million tonnes of lethal garbage and 10.3 *billion* pounds of toxic chemicals that are spewed each year into the air, discharged into public waters, and flushed into the sewers[8]—enough to fill 8,000 Love Canals. The countries of the European Economic Community turn out more than 30 million tonnes of toxic waste each year.[9]

The leaders of the corporate world maintain that they are not responsible for the environmental toll of industry. In truth, environmental accountability *is* complex, because "fault" is often shared by many and diffusely apportioned. But the underlying claim of innocent corporate impotence in determining our environmental fate holds no water at all. With even passing consideration, it is clear that industry-generated environmental destruction is seldom, if ever, an inevitability; it is *not* a necessary, no-choice-about-it side effect of industrial "progress." The fable is simply that, a fable.

This truth is thrown into sharp relief on those rare occasions when corporations are forced by legislation or public outcry to clean up their environmental act—only then does it become clear that, for example, simple pumps *can* easily replace aerosol-propelled sprays, that smoke-stack "scrubbers" *can* eliminate most sulphur dioxide emissions, that most of the toxins produced by industry *can* be recycled, reused, recaptured, neutralized, or eliminated altogether from the industrial process. It becomes clear that the decision to produce, test, and dispose of the products and by-products of industry in an environmentally negligent manner is, in fact, in most cases *a decision*.

Despite the vague awareness that environmental problems may derive from specific and identifiable business decisions, it is easy for environmental problems to appear to be orphans. The trail of decisions that leads from corporate boardrooms and drawingboards to toxic dumps and environmental catastrophe is often impossible to follow. By the time environmental damage becomes evident, it is often far removed from the decision-making source, both in time and geography, and so distant from the corporate parent that there appears to be little connection between the two. But this is not always the case, and the truth that environmental destruction is *not* an inevitability becomes most apparent when corporations are discovered to have acted with intentional and premeditated environmental recklessness.

Designer Tragedies

The leaders of industry claim innocence when confronted with the litany of environmental calamity. But they have a knack for destroying their own case.

• The use, or as some argue, the misuse, of agricultural pesticides causes widespread ecological disasters. Poisonous runoff from pesticides has almost destroyed Thailand's fish-breeding industry; DDT has contaminated much of the water table and the food chain in El Salvador; almost a million people a year are poisoned by pesticides, most in the Third World. Consumer advocates point out that chemical companies use the Third World as a dumping ground, routinely selling and manufacturing there highly dangerous and unnecessarily toxic chemicals that have been banned in industrial nations.[10] In reply, agrobusiness and chemical companies blame most of the pesticide problems on "ignorant farmers" whom, they say, misuse the chemicals, don't read or understand the directions, or apply and store chemicals improperly. Companies such as Ciba-Geigy, a Swiss-based chemical conglomerate and one of the largest chemical manufacturers in the world, have called for fairer treatment of the agrochemical industry.

This debate took a sharp turn when, in the early 1980s, a Swiss television crew uncovered evidence that Ciba-Geigy had intentionally sprayed unprotected Egyptian children with a new insecticide they were introducing, "Galecron." The purpose of the experiment apparently was to see how much of the insecticide was retained in the human body and how much was excreted.[11]

• In the 1970s "DBCP" was a popular pesticide, produced by several large American chemical companies for use on pineapples, bananas, and citrus fruits. In 1977, the three largest chemical companies stopped producing DBCP after the California legislature banned its use in the face of strong evidence that it caused sterility and was a probable carcinogen.

A smaller chemical company, Amvac, not daunted by the grim medical verdict, saw this as a profitable opportunity in the chemical market and jumped in to fill the marketing void. The annual report of Amvac for 1977 explained the company's decision to take up production of DBCP:

> Management believes that because of the extensive publicity and notoriety that has arisen over the sterility of workers and the suspected mutagenic and carcinogenic nature of DBCP, the principal manufacturers of the product have, temporarily at least, decided to remove themselves from the domestic marketplace and possibly from the world marketplace. . . . *Notwithstanding all the publicity and notoriety surrounding DBCP, it was [our] opinion that a vacuum existed in the marketplace that we could temporarily occupy. . . . [we] further believed that with the addition of DBCP, sales might be sufficient to reach a profitable level* [emphasis added].[12]

Examples of blatant and intentional disregard of environmental common sense, and resulting environmental horror stories, are easy to find in the annals of the *chemical* industry because these corporations deal almost exclusively in toxic materials—and produce 70 percent of all industrial toxic effluent in the US. But other industries do not lag far behind.

• Over the next 30 years, 240,000 people—8,000 per year, almost one every hour on average—will die from asbestos-related cancer. For fifty years, the Manville Corporation (formerly Johns Manville), the largest American manufacturer of asbestos knew of the mounting evidence that exposure to asbestos was a threat to human health. For fifty years, Manville executives suppressed documents, lied about health profiles, and even cut off workers from their own health records. "As long as the employee is not disabled," rationalized the company's medical director in 1963, "it is felt that he should not be told of his condition so that he can live and work in peace, and the company can benefit from his many years of experience."[13]

Now that asbestos production has been restricted throughout much of Europe and North America, multinationals have simply shifted production and markets to the Third World. The asbestos manufacturing industry thrives in countries such as India, South Africa, Mexico, and Brazil, where asbestos products are made in factories managed by, supplied by, and financed by the same American and British firms that have been run out of the home market.[14]

• In 1949, the General Motors car manufacturing company (GM) was convicted of "conspiracy to destroy American mass transit systems." This they did by pursuing an aggressive program of buying up electrical trolley systems in urban areas around the country—which they then dismantled, thereby forcing increased reliance on private automobiles.[15] The environmental consequences of this deliberate act of sabotage plague us today: Los Angeles, which in the 1930s boasted an efficient system of electrified public transit that linked together 56 communities, saw the system destroyed and replaced with a freeway system populated largely by GM cars. Los Angeles today has one of the worst air pollution problems in the US.

• Rockwell International Corporation, under contract to the Department of Energy (DOE), runs the Rocky Flats nuclear weapons plant near Denver, Colorado, a facility that produces the plutonium "triggers" for atomic bombs. In 1989, Rocky Flats was designated the most environmentally hazardous site in the nuclear weapons industry. The list of environmental violations at the plant is staggering: toxic chemicals were routinely discharged into creeks leading to drinking water supplies for Denver; radioactive wastes were routinely mixed with other oils and solvents; a wastes incinerator, which plant officials said had been shut down for safety reasons, was operated illicitly at nighttime; workplace safety violations included leaky "glove boxes" (used for manipulating plutonium), out-of-date fire alarm systems, and frequent accidents. An epidemiologist found eight times more brain cancers among Rocky Flats workers than would otherwise be expected;[16] there has been an estimated dispersion of plutonium and other radionuclides throughout the Denver area equivalent to 10 Nagasaki-type bombs; prospective homebuyers in the Rocky Flats area are provided with an "Advisory Notice" from the Department of Housing and Urban Development warning of "varying levels of plutonium contamination of the soil . . . in parts of Boulder and Jefferson Counties."

The plant had a history—known to DOE officials—of fires, faulty equipment, emissions accidents, and fraud. A lengthy investigation in the late 1980s established that managers at the plant had, for years, permitted the improper handling of wastes, and in fact had falsified health and safety records for the plant—a Justice Department affadavit in 1989 concluded that, "There was probable cause to believe that Rockwell and Energy Department officials knowingly and falsely stated Rocky Flats' compliance with environmental laws and regulations and concealed Rocky Flats' serious contamination."[17] In 1987, the DOE rewarded Rockwell with an $8.6 million bonus for "excellent management;" in 1989, executives at Rockwell were threatening to close the plant (considered "vital" to national security) unless the government guaranteed them immunity from criminal and civil prosecution of laws governing waste disposal.[18]

There are countless other examples of environmental outrages perpetrated knowingly and apparently with forethought by corporate decision-makers. The stories are, by now, generic tales in the popular culture of industrial countries, into which interchangeable corporate particulars can be plugged: The manager of a chemical factory who, acting with the tacit assent of his bosses, falsifies records that show routine dumping of chemicals directly into public sewers . . . The periodic, industry-wide "adjustments" in standards for recording worker-injury rates in electronics assembly plants . . . The officials at a privately contracted weapons production plant who deny for years the rumors of routine releases of radioactive materials into the air and water, and who then deny accountability when the truth of these releases is uncovered through the dogged persistence of a small watchdog group.

There is no denying that these egregious environmental abuses occur with alarming frequency. The problem is how to interpret these corporate excesses. Simply put: are they aberrations, exceptions to the rule, or are they in fact part of the normal "business as usual" continuum? If acts of flagrant disregard for environmental integrity are aberrations from the norm in corporate/industrial behavior, then there is little we can do to protect ourselves from the environmentally destructive acts of a few unethical factory owners and corporate managers. If, however, environmentally threatening acts lie on the continuum (perhaps on the extreme end, but albeit still on the continuum) of sanctioned "business as usual" practice, then the course we need to follow to protect ourselves from environmental deterioration is more pointedly clear.

Feminist Parables

Women have had cause to develop finely-honed instincts about when a behavior is an aberration and when it is an extension of the norm. Nowhere has this debate been more salient than in the formulation of explanations of male violence against women: is wife battering (or rape or incest) perpetrated by a few crazed, "abnormal" men, or is it part of a larger pattern of misogyny into which most men are socialized?

This debate was galvanized by the mass murder of 14 women engineering students in Montreal in December 1989. The horrible details flared briefly into national and international prominence: on a bright afternoon in December, Marc Lepine, a quiet and generally unremarkable young man, donned hunting gear and a semiautomatic rifle and stalked the halls of the University of Montreal, singling out women students and gunning them down wherever he found them, shouting epithets of "feminist" at his victims. When his half-hour rampage was over, 14 women were dead and dozens more were injured.

Within hours of the shooting, Lepine was described as a "madman," a "lunatic" and a "psychopath." There is no question that he was, by definition, all of those things. But feminist commentators, given a rare platform in the mainstream press, tried to break through the circular reasoning that dismissed Lepine as a lone psychopath, and that cast his mass murder as a tragic, random, single, unpredictable event. Rather, feminist analysts located his acts on the continuum of violence (and the threat of violence) that women face from men every day. Many women, reacting to the slayings, shared an understanding that it was dangerous to treat the massacre as an isolated incident: "The massacre may have been the isolated work of a single psychopath, but it is also a horrifying extension of something that goes on between men and women every day."[19]

Was the stalking murder of women by Marc Lepine the ephemeral, inexplicable act of a crazed individual? Or are his actions explicable within the context of the "ordinary" culture of violence against women that imbues our daily lives? Feminist analysis clearly suggests the latter. The debate over extreme acts of violence suggests an analytical context for understanding acts of extreme environmental destruction. Were the executives at Ciba-Geigy who approved the pesticide spraying of Egyptian children simply amoral individuals? Were the managers at the Rocky Flats weapons production plant who knowingly allowed radiation to leak from the plant for years on end, polluting the groundwater for miles around, simply "bad managers"? Are their actions to be understood as the unusual excesses of a few misguided individuals, actions that lie outside the norms of conventional business practice, or are their actions embedded within a corporate culture that nurtured and promoted and rewarded these individuals for their actions—up until the moment they were caught?

A radical analysis, and certainly a feminist analysis, of environmental calumny in the business world suggests the latter. (Jumping from a consideration of violence against women to "environmental violence" is not just an analytical convenience. As many ecofeminists point out, violation of the environment may be seen as analogous to and a continuation of violence against women. Many women point out the that the degradation of woman-identified Nature by male-dominated industrial enterprises stems from the same social structure that condones violence against women.)

PROFIT AT WORK

Many analysts have tried to explain why corporations act so consistently with flagrant environmental disregard. Most conventional analysis focuses on the workings of free market economics and "the profit motive" in explaining corporate environmental "villany." This is a necessary but not sufficient explanation. We *do* need to understand the workings of economic imperatives in corporate business. Having done so, though, we need to then extend our curiosity into the *cultural* heart of corporate life.

Individual corporate officers do not necessarily make decisions to intentionally cause environmental damage (although sometimes they do, as the record shows). The leaders of the corporate and industrial sectors who make the decisions that end up poisoning whole valleys and cities around the world do not necessarily start off with that intent. Rather, the decision-making process in industry typically has nothing whatsoever to do with environmental issues—and that is both the beginning and the end of the problem.

All modern corporations are driven by the profit motive.[20] Most decisions made in most corporate boardrooms focus *exclusively* on profit. A successful CEO ("chief executive officer") is defined as one who maximizes profit for the corporation, and the decisions he makes about manufacturing costs, manufacturing processes, labor relations, factory siting, and promotional strategy are all designed to increase profitability—if they aren't, he's out of a job, and the corporation may be out of business. An unspoken and often overlooked corollary of the singular focus on profitability is that the maximum-profit choice often carries a high environmental price tag. The "convenience" of using environmentally dubious practices often enhances industrial profitability in the first place, and corporate managers are well aware of the short term cost-benefits of evading environmental responsibility. It often appears to be cheaper (and therefore more profitable) to dump toxic effluent into the nearest river than it is to build purification plants or to design a cleaner manufacturing process in the first place. Both by omission and by commission, the corporate decision-making process favors environmentally destructive patterns of industrial enterprise.

The fact that profit is the primary mechanism for economic planning has many environmental consequences, some of which can be summarized quickly:

• It determines which technologies flourish and which die. In *The Closing Circle*, environmentalist Barry Commoner describes how natural organic products such as soap and cotton were replaced after World War II with unnatural synthetic ones such as detergents and plastics. In almost every case the new technology damaged the environment more than the old one, but in each case the new production process was also more profitable.[21]

• The drive for a fast return on investment leads to more rapid depletion and extraction of natural resources, even though a slower pace would cause less damage to both the environment and to worker health. Mining industries are particularly liable to favor quick extraction methods which would be hampered by stringent environmental safety regulations.

• The profit imperative, combined with lax regulation, leads to inadequate expenditures on pollution control. Toxic wastes are often disposed of in the least costly way possible; most of the major industrial companies, worldwide, have been caught using illicit disposal techniques and fly-by-night disposal firms.[22]

• Taking profit through environmentally risky behavior spawns secondary profit-making industries aimed at containing the damage of the primary ones; there is a self-sustaining circle of mutual profiteering. Thus, for example, huge "waste-management" and "cancer research" industries have emerged. Cancer is a serious, global problem. But, obviously, it is not because scientists have yet to find the cure for cancer that cancer proliferates. Rather, as one feminist notes, cancer proliferates because very little is done to reduce the burden of stress and poisons that modern industry produces:

> It is far more profitable to continue cancer research and violate more lives . . . than to eliminate the major carcinogenic materials that pour out from the vast industrial complex. . . . Scientists are said to have made great progress in "understanding" cancer by reducing it to genetic factors. They hope to devise ways to block it genetically, thus preventing the development of cancer or curing it once it is detected. This biological, cellular intervention completely begs the question of the environmental sources of cancer.[23]

Many of the largest polluters in the chemical industry own pharmaceutical and research divisions that are the major beneficiaries of cancer-research funds. Critics point out that the direction of cancer research is misguided—perhaps intentionally so. Most of the money in cancer research goes toward treatment; and yet, most of the major diseases of the modern world (malaria, typhus, tuberculosis, among others) have been controlled primarily through prevention, not treatment. Cancer-*prevention* strategies that would include cleaning up the workplace, cleaning up the environment, and changing consumer-product manufacturing are not in the interests of industry.[24]

• Corporations often evade the costs of cleaning up environmental waste. It is an expensive undertaking for local governments or regulatory agencies to identify and prosecute corporations who may be polluting towns and countries far removed from their headquarters. Even if corporate culprits are identified, it is often impossible to hold a company legally liable for cleanup—if the site has since been sold to another party, or if the company contracted with a waste disposal firm to haul away their toxins, then the company is often immune from legal liability. The Love Canal debacle offers a dramatic case in point: Hooker Chemical spent $1.7 million to dispose of its waste in an abandoned canal in upper New York State that the company then sold to the local school board; nearly 25 years later, when the Love Canal residents finally forced the state and federal governments to clean up the site, the cleanup bill which came out of public tax funds exceeded $61 million.

The difficulties of pursuing legal liability are compounded when dealing with multinational corporations, who often locate their most polluting facilities in Third World countries whose governments do not have the resources to enforce environmental accountability. The Bhopal disaster, discussed later, grimly illustrates this point.

• Profit-driven companies are constantly trying to stay one step ahead of the competition by introducing continuously "newer" and "improved" and "better" products. The petrochemical industry, for example, introduces hundreds of new pesticides and other chemical products into the marketplace each year, many of which are only cursorily tested for environmental safety. The rapid journey from laboratory to marketplace usually means that there is little time for realistic assessment of the environmental impact of the new product, let alone a realistic assessment of whether the new product is really needed in the first place.

Because problems in a corporation are subjected to the profits test, the profit imperative *can* be a force for positive environmental action. If the profitability of freewheeling pollution is undercut by stiff environmental regulations and enforcement, or by "boomerang" environmental effects that entail costs that directly impinge on the corporation, then pollution-causing behavior can become an economic liability. There is recent evidence that corporations do change their environmental behavior when the costs of pollution exceed the benefits. But this is rare—environmental and health effects, because they may be subtle and exert their influence on segments of society that are invisible to corporate leaders, seldom enter the balance sheet of cost/benefit analysis.

The primacy of the profit motive in guiding corporate policy explains much of the environmentally irresponsible behavior of major corporations. Simply put, every dollar not spent on environmental protection is a dollar of profit (or at least appears to be in the short run). This simple but potent observation is not new—it has been made by a number of social and environmental analysts, perhaps most poignantly by Greenpeace activists who hung a banner on a smokestack in West Germany that said, bluntly,

"After you have cut down the last tree, poisoned the last river, and fouled the remaining air, you will find that you cannot eat your money."

The profit imperative is compelled by "free-market" economics. While the environment and the economy are tightly interwoven in reality, they are almost completely divorced from one another in economic structures and institutions. Economic accounting based on a GNP bottom line makes resource destruction economically invisible—as factories, textile mills, office buildings, and other artifacts fall into disrepair, a subtraction is made from national accounts to reflect their depreciation in value; no similar subtraction is made, however, for the deterioration of soils, air quality, and other natural endowments. Natural wealth of all kinds is whittled away with no losses appearing on the national accounts.[25] Moreover, conventional economic accounting actually counts as "pluses" expenditures on pollution remediation: the Alaskan oil spill of 1989, for example, actually created a rise in the GNP, since much of the $2 billion spent on the cleanup was added to income; much of the $40 billion in health care expenses and other damages incurred by citizens as a result of air pollution is counted on the plus side of the national ledger. Although the nation certainly would be better off had the oil spill never occurred and if people didn't suffer respiratory ailments from air pollution, the GNP suggests otherwise.[26] "Progress," as defined by our modern economic system rewards and perpetuates environmental deterioration.

MEN AT WORK

But if we end our environmental analysis by saying simply that "the free market/profit motive" explains environmental excesses, then we end with an abstraction. The "profit motive" explanation for outrageous corporate environmental violations is a mechanistic explanation—it explains what fuels the corporate engine, but it doesn't address the behavioral dimension of the problem. Environmental degradation on the scale we encounter today raises questions, at the least, about morality and behavior and socialization. How can corporate officers approve the pesticide spraying of unprotected children? How can plant managers contract with fly-by-night waste haulers when they know with certainty that their toxins are going to end up in the backyards of innocent people in Georgia or Nigeria?

Proponents of explaining environmental degradation as a logical consequence of capitalism are not very interested in root structures, nor in whether masculinity and feminity have anything to do with those structures. The capitalism-as-cause theorists take for granted that it is men who are the corporate executives, government officials, and workers; that is, they don't believe that it is useful to investigate or try to explain this male predominance. For the purposes of economic analysis, the CEO of General Dynamics or the defense minister of Kenya are not men; they are simply capitalists or the representatives of capitalists.[27] This sort of theorizing is disempowering and parochial.

Analyses of corporate structure that focus exclusively on economic explanations of corporate malfeasance (the "imperatives of capitalism") don't take into account the rich feminist literature on the *mutual accommodation* of capitalism and patriarchy. It is an inadequate analysis to say simply that "capitalism" compels environmental irresponsibility. We need to be more curious than that—"who are the capitalists?" and "what are the linkages between the power of men as leaders of capitalist enterprises and the power that derives from the broader exercise of male privilege?" Indeed, many feminists argue that capitalism is itself a project of patriarchy—that it is men who have created this particular system of economic behavior, one that protects and sustains their power *as* men.[28]

More simply put, an analysis wedded to purely economic explanation overlooks the fact that behind the corporate anonymity, behind the translucent screen of "the profit motive," and behind the staggering statistics about tons of lethal waste, miles of polluted rivers, and acres of blighted land, are real men (and a few women). There are real people, and not very many of them—after all, there are only about 9,000 corporate officers among the largest 500 American corporations—who make concrete day-to-day decisions that determine our collective environmental fate. The decision-making echelons of major corporations are almost entirely male. In 1987, only three of the American "Fortune 1,000" companies (the 1,000 largest industrial corporations) had a woman as chief executive, two of whom were widows of former corporate managers. Predictions are that by the late 1990s, women will hold no more than 2 percent of the top jobs at major corporations. Among Fortune 500 firms in a 1986 study, there were only 217 women (versus 5,889 men) on corporate boards, and 152 women corporate officers versus 9,048 men (1.7%). White males still hold 95 percent of the top management jobs at the country's largest corporations. The situation is remarkably similar around the world: women hold only 2 percent of senior executive posts in most West European countries, and a discouraging .9 percent in Japan.[29]

To understand our environmental crisis, we must understand the culture that nurtures and rewards these men; we must probe the culture that protects (and numbs?) them from the horrible reality that the price of their profit-maximizing success is the degradation of the air that we all need to breathe, the water that we all need to drink, and the planet that we all need to live on. And we must ask whether it is significant that the world of industry and the world of large-corporation policy makers is one noticeably populated almost entirely by men . . . that it is men who are the engineers of the train of progress.

Stalking the Corporate Culture

There is a large and profitable publishing subculture devoted entirely to books on business theory, managerial strategy, and the general theme of "what it takes to make it to the top in big business." Those of us who are not on the corporate ladder are aware of this literature only when a particularly fanciful title tweaks the public imagination, such as the best-selling *Leadership Secrets of Attila the Hun*.[30] As a genre, these books tend to be tedious and self-congratulatory, slow reading at the best of times, but they offer an intriguing glimpse of corporate culture, a world that is generally closed to outsiders. What emerges is a picture of an insular, self-referential corporate system.

Moral Bracketing   Large corporations are bureaucracies, and it is a commonplace observation that bureaucratic work shapes the consciousness of the people who work in them. Particularly, bureaucratic work necessitates people to bracket, while at work, the moralities that they might hold outside the workplace. Within the corporate world, strong convictions of any sort are suspect. As one manager says:

> If you meet a guy who hates red-haired persons, well, you're going to wonder about whether that person has other weird perceptions as well. You've got to have a degree of interchangeability in business. To me, a person can have any beliefs they want, as long as they leave them at home.[31]

The successful corporate "team player" is one who sublimates private morality to the shared value system of the corporation. Unless concern for environmental issues is explicitly structured into the shared value system of a corporation, questions of environmental efficacy will be interpreted as inappropriate intrusions of "personal values." And certainly this is the case if a worker expresses strong opinions, perhaps even emotional sentiment, about environmental issues.

Emotional Bracketing   One observer of corporate culture points out that "a sure sign of a company in trouble is emotional outbursts . . . such as denouncement of a company policy at a meeting."[32] In a successful

IT WON'T BE YOU, BUT THESE EIGHT MEN, WHO WILL DECIDE THE FATE OF YOUR CHILDREN.

EVERY DAY, we hear something new about the vital significance of the rainforest to the future well-being of our planet.

News stories describe the crucial role it plays in supplying the world with water and oxygen. Medical experts tell us about the recently-discovered, cancer-curing properties of its indigenous plantlife. Scientists explain the unique defense it provides against the "Greenhouse Effect," which may already be raising the earth's temperature to dangerous new levels.

We also hear that this irreplaceable ecosystem is being destroyed at the alarming rate of 75,000 acres a day. And we fear for the future of our planet. And our children.

And yet, all too often, we feel powerless to do anything about it.

The same, however, cannot be said of the eight men pictured above. In the next few months, these men will determine future actions on three separate fronts that, together, significantly affect not only the rainforest, but, ultimately, what kind of world future generations will inherit.

It is important that we understand the issues they face. And equally important that they know what we think about those issues. While there is still time.

I. WORLD BANK LOAN TO BRAZIL

The World Bank is presently arranging a series of loans to Brazil for up to $500 million. When the loan was originally proposed, much of the funding was slotted for the building of over 50 dams. If these dams are built, 16 million acres of rainforest will be trapped under water forever.

Earlier this year, public opinion (much of it from U.S. taxpayers, the largest contributors to the World Bank) forced postponement of this proposal. At this time, it appears that the loan might well become reality. What hasn't yet been resolved is whether or not the dams will be built, or how many.

Four of the men pictured above will wield considerable influence on what happens next. They include World Bank President Barber Conable and Brazilian President Jose Sarney, as well as President George Bush and Secretary of Treasury Nicholas Brady, who not only determine our country's vote, but set the tone for many others as well.

Voting on the final loan proposal could easily take place within the next few months, if not sooner.

II. THE PENAN RAINFOREST OF BORNEO.

The island of Borneo is divided into two countries which, together, share the vast Penan Rainforest.

In Malaysia, that rainforest is being destroyed, literally almost overnight, by a burgeoning logging industry. The Penan tribespeople, among the last hunter-gatherer societies left on the planet, have bodily blockaded logging trucks. But the Malaysian government is arresting the protesters, and Japan continues to import 60% of the forest's yield, in the face of over thousands of letters of protest from its own citizens.

In Indonesia, the Penan Rainforest is also being systematically stripped, at the rate of hundreds of trees a day. The raw timber is then converted to plywood and shipped to lumber companies in the United States.

Three of the men pictured above have great influence on the future of the Penan Rainforest. They are Shinroku Morohashi, President of Mitsubishi Corporation, one of the largest Japanese importers of Penan timber; T. Marshall Hahn Jr., Chairman and CEO of Georgia-Pacific Corporation, the principal importer of Penan plywood to the United States; and the Malaysian Chief Minister of Sarawak, YAB Patinggi Haji Abdul Taib Mahmud.

In a similar situation in Thailand, after the loss of much of their rainforest, a total ban on logging was enacted earlier this year. Without the same kind of ban in Borneo, the Penan Rainforest will almost certainly disappear within five years.

III. THE SCOTT PAPER COMPANY.

In the near future, this American-based company is planning to convert over 2,000,000 acres of Indonesian rainforest into a pulp and woodchip plantation for the manufacture of paper products. Future plans reportedly include a similar venture in Papua New Guinea.

The loss of this rainforest, like all rainforests, will be irreversible. In this case, at least 15,000 tribespeople — people who for centuries have relied on the rainforest for survival — will also lose their only home and way of life.

One of the gentlemen pictured above, Scott Paper CEO Philip Lippincott, will have great influence on how this project is finally enacted.

At present, Scott Paper Company says the final decision is two years away; but it has already established the first nursery for seedlings that would replace the existing forest. Concurrently, it has begun negotiations for an imported work force and machinery. An area of rainforest larger than Delaware hangs in the balance.

Between them, the eight men pictured above share the awesome responsibility of determining the future of much of the remaining rainforest on earth, and, consequently, the future of our planet and children.

Between all of us, we share the responsibility of expressing our opinions of what should happen next. And so, we urge you to write these men and speak our mind now.

We recognize that they find themselves under enormous pressure, facing the most complex and delicate of situations. But as men of reason and intelligence, they will listen.

And, at this important time, they deserve nothing less than full knowledge of what you think.

For more information about participation and contributions to the battle to save our rainforests, please contact:

Save The Rainforest/
Rainforest Action Network,
301 Broadway, Suite A,
San Francisco, CA 94133

1 BARBER CONABLE, President, World Bank, 1818 H Street N.W., Washington, D.C. 20433

2 NICHOLAS F. BRADY, Secretary of Treasury, U.S. Treasury Department, 1500 Pennsylvania Ave. N.W., Washington, D.C. 20220

3 PRESIDENTE JOSE SARNEY, Presidencia da Republica, Gabinete Civil, Placio do Planalto, 70150 Brasilia-DF, Brazil

4 GEORGE BUSH, President, The White House, 1600 Pennsylvania Ave. N.W., Washington D.C. 20500

5 YAB DATUK PATINGGI HAJI ABDUL TAIB MAHMUD, Chief Minister of Sarawak, and Minister of Resource Planning, Chief Minister's Office, Petra Jaya, Kuching Sarawak, Malaysia

6 T. MARSHALL HAHN JR., Chairman and CEO, Georgia Pacific Corporation, 133 Peachtree St. NE, 31st Floor, Atlanta, GA 30303

7 SHINROKU MOROHASHI, President and Director, Mitsubishi Corporation, 6-3 Marunouchi 2-chome, Chiyoda-ku, Tokyo 100-86

8 PHILIP LIPPINCOTT, Chief Executive Officer, Scott Paper Company, 1 Scott Plaza, Philadelphia, PA 19113

Figure 2.1. The emphasis here perhaps should be on the word *men*. (© Rainforest Action Network, by permission.)

company, the harmony of a shared corporate ethos is always perceived to be under imminent threat by "subcultural" values. There may be appropriate channels for dissent, but they are highly ritualized.

Personal concerns of any sort are generally not considered appropriate workplace issues. Researchers such as Lotte Bailyn at MIT, and others studying day-care and workplace-quality-of-life issues, conclude that, "Despite evidence pointing to the imperative nature of workers' domestic concerns, executives still behave as if caring for children or aging parents, or simply the desire for a more balanced life, represents 'special needs' that are marginal or even antithetical to the central goals of business."[33]

Distancing   The sheer impersonality of the vast markets that corporations serve also helps managers achieve the emotional distance and abstractness necessary for their roles. A chemical executive reflects on the possible harm that one of his chemicals might produce:

> It gets hard. Now suppose that the ozone depletion theory were correct and you knew that these specific fifty people were going to get skin cancer because you produced chlorofluorocarbons (CFCs). Well, there would be no question. You would just stop production. But suppose that you didn't know the fifty people and it wasn't at all clear that CFCs were at fault, or entirely at fault. What would you do then?[34]

Another executive at the same company echoed a similar sentiment:

> Certainly no one wants to significantly damage the environment or the health of individuals. But it's a different thing to sit and say that it's OK for twenty people out of one million to die because of chlorinated water in the drinking water supply when the cost of warding off those deaths is $25 million to remove the halogenated hydrocarbon from the water. Is it worth it to spend that much money? I don't know how to answer that question as long as I'm not one of those twenty people. As long as these people can't be *identified*, as long as they're not *specific* people, it's OK. So you put a filter on your own house and try to protect yourself.

Managerial distancing from the consequences of decisions is also achieved through the elaborate use of a linguistic code marked by emotional neutrality. Instead of talking about firing people, managers typically talk about "housecleaning," "redundancies," or "census reductions." It is widely remarked that sports metaphors, especially football metaphors, sprinkle the business lexicon—managers "take the ball and run with it," perhaps while someone "runs interference" for them, while others get "sidelined." Euphemisms dominate business discourse. In the textile industry, cotton dust becomes an "airborne particulate," and brown lung disease a "symptom complex." In the chemical industry, spewing highly toxic hydrogen fluoride into a neighboring community's air is characterized as "a release beyond the fence line." Environmental disasters become "incidents." For a time, the EPA called acid rain "poorly buffered precipitation." These abstractions and euphemisms, while not necessarily used with the intent to deceive, do in fact obfuscate issues, deflect criticisms, and allow managers to grapple dispassionately with problems that otherwise could generate high emotions.[35]

Hierarchies   Most corporations are tightly structured as a series of hierarchical plateaus. To succeed in this system, to climb from one plateau to the next, a subordinate must *be* subordinate—always follow the boss's lead, and never contradict the boss in public forums. The higher one climbs on the corporate ladder, the more crucial it is to establish oneself as a reliable defender of the corporate good:

> High-ranking executives "go to the mat" with one another striving for the CEO's approval and a coveted shot at the top. Bureaucratic hierarchies, simply by offering ascertainable—but scarce—rewards for certain behavior, fuel the ambition of those who are ready to subject themselves to the discipline of . . . their organization's institutional logic, the socially constructed, shared understanding of how their world works.[36]

A major MIT study of "industrial productivity" concluded that one of the major weaknesses of the American corporate system is that "steep hierarchical ladders and organizational walls" inhibit communication and cooperation—both within organizations, and between corporations and their consumer base.[37]

Enemies Everywhere   A primary explanation of why corporate managers can tolerate so little internal dissent is because they typically feel beleaguered by a wide array of external adversaries who, it is thought, want to disrupt or impede their attempts to further the economic interest of their companies. In a far-reaching study of corporate culture, one researcher found that, consistently, managers see themselves as being under siege from a host of outsiders, including government regulators, consumer activists, affirmative action activists, workmen's compensation lawyers, labor activists, academic critics, environmental activists, and the news media.[38] Managers see themselves as constantly fending off nosy outsiders, particularly those who try to make political issues out of what are seen as internal business prerogatives.

"Managing" Problems   When political issues *do* intrude, the typical corporate response is to transform them into administrative problems, and as problems to be defused through aggressive public relations campaigns. The response of the textile industry to the cotton dust issue illustrates this strategy. Prolonged exposure to cotton dust causes in many workers a lung disease called byssinosis, or "brown lung." The brown lung problem in the textile industry was recognized in the early years of this century, but it wasn't until the late 1970s that the American government proposed a ruling to require textile companies to invest large amounts of money to reduce worker exposure to dust. The industry fought the regulation fiercely on many fronts, including an intense lobbying campaign directed at legislators and the general public. With public relations experts at their elbow, industry executives hammered out an ideology to explain the problem. It went like this:

> There is probably no such thing as byssinosis; textile workers suffering from pulmonary problems are all heavy smokers and the real culprit is the government subsidized tobacco industry; besides, 20 to 25% of the general adult population has some form of chronic breathing problems with symptoms similar to byssinosis; therefore if there is a problem it is one almost impossible to diagnose accurately and fairly, since the pool of workers from which the textile industry draws already have lung disorders; the textile industry should not have to bear the burden of medical problems unrelated to work; moreover, if there is a problem in the workplace, only a few workers are affected and must be particularly susceptible; rather than engineering controls, the more reasonable solution then would be increased medical screening of the work force to weed out those with a predisposition to lung problems.[39]

Similar arguments to deflect culpability—and put blame on the victim—are commonplace in industry. In early 1990, a member of the Texas pesticide regulatory board, referring to a controversial household pesticide, Chlordane, said, "Sure it's going to kill a lot of people, but they may be dying of something else anyway."[40]

Environmental problems are often cast as failures of public relations; and, by corollary, then, it is assumed that "solutions" lie with aggressive public-relations management. A speaker at the 1990 conference of the Public Relations Society of America lamented the failure of public relations in the Exxon Valdez oil spill:

> We are not here today to debate environmental or ethical questions. We are, at least for today, not concerned with the fate of sea otters, but with how a huge American corporation spent $2 billion on the cleanup of what was not the worst oil spill ever, yet lost the battle of public relations and more than a year later is still struggling with one of the worst tarnishings of its corporate image in American history. . . . Well, let's be realistic. Nobody died at Prince William Sound— no humans, anyway. Exxon failed to tell the best sides of its story.[41]

This lament aside, the truth of the matter is that the mass media is a powerful ally of corporate efforts to obfuscate environmental realities. The news media is closely intertwined with corporate America through a closed-loop network of interlocking directorates and corporate ownership of media outlets. The consolidation of media control by large corporate conglomerates is such that by the mid 1990s, five to ten corporations will control most of the world's broadcast stations, magazines, and newspapers.

"Public relations" management strategies mark industry response to most regulatory threats. Rather than throwing their resources behind solving the problem at hand, industry executives often spend millions of dollars to cloud and divert the issues. One favored technique is to form a fictitiously neutral "scientific institute" to study the particular problem at hand. For instance, the Formaldehyde Institute, formed in 1979 with the stated purpose of furthering "the sound science of formaldehyde and formaldehyde-based products and to ensure that the data are used and interpreted properly," was a creation of the major formaldehyde producers who were coming under increased scrutiny because of serious health problems found among many people living in houses insulated with urea formaldehyde foam. Similarly, in 1980, the major chemical companies producing CFCs (which are the main culprits in ozone depletion), seeing the regulatory handwriting on the wall, created a front-group called the "Alliance for Responsible CFC Policy." The Alliance organized speaking tours, public forums, and staged media events. They organized a national campaign of sober scientific debate, and coordinated a write-in campaign aimed at legislators and regulators. The EPA, it was said, was buried with a mountain of paper generated by the campaign. In the end, the new Reagan administration quietly buried the proposed regulations to restrict CFC production, and the Alliance claimed a victory. In the 1990s, industry-based "associations" have proliferated to fight environmental regulation: the "Clean Air Working Group" and "Citizens for Sensible Control of Acid Rain" both work to defeat clean air regulation; the "US Council on Energy Awareness" is a front group for the nuclear industry; the "National Wetlands Coalition" works to narrow the definition of what constitutes a wetland.[42] At the same time that industry is trying to sell a "green" image, industry trade groups are engaged in aggressive public relations subterfuge.

Bureaucrats transform moral issues into pragmatics. In so doing, corporate executives insulate themselves from messy political intrusions. If a problem is transformed into a public relations campaign, then it's a problem for middle management. Despite the rigid leadership hierarchy of most corporations, responsibility for working out logistics of any policy is pushed down the corporate ladder. Executives distance themselves from detail work; they make policy, and then leave it to others to work out how the policy is to be implemented. This results in an amorphous system of accountability when things go wrong. It is almost impossible to track any particular decision back through the corporate maze, and this allows individual managers to wriggle out of responsibility when confronted with, for example, an environmental disaster. Thus, the finger of blame typically ends up being pointed at the lowly "factory operative," at the very bottom of the chain of accountability, when radiation leaks from a cooling tank or when toxic vapors escape up the smokestack.[43]

Short-Term Planning   Another noted characteristic of life in the Euro-American corporate world is that managers of large companies seldom have time to engage in long-range planning. Managers must think in the short term because they are judged on short-term results.[44] Their days are filled with serial, short meetings, terse one-page memos, and summary reports prepared by subordinates. Managers are trained to unload ailing projects quickly, to hitch their wagon to fast-rising stars, and to turn in fast results. Despite best intentions, most managers turn out to be "satisficers"[45]—that is, without the time to search out the "best" course of action, managers turn to decisions that are "good enough," those that meet a minimum set of requirements. Environmental integrity is not often on the short list of "minimum" requirements.

A Buffered Elite   Executives may not include environmental efficacy on their list of minimum requirements for the simple reason that in their everyday lives they are buffered from the effects of environmental degradation. By virtue of their place in the economic elite, most corporate officials are unlikely to live near a toxic waste dump. They can afford to buy healthy food, pure air and clean water. They have access to the best health care. Neither they nor anyone they socialize with have to work in their own factories.

Militarized Models   Militarized metaphors and military analogues run deep throughout the corporate world. This is partially due to the fact that a sizable proportion of the current crop of CEOs in American, European, and Japanese corporations have military experience in their backgrounds. In a recent collection of profiles of 16 CEOs of major corporations (mostly US, with a few foreign companies thrown in), virtually all of the men, and one of the three women in the group, listed their military service in the one-page summations they provided of their career.[46]

Increasingly, CEOs use militaristic metaphors to describe competition in business:

> In the last 10 years, mainly in the last five, US industry has moved from an essentially secure domestic domain to one facing international competition. . . . the emerging nations of the Far East have amassed a formidable capability to compete in the arena we used to call our own. . . . We compare the cost, quality and delivery capability of our Japanese competition against our standards. . . . they have less product proliferation than we do. . . .[47]

Some of the hottest-selling management books in the early 1990s have been primers that translate military theory into management practice— books such as *Sun Tzu—The Art of War*, a text that promises business executives "if you look upon the marketplace as the equivalent of a battleground, you can readily understand how the principles of a success-ful military operation, as gleaned from this classic 'bible,' can be effectively applied to today's domestic and global business environment." Or *The Book of Stratagems*, a 1991 bestseller that promises an extraordinary exercise of power usually associated with military might: "He who is versed in the application of stratagems can plunge an orderly world into chaos or bring order to a chaotic world; he can produce thunder and lightning from a clear sky; can transform poverty into riches, insignificance into prestige, the most hopeless situation into a promising one."[48] A recent report on American business in a popular economic periodical characterized the trends in industry in highly masculinized and militarized terms: "Domestic corporations have put on new muscle and become industrial warriors capable of blowing away even the most intimidating global competitors."[49]

The militarized subtext of modern corporate life has a number of ramifications: • It adds hostility to competition. • It heightens the value of secretive protectiveness in the corporate world. • For employees, it raises the stakes around loyalty to a firm—to be an environmental whistleblower, for instance, in a quasi-militarized corporation is to be a traitor, not just an unenthusiastic company worker, a pressure that will obviously be greater in firms that are actually military contractors. A militarized ethos may also strengthen the sense among CEOs that they are beleaguered by "enemies," and it hardens the line between "us" and "them." In this dichotomous world, environmentalists are often cast as "them." • Using such a highly masculinized organization as the military as a corporate exemplar reinforces the "men's club" ethos of the corporate enterprise.

Faith in Science   An even stronger guiding paradigm in the corporate world is a ritualistic dependence on scientific rationality. If an environmental problem cannot be absolutely "scientifically proven" to be the responsibility of a particular industrial activity, then corporations feel justified in continuing business as usual (and are usually legally vindicated in doing so). "Common sense" may tell us that toxins dumped into an unlined landfill near a residential area will pollute drinking water supplies. "Strong evidence" may point to the connection between CFC production and ozone depletion. And "the likelihood" may be strong that toxic chemicals stored in barrels at a dump site *will* leak. But without incontrovertible Scientific Evidence, corporations evade responsibility.

The problem with reliance on "scientific certainty" is that science is seldom certain about anything. To the contrary, *uncertainty* is central to the nature of science. For every theory, there is a countering theory; for every piece of evidence there is an opposing piece of evidence. To achieve absolute scientific certainty takes years (sometimes centuries) of study, debate, consideration, and reconsideration—which may be appropriate if the problem that scientists are debating is whether it is the sun or the earth at the center of our universe, but waiting for this approach to provide certainty is deadly if a pollution time-bomb is ticking.

Industry's response to the ozone depletion crisis is a classic study of the corporate retreat behind the screen of scientific uncertainty.[50] As early as 1974, atmospheric scientists detected a loss of ozone—the protective shield that sustains all life on earth. Suspicion soon focused on a class of chlorine chemicals called chlorofluorocarbons (CFCs), manufactured, very profitably, by a handful of large corporations in the US and Western Europe. The CFC industry at first simply ignored the reports. Then, as scientists proved in lab tests during the late 1970s that chlorine chemicals were a threat to ozone, the chemical manufacturers refused to accept the laboratory results as sufficient evidence. As the clamor of anxiety grew, DuPont (the largest manufacturer) and other producers promised that they would take remedial action, but only if the theory of CFC-caused ozone depletion could be scientifically proven—which implied years of monitoring to gather evidence of actual damage to the ozone layer. The voices of industry executives were unanimous:

> Richard Barnett, Vice President of York Air Conditioning (1987) : "We don't know [that CFCs don't disperse in the lower atmosphere], and we don't know that the scientists know that. There is growing evidence that seems to suggest that. . . . The most recent data from the last Antarctic expedition was not specific in terms of the cause of the hole over Antarctica. There've been several theories professed, and quite frankly, I don't think anyone has positive proof."

> Igor Sobolev, Kaiser Chemical Company (1987): "The theory, in our view, is full of uncertainties."

> John Roberts, DuPont Company (1987): "It's going to take about three years to find out whether we have a problem, and if there is one, to have an idea of what its magnitude is."

> Richard Leader, Phillips Petroleum Company (1987): "The problem, I believe has been proven, is not that great."

> E.P Blanchard, vice-president, DuPont Company (1987): "I feel very comfortable that based on our projections of the effect on the ozone layer, of current levels of chlorofluorocarbons emissions and modest growth in these levels, does not represent a threat to the health of the environment, human race on out 50 to 100 years from now."

During the 1970s, DuPont began quietly researching CFC substitutes, but in 1980, when Reagan was elected and the threat of federal regulation in the US waned, DuPont stopped its research: "There wasn't scientific or economic justification to proceed," a DuPont division manager said in 1987.[51]

Industry uses scientific uncertainty as an argument for delaying regulatory action, and as a tool to confuse a concerned public by bombarding us with impenetrable scientific debate. The claim of scientific uncertainty is usually not untrue (there *is* scientific uncertainty about most things), but it is used disingenuously. How does one respond to uncertainty? In the face of uncertain scientific evidence about chemicals that might very well be altering the climate on a global scale and threatening the web of life on the planet, the prudent course of action is to *cease* production of the suspect chemicals—because waiting for absolute results might raise the problem to a magnitude beyond solution. But common environmental sense has little place in the corporate lexicon.

Culture Cops   The weight of corporate culture appears to sustain its own inertia. But it is not just because of "inertia" that change is slow to come to the corporate world: progressive or visionary corporate managers are "policed" by their peer group. Studies of companies that have made beneficial environmental changes suggest that a change in corporate philosophy and business practice nearly always starts with a visionary individual.[52] But the network of corporate leaders is a small, close-knit one, and mavericks who don't toe the line have a short tenure. Whistle-blowers are routinely fired; managers who even express doubts about prevailing "business as usual" practices are edged out of power. Work-place-trends observers in both the US and the UK believe that suppression of whistle-blowers is on the rise in both the public and private sectors, and is increasingly supported by the courts and government.[53] Workers at the bottom of the chain of command have even less protection: a recent survey concluded that 96 percent of workers in the US who file OSHA workplace-safety complaints are subsequently fired.[54]

Recent examples from two very different industrial trade groups illustrate the pressures that executives face from their own circle—their clients, boards of directors, trade associations, and in the US, the Republican party:[55] • In the mid 1980s, before defense contract scandals in the US blew up in public, executives at the major defense firms refrained from speaking out about industry ethics because of pressure from their own trade groups. In 1986, when a vice-president at Raytheon did speak out, publicly challenging the Pentagon's priorities on defense spending, he was summarily fired. The courts later upheld his firing, agreeing with the company that the vice-president was "no longer an effective representative" for them. • In 1990, the Business Roundtable, a national group of business executives, publicly opposed the civil-rights bill then under consideration by the American Congress; the Roundtable supported the Bush administration's arguments that the bill would hobble business. But in 1991, a small group of executives, lead by managers from AT&T, sat down with civil-rights leaders, "for the first time ever," to exchange views. Almost immediately, two fatal blows were dealt to AT&T's efforts, first from the Bush administration, and then from the business establishment. White House Counsel Boyden Gray demanded a meeting with the executives in which he told them that they were "complicating Bush's efforts to defeat the bill"; then, a month later, the *Wall Street Journal* ran an article accusing the AT&T executives of "opportunism" and suggesting that they were dupes of civil-rights leaders who were using the bill to promote quotas in business hiring. That same day, the ongoing talks were canceled.

The conservative press plays a crucial "policing" role. For example, when Heinz announced in 1990 that Star Kist tuna would be harvested in a way that protects dolphins, *CEO Magazine* denounced the chief executives for "capitulating to ecoterrorists." The Capital Research Center in Washington, DC, publishes an annual volume of loyalists and traitors to the free-market cause. The 1990 volume, "The Suicidal Impulse," criticized such firms as General Mills and Aetna Life for giving in to consumer pressure groups on issues of health care and the environment. The Bank of Boston, for example, was slammed for donating money to the League of Women Voters' education fund, which the book says "advocates school busing, quotas, and 'forced' energy conservation." The Center reports that it gets nervous inquiries from companies every year that don't want to be portrayed in the book as "too liberal."

NORMAL ACCIDENTS

Public environmental consciousness is often galvanized by spectacular environmental "accidents"—the explosion at Bhopal, the Exxon Valdez oil spill, the Chernobyl disaster. Accidents such as these are usually construed to be unpredictable events—the work of a moment's slip in judgment or unforseeable circumstance. In fact, this is seldom the case. As one author points out, given the nature of corporate enterprise, accidents are "normal."[56]

The Exxon Valdez Late in the evening of March 24, 1989, the Exxon Valdez, a supertanker loaded with millions of gallons of crude oil left the port of Valdez at the end of the Alaska Pipeline, under the direction of Captain Joseph Hazelwood, sailing with a crew of 20. Just outside port, the ship rammed a clearly marked reef, reportedly because of confusion on its bridge in the temporary absence of the captain. The tanker's hull was ripped open and its cargo gushed into Prince William Sound, a pristine Alaskan wilderness, causing the worst oil spill in American history and one of the worst anywhere in the world. Hazelwood, it turns out, had been drinking before the accident.[57]

One of the first public messages from Exxon after the spill was an apology. Running full-page ads in a number of newspapers across the country, the Chairman of Exxon, Lawrence Rawl, wrote an open letter to the public, saying "how sorry I am that this accident took place," and committing the full resources of Exxon to cleanup operations. (Not incidentally, under American tax law, spill cleanup costs are deductible from company income taxes).[58] Within a few months, it would turn out, Exxon was backtracking from its "full commitment" and in fact withdrew its cleanup operation with less than half the job done.

Legal action followed quickly after the disaster. In a dramatization of the business maxim that "credit is pushed up, blame is pushed down," Exxon denied corporate responsibility, pointing the finger at the allegedly drunk Captain Hazelwood. The captain pointed the finger at the third mate, an unqualified subordinate whom he left in charge of the tanker when he retired to his cabin.

In the most immediate sense, Hazelwood clearly was responsible for the accident—captains always are. He was piloting a ship (or, rather, not piloting it) through a tricky channel sailing at night, while possibly under the influence of alcohol. But Hazelwood's culpability is only the tip of the iceberg. Lurking below the surface lies a corporate credo of cost-cutting, personnel-cutting, profit-maximizing and managerial-interference that directly bear on the grounding of the Exxon Valdez.

It costs a lot to run a supertanker—upwards of $30,000 a day. In the 1970s, oil companies realized that they could save a lot of money by sailing fewer, but bigger, ships. In 1969, the average length of an oil tanker was 641 feet; by 1989, the average had crept up to almost 1,000 feet. These supertankers are underpowered and extremely difficult to maneuver. It takes them 3 miles, and almost a half-hour, to stop. They need two miles in which to make a turn. When on the bridge, the captain is approximately 100 feet above the water. The size of these ships makes them unforgiving of error.

As a further cost-cutting measure, most of these oil supertankers, built in the 1970s, were constructed with only a single steel hull—unlike ships that carry other hazardous materials which have double hulls. In fact, Exxon was almost single-handedly responsible for defeating government legislation proposed in the 1980s that would have required double hulls on oil tankers: "They successfully lobbied, and engineered, and conspired, to get the requirement for double-bottoms out of the regulations."[59] The American Coast Guard estimates that a second hull on the Valdez would have reduced the amount of oil spilled by anywhere from 25 to 60 percent.

At the same time that the ships have become bigger and more complex, oil companies have implemented dramatic reductions in personnel. In 1989, as the Valdez was ramming the reef, Exxon was applying to the Coast Guard for approval to cut personnel on the Valdez, and six other ships, from a crew of 20 (on other ships, 25) to 15. Exxon had instructed its captains to eliminate records of overtime, in order to "prove" to the Coast Guard that personnel cuts were justified—an action that the seaman's union has characterized as "knowing falsification" of records.[60]

Personnel reductions are increasing stress in what is already widely acknowledged as a high-pressure life. Sailors on supertankers often work up to 100 hours overtime in a 15-day period. They are at sea for three months at a time. Their daily lives are comprised of long periods of boredom, interspersed by frantic moments of activity as the ship docks, loads, or unloads. Alcoholism is known to be a persistent problem on supertankers, and as one sailor commented sardonically, "Things have changed. When you had 48 men you could carry a drunk or two, but that's not the case any more."[61] In the specific case of Hazelwood, Exxon officials knew that he had a history of alcohol abuse. At the time he was piloting the Valdez, his driver's license was under suspension in his home state for drinking violations.

In the aftermath of the accident, three managers at Exxon (who remain anonymous) charged that the company was concerned with cost-cutting to the point that it endangered ship safety. One of the biggest problems, they said, is constant interference with shipboard decisions by shore-based middle managers. In the aftermath of the accident, maritime observers suggested that Hazelwood had showed poor judgment by sailing the ship at night in a channel known for icebergs. But the Exxon managers point out: "If Joe [Hazelwood] hadn't pulled out of Valdez at 10, after he was finished loading at 8, in a few minutes the ship's satellite phone would have rung, telling Joe to get his ship out of there. He'd have to come up with a real good reason to stay."[62]

The same three Exxon whistle-blowers identified a string of problems that derived from Exxon's management philosophy, including: forcing automation on older ships that are not designed for it; long working shifts for mates—up to 100 hours a week—in violation of federal regulations; cost-cutting for ship repairs and crew expenses; and constant performance pressure that compels captains to operate in risky conditions, sometimes against their better judgment.

In 1991, the corporate owners of the trans-Alaska pipeline, including Exxon, Arco, Mobil and Phillips, were charged with illegal surveillance of employees suspected of providing oil company documents to congressional investigators after the Valdez oil spill.[63]

Bhopal   The environmental disaster in Bhopal offers another case study of the corporate ethos in action. Following a script similar to the textile industry's response to concerns about byssinosis, Union Carbide (the owner of the Bhopal plant) hammered out an explanation of the disaster that, in turn, minimized the magnitude of the problem, denied there was an accident, turned the blame on individual factory operators, and denied corporate culpability.

The chronological facts of the chemical explosion in Bhopal, India, are mostly clear. Union Carbide, an American multinational chemical company, opened a plant in Bhopal in 1970 to produce pesticides. It was never a very profitable operation, and continued to run at a loss. Rumors were afoot in the early 1980s that Carbide was going to close the plant and move its pesticide operations to Indonesia. In late 1984, pesticide operations were temporarily halted, and large amounts of a chemical called methyl isocyanate (MIC) sat in underground tanks, waiting to be used in the next manufacturing cycle. Late in the night on December 3, 1984, those tanks exploded, spewing 40 tons of lethal gas over a densely populated, sleeping neighborhood. Official records count 1,700 people killed immediately after the leak, and an equal number dying in the next few months. Local relief workers put the death count at 5,000 to 10,000. Another 200,000 people suffered health damage from the gas leak, primarily respiratory problems, blindness, and reproductive disorders. Seven years later, people in Bhopal are still dying at a rate of one a day as a result of illnesses caused by exposure to MIC.[64]

In the confusing days immediately following the explosion, Carbide's American president, Warren Anderson, issued a public expression of regret, and said that Union Carbide assumed full "moral responsibility" for the accident.[65] In the month after the accident, Anderson said that "quick and fair relief was the company's first objective" and that he would devote the rest of his career to resolving the accident problem. Personal expressions of regret, however, were soon swept away by the corporate machinery. Within a year, Anderson reversed his earlier statements: "I overreacted. . . . Maybe early on they thought we'd give the store away. [Now] we're in a litigation mode, I'm not going to roll over and play dead."[66]

The corporate strategy of Union Carbide emerged over the next six years, as legal battles volleyed between the US and India. The main thrust of Carbide's approach was to wear down (and, in many cases, outlive) its litigants with costly and baroque legal maneuvering, and to pin the blame onto everyone but themselves.

Carbide produces batteries, plastics, pesticides, and chemicals in 40 countries, and markets products in 125 more, most of which are in the Third World, and 75 of which have smaller economies than the corporation itself.[67] Multinationals seek out production sites in poor countries where labor is cheap, health and safety standards are lax, environmental regulations are often lacking and are virtually unenforceable even if present, and governments are at a tremendous disadvantage in negotiating with corporations sometimes twice their size. Efforts in the United States to pass laws that would force American companies to meet the same standards in domestic and foreign operations have been consistently opposed by multinational interests. In 1981, Jack Early, president of the National Agricultural Chemicals Association called such efforts "regulatory imperalism."[68]

Despite double-standard operating practices that are commonplace throughout the chemical industry, Carbide vigorously denied criticisms that the Bhopal plant was designed and maintained poorly. It insisted that conditions in the Bhopal plant were on a par with conditions in its home factory in Institute, West Virginia. (This was not necessarily reassuring to people in Institute). Carbide executives blamed factory operatives in Bhopal for not paying attention to rising pressure guages; workers responded by saying that pressure dials frequently malfunctioned, valves were frequently in disrepair, and the way that operatives typically detected chemical leaks at the plant was not through guage readings but when their eyes started to tear up. On the night of the accident, all of the built-in safety systems were inoperative. An internally-commissioned Carbide report on the Bhopal plant two years earlier had warned that, "a number of factors make the MIC feed tank a source of concern. . . . there were several conditions or operations in the unit that presented serious potential for a sizable release of toxic materials."[69]

A 1990 study concluded that Union Carbide's problems with health and safety could be traced as far back as the 1930s—and that "Carbide persistently shows wanton disregard for the health and safety of its workers and the communities in which it operates."[70] In the US, Union Carbide continues to be a major discharger of toxic substances into the environment and a major generator of hazardous waste. In 1988, the company generated more than 300 million pounds of hazardous waste—an increase of 70 million compared with 1987.[71]

Because of ongoing speculation about the imminent closing of the Bhopal plant, many of the most skilled workers had quit their jobs in the preceding year, leaving the plant seriously understaffed. Further, Carbide had introduced cost-cutting measures, including a realignment of the blue-collar pay scale that resulted in wage cuts for many workers, and left a generally demoralized staff. In the year before the accident, the work force had been reduced by one-third.[72]

Carbide denied that there was an accident. They circulated a sabotage theory, casting blame for deliberate destruction of the plant on an a never-identified "disgruntled Indian employee."

Facing huge lawsuits, Carbide argued that the site of litigation should not be in the US as the Indian government wanted (where settlements might be higher, measured in US dollars), but that the case should be tried in the Indian courts. Once they succeeded in having the legal battle shifted to India, Carbide argued that Indian courts had no jurisdiction over an American company. An Indian order that Carbide officials appear in India to answer charges was dismissed by a Carbide official as "one of the silliest things the Indian courts have done."[73] They engaged in a low-intensity campaign of contemptuously discrediting the Indian courts, arguing that the judges were prejudiced and that the legal system was antiquated and unable to deal with such a complex issue.

In the year following the disaster, Union Carbide reorganized the corporation, made extraordinary payouts to shareholders, and reduced its equity from about $5 billion before the disaster to less than $700 million on the books by the end of 1985.[74] By liquidating its assets, Carbide reduced the chances of ever being forced to pay large damages.

Doctors at the scene of the disaster in December said that there was no medical information available to them on how to treat MIC poisoning. All of the Carbide information said that MIC was "an irritant," not lethal. Carbide continued to maintain that there were not long-lasting effects from MIC exposure, even as hundreds of people died in the weeks after the accident. Carbide is still not prepared to say that long-term health effects from MIC exposure exist, a point that disregards the ongoing suffering of the people of Bhopal.[75]

Paying the Price for Accidents   The impacts of environmental degradation—including spectacular environmental "accidents"—are never gender neutral. Women's social roles as family caretakers, and in agrarian economies as primary subsistence providers, situate them in the environmental front lines.

No people died in the Valdez disaster (although in early 1992, reports were beginning to circulate about serious health problems from prolonged exposure to oil and oil by-products among people employed in the Valdez cleanup). The marine food-chain was poisoned, but the human health effects of this are assumed to be minimal. Despite the smug presumption that the human costs of the oil spill are negligible, this is not necessarily the case when the impacts on women's lives are assessed. With the closing of the fisheries in the aftermath of the oil spill, unemployment—especially male unemployment—skyrocketed in the dozens of small communities on the edge of Prince William Sound. At the same time, Exxon pumped into these villages hundreds of thousands of dollars in damage cleanup payments. The combination destabilized social dynamics; as one teenager reported, "Things have gotten really weird around here since the oil spill. A lot of people are getting drunk again. There's just too much money and too much stress."[76] The combination of stress, alcohol, and anger took a particular toll on women: in the aftermath of the spill, social workers in the region reported a dramatic rise in rates of alcoholism, "hooliganism" and violence—especially rape, sexual abuse, and wife-battering.

The human costs of the Bhopal disaster are more immediately evident, and once again it is clear that women suffered disproportionate effects. Pregnant women living in Bhopal at the time of the disaster were not given any health information about the possible effects of gas exposure on their reproductive health. Despite demands by a few women's groups, no facilities for ultrasound monitoring were brought to Bhopal, no additional abortion services were set up, and women who did venture to inquire about abortions were told by government officials that they were being alarmist. Of the 2,210 live births recorded to women living in Bhopal at the time of the gas leak, 150 of the infants were dead within six months.[77] Despite Carbide claims that the MIC exposure did not cause genetic or reproductive disorders, a study by a Bombay women's medical group found that the rate of spontaneous abortions among pregnant women in Bhopal shot up from 9 percent in 1984 to 31 percent in 1985. Bhopal women were found to be suffering high rates of pelvic infections, menstrual irregularities, and fertility disorders.

Investigators with the official medical team sent to Bhopal examined the effects of the gas on eyes, lungs, the gastric system, and the skin, but completely overlooked gynecological and reproductive problems. It was generally assumed that the consequences of the disaster were the same in men and women; one Indian feminist notes cynically, "It is not surprising that a system that has long been geared to dismiss women's problems as inconsequential should hardly take note of signs of deteriorating health that women particularly are presenting."[78] Even now, the full extent of the impact of the Bhopal disaster on women's health is unknown.

The people most affected by the gas leak were the poor, who lived in colonies close to the factory. The Indian government estimates that 30,000 people have suffered such severe respiratory damage that they cannot now earn a living.[79] Universally, women are the poorest of the poor, and can least survive any reduction in their circumstances. Thousands of women became the sole wage earners for their families after the accident left them widowed or caring for crippled husbands.[80]

Indian feminists report increasing rates of violence suffered by poor women in Bhopal, a side-effect of the mounting frustrations of increased poverty and ill-health. Women maimed and blinded by the accident are especially vulnerable to abandonment and abuse. A high proportion of women lost all or most of their children—there will be increased pressure on them, including sexual coercion, to bear children again.[81] Women who cannot or will not bear more children may face ostracization and a loss of their social stature.

Figure 2.2.   Waiting for justice. A women whose sight was damaged in the Union Carbide gas leak in 1984 sits with other victims outside a courthouse in Bhopal, India. (© Reuters, by permission.)

For Union Carbide, the Bhopal issue is closed. They settled with the Indian government for $460 million in 1989, a deal that many decry as a "sellout," and considerably short of the $3.3 billion originally sought. Carbide's status on the stock markets has rebounded, the company has sold off some of its least-profitable divisions, and, as one official says, "as far as the company is concerned, the matter is over."[82] For the people of Bhopal, who have still not received any compensation, the Union Carbide legacy is far from over. For women, there is perhaps a glimmer of a silver lining in the Bhopal cloud: in predominantly Muslim Bhopal, women's groups were virtually unheard of before the accident. Now there are several, and it is women are at the forefront of the campaign for justice.[83]

WOMEN AS "OTHER"

Culture Clash

The corporate world is a man's world. More than this, it is a "masculinist" world. The fact that individual women may succeed in corporate life, and that some men do not, does not undercut the saliency of this argument. It *is* clear that men suffer the predations of corporate life—men who try to structure their work life to incorporate personal and family concerns, for example, often sacrifice benefits and advancement to do so.[84] Nonetheless, corporate values and propensities are shaped by the dominant culture, and they reflect the values and propensities of men's socialization, not of women's. Attributes for success in the corporate world—a privileging of emotional neutrality, of rationality, of personal distancing, loyalty to impersonal authority, team playing, scientific rationality, and militarized paradigms—reflect characteristics that define "manliness" in our culture.

An American business organizational consultant, a man, describes the corporate workplace in these terms:

> No matter how many women there are in the workforce, the workplace is overwhelmingly male in its code of conduct, its unwritten process, its governing principles. It is a male system, determined by male modes of thinking and male communications patterns. This is not just an underground subculture. The official policies and procedures that guide or regulate work practices are slanted toward favoring men's convenience, comfort, and familiarity. The dice are loaded in ways women may only barely suspect.[85]

Women who are making small inroads into the corporate world may be successful by material measures, but they are not necessarily comfortable in the culture. A number of business analysts have detailed the alienation of women from the corporate world,[86] an alienation that appears not to be diminished by the fact that there are more women now in business than ever before.

Like boot-camp inductees who need to be socialized into military manliness, most women executives are socialized into the corporate world through the now-ubiquitous plethora of training seminars designed expressly for them. These seminars are designed primarily to acclimate women to overcome their reluctance about a single-minded preoccupation with corporate success:

> Women see a career as personal growth, as self-fulfillment, as satisfaction, as making a contribution to others, as doing what one wants to do. While men indubitably want those things too, when they visualize a career they see it as a series of jobs, as a path leading upward with recognition and reward implied. Women must learn to view their careers in terms of movement up through the hierarchy, not in terms of the intrinsic value of their actions; they must develop a personal strategy that always asks 'what's in it for me?' The cultivation of these organizational values and skills should enable women to overcome

the obstacles to succeed posed by their traditional feminine socializa-tion, so that they can present the appropriate image to male colleagues and superiors, emulate the appropriate conduct, and gain access to the field of bureaucratic discourse.[87]

One of the most difficult transitions for women is the necessity of bracketing personal ethics from work morality. In a recent survey of women executives, it was found that most women "demand that their careers be not only lucrative, but a means of self-expression, a meaningful experience that meshes seamlessly with their personal life. . . . younger women tend to view their careers wholistically, ever seeking to merge the personal and professional, whereas the old guard separates the two."[88] Since women today, in much higher proportions than men, express con-cern for environmental safety and are more likely than men to be active in local environmental politics, conflicts over environmental issues are likely to be central to the tension between private values and the corporate good.

It is still the case that the most successful women in big corporations are those who deny their difference from the men they work with, "eyes fixed firmly on the goal [of becoming CEO], their credo seems to be to fit in with as little noise as possible. A few have gone so far as to refuse all interviews on the subject of women in business."[89] As women entered the corporate work force in large numbers in the 1970s, they veered sharply away from any agenda of "difference" or any suggestion of special consid-erations, fearing, with justification, that it would be used against them. Male-assimilated women are still the most successful corporate leaders—otolaryngologists even report that women coming up through the ranks of male-dominated organizations tend to force their voices into lower octaves as their status rises.[90]

It is wrong, then, to speak of "women" in corporations as bringing a unified philosophy of change to their work. Many women are as conven-tional in their view of hierarchy, bureaucracy, and management as the men they join. Many women harbor suspicions about or profess disdain for feminism. And, yet, as a British researcher recently found, "there is a strong voice *among* women that is characteristically *of* women, speaking for 'a different way of doing things.'"[91] This British study echoes the conclusions of studies of women in American corporations: women priori-tize a social orientation over a narrow task-orientation in their work:

> For me it's about changing the workplace. Breaking down the division between work and home. Recognizing a whole variety of different management styles. The expectation that people will work in different

ways. . . . It seems to me, working with women, that they work in
different ways. I don't think it's innate. I think it has to do with
socialization . . . the ways girls are brought up, women have better
interpersonal skills, often they don't have the same sense of hierarchy.
They are less competitive, more cooperative. And the work environment hasn't valued those things in the past.[92]

As the women quoted above suggest, one need not argue (and I do not)
that women are *inherently* more inclined to egalitarian, nonhierarchical,
more expressive ways of knowing. To sustain a critique of corporate
culture, it is sufficient to agree that, at this point in our collective socialization, the values that are structured into women's and men's experiences
are very different.[93] And, in general, the values that structure the world of
corporate policymakers are not those that structure the world of women's
experience. Carol Gilligan's research suggests that one of the major differences between men's and women's intellectual and emotional development arises in the formulation of moral ideologies: men develop an
ideology based on an ethic of rights, women develop an ideology based on
an ethic of responsibility.[94] Gilligan consistently found that men describe
themselves in terms of great ideas, distinctive activity, and individual
achievement. Their moral development is predicated on a hierarchy of
rules and rights. They perceive their relationships through positioning
and distribution of power and authority. The (women-identified) ethic of
responsibility rests on an understanding that gives rise to compassion and
care. Perhaps this is a crucial difference that explains why men and
women often express quite different environmental sensibilities. One local
journalist, a woman, echoed many women's sentiments when she voiced
her environmental concern in this way:

> How did we ever reach this state of affairs? Why haven't scientists put
> as much energy into disposing of poisonous substances as they have
> into inventing them? This isn't an idle question. I can't imagine myself
> bringing something new and dangerous into being without some
> thoughts about its eventual disposal. And I certainly can't imagine
> making a mess without thinking it was my responsibility to clean it
> up, or to enable someone else to.[95]

If women don't make it in the man's corporate world, or don't want
to try, there are few professional options available to them. Feminist
alternatives to technocratic institutions are few and far between. Recently,
feminist theorists such as Jean Elshtain and Sara Ruddick[96] have initiated
a discourse that calls on female values for inspiration rather than on
technocratic values, challenging the dualism of private/expressive versus

public/instrumental, but at this point in time their program for change remains only visionary.

If women are uncomfortable in the corporate world, men are equally uncomfortable having them there. Businessmen are accustomed to the privileges that accrue when the economic elite of a country is also a male elite. The corporate "power world" of high finance, high prestige, and high salaries not only derives from patriarchy, it protects and extends male power. Businessmen are notorious for resisting women's intrusions into their comfortable men's clubs. They fight to preserve men-only spaces, they complain that having women around cramps their style—which mostly means that women might interfere with the male bonding that comes in some measure through shared sexist jokes, shared prostitutes on business jaunts, or trivializing treatment of secretaries. And men are not comfortable with women who earn high salaries, perhaps higher than their own. The crucial role played by adherence to "shared values" in a corporation is threatened by heterogeneity. Heterogeneity in class, ethnic background, or gender makes it much more difficult to instill in workers a sense of a common corporate good.

Secretaries   Most businessmen simply do not have experience dealing with women as peers. The women that businessmen deal with the most are secretaries. Secretaries play a crucial role in the corporate culture, most noticeably as a currency of status. As men climb the corporate ladder, they move from the services of the typing pool into the privilege of having a private secretary, or even, near the top of the ladder, a bank of private calendar-keepers, appointment-makers, and receptionists. Secretaries are the least entitled to voice opinions or to sway corporate decisions, even at the same time that many of them work close to the locus of corporate power. If secretaries have opinions on corporate policy, they are well advised to keep them to themselves.[97]

Wives   Women who are corporate wives pose yet another kind of dilemma for the men who rule the world of business. CEOs need the wives of their managers to be supportive of their husbands' work, but wives are beyond the reach of corporate normative socialization. Wives often do not share the values of the "corporate common good," and can be a countervailing influence on their husbands. Environmental issues are increasingly appearing at the hub of corporate/wife clashes, in large part because as women become increasingly active in grassroots environmental work they often find themselves confronting the very corporations for which their husbands work. In a story about the dissolution of a marriage, Jamaica Kincaid captures the poignancy of the clash of these worlds:

Mariah and Dinah and other people they knew had become upset by what they said was the destruction of the surrounding countryside. . . . Mariah decided to write and illustrate a book on these vanishing things and give any money made to an organization devoted to this point of view. Some days she would go out from early morning until late afternoon sketching specimens of all sorts in their various habitats; she gave me the impression that everything was on its last legs and any day now would vanish from the face of the earth. Mariah was the kindest person I had ever known. . . . And that was the reason I couldn't bring myself to point out to her that if all the things she wanted to save in the world were saved, she might find herself in reduced circumstances. I couldn't bring myself to ask her to examine Lewis' daily conversations with his stockbroker, to see if they bore any relation to the things she saw passing away forever before her eyes. . . .[98]

An executive in a chemical firm posed the problem more bluntly. In reference to a manager whose wife was known to be active in environmental action groups, lobbying in fact for legislation on chemical waste disposal, he said, "If a guy can't even manage his own wife, how can he be expected to manage other people?"[99]

Workers   Women who work in the factories of industry are especially prone to being defined as problematic. The manufacture of chemicals and poisons pollutes the air and water for everyone, but takes its first toll on industrial workers. According to federal statistics, each year 100,000 American workers die from occupational illnesses, and each year there are 400,000 new cases of work-related disease.[100] Workers at the bottom of the occupational hierarchy are the most vulnerable to workplace hazards—a recent American study established that the employees who are most likely to be placed in the most dangerous situations are contract workers who are poorly educated, inadequately trained, and unprepared to confront emergency situations.[101]

Of all the occupational hazards, reproductive hazards are the most invidious, often the least conspicuous, and quite possibly the most widespread—more than 14 million workers in the US alone are exposed to substances that are known or suspected toxins to the reproductive system.[102] In Canada, 50 percent of all pregnancies fail to produce a child; many researchers think that workplace hazards contribute to a significant proportion of these reproductive failures.[103] High rates of fetal deformity, stillbirths, and childhood disorders are commonly reported among pools of women who work—or whose husbands work—with toxic materials. Poor women and women of color face disproportionately high risks because of their concentration in the lowest paying, most hazardous, and least protected occupations.

A startling number of corporations that operate facilities that expose workers to possible reproductive toxins have responded to pressure to reduce hazards *not* by eliminating workplace hazards, but by attempting to eliminate women as workers.[104] On the grounds of preventing risk to fetal health, (but primarily to avoid potential lawsuits), a wide range of companies has implemented policies that exclude women from particular jobs, but that permit men to perform the same jobs. Not coincidentally, the jobs from which women are excluded are often the higher-paying skilled jobs in a factory. One 50-year old woman who has worked in an automobile battery factory for years was recently demoted "because the level of lead in her blood posed a potential risk to a fetus she might conceive"—she now has a lower-paying job washing the protective face masks of the men who do her old job.[105] Observers note that exclusion policies are rarely implemented in traditionally women's jobs that may also pose reproductive hazards, such as in x-ray clinics, where there are few men waiting to take over the work.[106]

Corporate exclusion policies typically take one of three forms.[107] Some companies have required women to prove their sterility before employing them in certain areas of work; in 1978, five women in West Virginia who faced an exclusion policy at the American Cyanamid Company (a large chemical conglomerate, and one of the top ten polluters in the US) actually underwent surgical sterilization in order to keep their jobs.[108] More common exclusionary policies are those that restrict "women of childbearing age" from work that the company deems hazardous; the definition of "childbearing age" is left up to corporate whim, and in some cases is defined as including all women aged 5 to 55. Finally, some companies exclude pregnant women from particular jobs.

A 1991 US Supreme Court ruling deemed these policies unconstitutional, but the effects of 'fetal rights' initiatives are invidious. In the late 1980s, these exclusion policies added vigor to the rising right-wing "pro life" movement, which values women primarily as vessels of pregnancy, and the presumption that women are "special" workers will not be easily dislodged by judicial findings to the contrary.

◑

While many corporations do have environmental safety regulations in place, their saliency is undercut by the larger culture of corporate enterprise, which more typically casts environmental issues as the ushers of outside chaotic political forces that threaten the protected world of the corporation, as costly externalities, and as public relations problems. Individual managers are distanced from environmental consequences by virtue of their economic and social privilege, by the anonymity of the people their decisions effect, by maze-like channels of responsibility that push "details" down the corporate ladder, and by elaborate linguistic buffers. Internal questioning of priorities is stifled by the hierarchical system of reward and promotion, by the importance placed on a sense of "shared values," by the priority given to emotional neutrality, by the partitioning of private values from work values, and by the frustrations of embattled executives who often see the world in terms of "us" versus "them." The reluctance to embrace heterogeneity, which means among other things that women (and minority cultures) may exist *in* corporations but are never *of* them, means that no alternative moral voice is heard.

Conventional environmental analysis focuses on "the profit motive" as the driving force behind environmentally destructive corporate behavior. But it is the larger corporate ethos, more than just the profit motive, that subverts environmental consciousness in industry. It is an ethos that is thoroughly masculinist—it is designed by men, for an almost exclusively male enterprise of running Big Business, and it acts to prop up male privilege in the larger culture.

Chapter 3

On the Coattails
of Men in Government

We are increasingly pinning our environmental hopes on the coattails of men in government. It is becoming more evident that solutions to the largest global environmental problems *must* come from governments. As we grapple with truly transnational issues, such as ozone depletion, global warming, or tropical deforestation, we turn, expectantly, to the governments of the world for environmental clarity, for leadership in forging both national and international policies that will avert global catastrophe. The environmental hope for the future is the expectation that governments will act as neutral players, arbiters of the greatest environmental good.

This expectation meshes with the broader and much-cherished perception that governments, especially "democratic" governments, act to protect the best interests of "all"—or, at least, that they act in the best interests of the majority of the citizenry. In principle, a democratic government is one that weighs the considerations of competing interest groups, and then, in a mediating role, arrives at objective policy conclusions that offer the greatest good to the greatest population possible. Governments, especially those that presume to be "of the people and for the people," are presumed to not have vested agendas of their own.

Nothing could be further from the truth. Governments are self-perpetuating bureaucracies that act in the first place to protect their own interests. Those interests are often defined by military priorities, and by officials within government who use their standing to protect private wealth, social status, and the interlocking interests of industry and the military that keep them in power.

GOVERNMENTS IN KHAKI

The wheels of government are greased by the influence of large clusters of vested interest groups, most noticeably and most ubiquitously, the military. Militaries wield enormous influence, and have established an intractable political base, throughout the governments of the world's 160-plus countries, a fact that has changed little even with the trend to the "democratization" of many countries in the 1980s. Even in newly democratizing countries, a change in government does not necessarily signal a change in regime; in countries such as Chile, Argentina, Uruguay, and Brazil, "new democracies" are still beholden to the "old military."[1]

Ruth Sivard, one of the most persistent analysts of military influence, outlines the basis for military influence in government:

> They [militaries] represent the largest single element in most government bureaucracies, the largest financial resources and, of course, the power of the sword. They also have several other unique advantages. They provide the visible trappings of prestige for political leaders, civilian or military: the requisite honor guards, jet aircraft, helicopters. They have a direct line to the world of wealth and business, the arms-producing corporations that are both beneficiaries of government largesse and contributors to political power. And they deal in matters of national security that can, to an extent largely within their own control, be made secret and inaccessible both to the public and to any of the usual checks and balances within the government.[2]

The fact that all militaries and virtually all governments are all-male clubs is not unimportant. As we have seen earlier in this book, the collusion of governments and militaries is held together with the glue of male bonding. Militaries and governments support each other in sustaining the power of men within their own institutions and it is a collusion that extends male power throughout the larger spheres of civic and private life.

There are very few governments in the world in which the military does not wield enormous influence.[3] Moreover, in an astonishingly high number of other countries, the military *is* the government—a pattern that is especially pronounced in the developing world. Sivard suggests that the decolonization of the "Third World" in the past three decades provided "exceptional opportunities for military entree into the political field. Where there was limited experience in self-government and administration, the military came to represent the strongest sources of leadership."[4] In her 1991 analysis of government/military relations, Sivard identifies 64 out of the 113 countries in the "Third World" as being under the control of militaries. The trend to militarized government is increasing: in 1960, 26 percent of the developing states that were then independent were under military domination in some form; by 1989–90, the proportion was up to 57 percent.[5]

Sivard identifies several features that countries in which the military play a political role have in common: • they tend to be more heavily militarized, in terms of expenditures on military arms and personnel • militarized governments stay in power longer than nonmilitarized governments • militarized governments tend to use force and repression against their citizenry, often summarized as "human rights" abuses • and those countries under military control have suffered more wars.[6] I would add to this list: • countries with heavily militarized governments are more likely to suffer severe domestic environmental degradation; • and, in such countries, environmental regulation is least likely to be effective, if not absent altogether. (In passing, one can also note the proclivity of self-aggrandizing male leaders to want to leave their mark—literally—on the land. Environmental horrors are often perpetuated for the sole purpose of building monuments to memorialize authoritarian leaders. In Chile, President Augusto Pinochet's pet project was the construction of an 800-mile highway into Chile's deep south, one of Latin America's last remaining unexploited frontiers.[7] In North Korea, a monumental dam project, completed in 1988 at a cost of $1.8 billion, was built primarily to honor the president, Kim Il Sung. "The Great Leader personally came in 1981 to point out where to build," a project manager explains. "At that time some scientists said it would be better to build the dam further upriver, where it would be narrower and easier to construct. But the Great Leader said that it would not be so useful. Later, during construction, the Great Leader offered important hints on construction that even the engineers would never have been able to think of.")[8]

Two features of military governance—the longevity of military governments and the tendency to war or continuous preparation for war—escalate the likelihood of severe environmental degradation in countries under military and quasi-military governance. Case studies of the USSR and of Myanmar (Burma) illustrate the environmental costs of military rule.

USSR/Eastern Europe   The lifting of the Iron Curtain from the combined territories of the former USSR and its allied states in Eastern Europe in the late 1980s exposed an unprecedented and unparalleled environmental catastrophe.[9] The landscape of the former USSR and Eastern Europe is scarred with industrial wastelands and dotted with valleys of death. In Czechoslovakia, the environment minister recently announced that the environment is so polluted with carcinogenic chemicals that compulsory testing of mothers' milk may be necessary. In heavily industrialized northern Czechoslovakia, infant mortality is 12 percent higher than in the rest of the country, which, in turn, has infant mortality rates 60 percent higher than averages in the West. Sixty percent of Bulgaria's farmland is classified as severely damaged by excessive use of agrochemicals and industrial fallout, and Bulgaria, which used to be part of the Eastern European food-exporting bread-basket region, started to import food in the late 1980s. Only 3 percent of river water in the former East Germany is drinkable, and air pollution in most urban and industrial centres poses severe health risks. The National Hygiene Institute in Budapest recently estimated that over 400,000 people in Hungary drink water with high arsenic content. One death out of 17 in Hungary, and one disability out of 24, is attributed directly or indirectly to air pollution. In the industrial area of the Ukraine, the air is so polluted that it is causing permanent genetic damage to the population. A Soviet biologist estimated in 1989 that 20 percent of the Soviet population lived in what he called "ecological disaster zones," and another 35 to 40 percent in badly polluted areas.[10]

Poland's pollution is widely described as the worst in the region: the Polish Academy of Sciences released a report on the environment in 1990 which concluded, among other startling findings, that 80 percent of the water from lakes and rivers was undrinkable, that the Vistula River is so polluted that its waters are unfit even for industrial use along 80 percent of its total length, and that one-third of the Polish population lives in "areas of ecological disaster"; in the late 1980s, the Polish government declared five villages in the industrial region of Silesia unfit for habitation because of high levels of heavy metals in the soil—and this was considered to be just the tip of the iceberg.[11] Governments in the newly reconfigured states of the former Communist bloc are facing the haunting possibility that the Chernobyl nuclear accident, far from being a singular event, could be just the first of several disasters to come, as the dozens of nuclear power plants across Eastern Europe, built with dubious technology and substandard safeguards in the first place, become even more precarious and dangerous as they age.

Still reeling from the enormity of these revelations, pundits in both the West and the East are casting around to explain the causes of such appalling environmental deterioration—and to understand why it didn't provoke a major state crisis until the late 1980s. Conventional wisdom is jelling around two explanations for the ecological catastrophe in the former Communist states: the role of Marxist economics, and the imperatives of Stalinist industrial development. The development of heavy industry on a vast scale was key to Stalin's grand rebuilding of the Soviet empire. In the post-World War II reconstruction of the USSR, industry and technology were represented as the primary vehicles for achieving socialist goals—they were key not only to "progress," but to the far loftier goal of the heroic restructuring of society into an ideal workers' state. In the aftermath of World War II, industrial complexes for the extraction of raw materials, especially steel and coal, and for the production of heavy machinery were built throughout the Soviet sphere on a vast scale, in keeping with the prevailing spirit of gigantism. As in the West, environmental considerations were largely absent in the rebuilding of the 1940s and 1950s new world order, and it is this postwar industrial sector that is responsible for the poisoning of much of Euro-Asia.

Post-Stalinist governments in the USSR and Eastern Europe retained the commitment to industrial socialism and continued to expand the industrial sector at breakneck pace. But it was fundamental economic principles incorporated into the Soviet system that prevented the environmental costs of the industrial regime from being acknowledged or ameliorated. In classic Marxist economics, nature is portrayed as the handmaiden of social progress, existing to serve "man's" needs. Indeed, in the Stalinist rush for social reordering, the environment was often portrayed as a recalcitrant force; one Stalinist slogan proclaimed, "We cannot expect charity from Nature. We must tear it from her."[12] In Marxist economics, as in capitalist economic theory, natural resources are assigned no economic value or cost—they are "free." However, in market economies, the costs of environmental deterioration do eventually impinge on the profitability of the market; in the centralized Soviet state economic system, the feedback loop between the continuity of production and the discontinuities produced by environmental deterioration was severed, and the ubiquitous subsidies from the central state blunted the effects of spiralling ecological costs.

Many Western economists and government officials take some satisfaction from explanations that point the finger at Marxist economics and Communist ideology for the environmental disasters in the Soviet sphere. However, there is little reson for ideological smugness. At root, the environmental disaster in Eastern Europe is the product of the same institutions that have caused environmental crises in the rest of the world; specifically, it is the demands of *militarized governance* in the Soviet sphere, in tandem with the prevailing industrial economic order, that is responsible for the environmental catastrophes of the USSR and Eastern

Europe. Since World War II, the USSR and most Communist states in Eastern Europe have been ruled by nominally civilian governments—however, they have been heavily militarized civilian governments over which the Soviet military exerted extraordinary influence. The priorities of the governments in the Soviet bloc were inextricable from the priorities of the military; domestic priorities were dictated by the national security interests of the Soviet Union's military.

Thus, the frenzied pace of development of industry and technology—responsible for polluting so much of Eastern Europe—was central not only to achieving socialist goals, but to achieving military goals, goals largely dictated by the escalating USA/USSR military competition. The state owned most industrial enterprises, a fact that in itself deflected any independent environmental oversight; but, more specifically, it was the military that controlled much of the industrial might throughout the Communist bloc. Defense industries, including resource extraction sites, were in a privileged position—and, indeed, remain so in the aftermath of "democratization." As recently as 1991, the chairperson of the Environmental Committee of St. Petersburg (formerly Leningrad) observed that 70 percent of the factories around St. Petersburg belong to the Department of Defense and existing laws make it possible for plant managers to refuse access to environmental inspectors.[13]

State intolerance of dissent, opposition, and free debate, which created an atmosphere in which there was a lack of public accountability and no viable venue for exposing the unfolding environmental crisis, was not just a product of Stalinist paranoia—it was an essential prop of military control. Throughout much of Eastern Europe, it was a crime against the state to question the government's environmental policies. Governments hid, falsified, or ignored medical and environmental statistics that exposed the severity of environmental degradation. Environmentalists were cast as enemies of the state, and expressing environmental concern was considered a subversive act—indeed, a ploy of the West—that threatened national security throughout the Communist bloc; one Czechoslovakian journalist recalls that, "We were told we have to fight the efforts of the West to bring us to our knees through ecology."[14]

The Soviet military, with the USSR government in its pocket and client governments under its thumb, is directly responsible for many of the most outrageous and dangerous environmental violations in the former Communist bloc. The Soviet military established thousands of bases and installations across Europe and Asia—approximately 150 military bases in Czechoslovakia, occupied since 1968; more than 1,000 military facilities in the former East Germany; more than 600 in Latvia.[15] The legacy of this massive military presence is a pancontinental trail of poisoned groundwater, chemical and oil spills, toxic waste, and vast tracts of agricultural and forest land degraded by years of target bombing and war games.

Like its counterpart in the US, the Soviet nuclear weapons industry existed for decades in secrecy, protected by barbed wire, miles of empty land, and an atomic brotherhood of scientists, engineers, and technicians.[16] The Soviet nuclear-weapons industry consisted of dozens of production facilities scattered across the USSR, and hundreds of deployment sites across the USSR and Eastern Europe. Environmental violations were rife throughout the nuclear system. Radioactive waste was systematically dumped into lakes and rivers; airborne radioactive emissions were routine; the nuclear-powered Soviet Navy and the icebreaking fleet dumped much of their radioactive waste in the shallow Arctic waters of the Barents and Kara Seas; nuclear weapons tests conducted for decades at Semipalatinsk, in the Asian republic of Kazakhstan, exposed thousands of local residents to highly radioactive fallout.[17]

Ironies abound in the new Eastern European world order.

Anxieties and anger over the state of the environment boiled over in the 1980s, despite the tight lid that Eastern European governments tried to keep on dissent. Citizens' ecology groups and nascent Green parties were in the foreground of the movement for democratization in Hungary, the Ukraine, Bulgaria, East Germany, Czechoslovakia, and Estonia, and then again in the lead of the revolution that led to the dissolution of the Soviet empire. But as one observer notes, "No sooner were the mechanics of Western democracy and market capitalism set in motion than popular ecological demands faded into the background."[18] The environment is rapidly receding as a popular rallying point, overshadowded by concerns about ethnic conflicts, soaring prices, Western-driven "economic reform," and unemployment.

The Chernobyl disaster was a catalyzing moment in the political changes that swept the Soviet territories in the late 1980s. Yet in the early 1990s, in their eagerness to throw off dependence on Soviet oil, the new governments of Hungary and Czechoslovakia are rushing into the arms of the nuclear club, and Western companies are anticipating a booming business in nuclear power plant construction in the former Communist bloc.

Perhaps the greatest irony of all is that the contraction of the Soviet military is causing its own environmental crisis. As the military retreats from previously occupied territories, it leaves blighted lands, toxic trails, nuclear wastes, and even nuclear weapons in the hands of inexperienced new governments. The extent to which resources in the civil sector were bankrupted and scavenged by the cumulative demands of maintaining militarized governance over several decades leaves the new governments under-funded and ill-equipped to cope with the environmental legacy of the previous militarized state. At the same time, most of the new Eastern European states remain highly militarized, but with such a fluid structure of authority that there is virtually no hope of ensuring any greater public oversight or environmental accountability in the newly democratized states than under the previous authoritarian regimes; indeed, in the midst of such sweeping political changes, the new 'civilian' governments may have considerably less control over the military and its environmental record.

Burma (Myanmar)   In many countries, most notably a cluster of those under military rule in Central America and Southeast Asia, government repression of indigenous groups, guerilla insurgency forces, and political opponents has reached the pitch of internal civil war. In many of these cases, the natural environment has become a pawn of such war, caught, literally, in the cross-fire. The wholesale and deliberate destruction of natural resources has, in many countries, become a deliberate tool of government (or, more accurately, government-cum-military) policy and power. Large-scale deforestation, in particular, has become a trademark of government exercise of power in many of these countries.

Recent events in Burma[19] illustrate the alarming exercise of government deforestation policy. Since 1988, the Burmese military government has engaged in the systematic destruction of the dense rainforest along its border with Thailand as a means of extending control over insurgency groups that are located there. In this case, deforestation is actually the result of tacit cooperation between *two* governments and two militaries, those of Burma and Thailand. In 1988, the Burmese military government, financially drained by years of internal civil war and facing an international embargo of foreign aid, opened up the northern forests to commercial exploitation. The government has since sold scores of logging concessions, most to Thai-based logging firms, most of which have direct links to the Thai military, and many of which have links to Thailand's ruling coalition.[20] Logging in Thailand, which reduced the extent of tropical rainforest in that country by approximately 85 percent, was banned in 1988;[21] Thai logging consortiums, desperate for new sources of timber, are aggressively seeking out logging operations outside their own country, and are spearheading the Burmese operations. The commercial allure of the Burmese forests is teak—Burma has an estimated 70 to 80 percent of the world's remaining teak forests. Thai businessmen refer to teak as "brown gold"; a century-old tree can be worth $200,000. In 1989 alone, the sale of logging and mineral concessions brought in $1 billion to the Burmese regime, eager for the hard currency it needs to buy weapons.[22]

Until the new logging policy in 1988, the government crackdown on opponents in Burma had only a minor effect on the forests. While the Burmese army destroyed villages, which forced refugees to clear new land for farming wherever they resettled, the fighting also discouraged large logging operations. Insurgent groups themselves relied on their own exports of teak and other tropical hardwoods as a source of income. But like the British and Burmese loggers who worked these forests before the current regime assumed power in 1962, they adhered to strict limits on the minimum size of trees to be cut, and preserved the teak growing stock by cutting on a 30-year cycle. In addition, the small-scale rebel operations used elephants instead of bulldozers and trucks, eliminating the need for roads. The ubiquitous logging access roads built by the Thais in the past four years for their heavy equipment have accelerated the destruction of forest ecosystems.

The reckless pace of deforestation is staggering: logging is proceeding so quickly that one reporter, writing in October of 1990 said, "whatever's left within 50 or so miles of the Thai border should be cleared out by December"; environmentalists and United Nations officials estimate that 500,000 hectares of tree-cover a year have disappeared during the past five years from Burma. Forest destruction has accelerated to the point where Burma is now losing at least 2 million acres (3.3% of its total forest cover) a year. Experts say that unless concessions are canceled, the country will be denuded within a few years.[23]

In a desperate grasp to retain power, the men in the Burmese military government have forged a fiscal policy that will, within a few short years, lead to the destruction of the world's last great teak forest. In this case, a change in government may be the only hope for halting this wholesale destruction. But it is erroneous to assume that "democratic" governments, per se, have an inherently better stance of environmental protection.

LOOKING OUT FOR NUMBER ONE

Most governments are composed mostly of men (and sometimes a few women) from the wealthy elite of their country. Even for the rare official who does not come from the elite, government standing propels them into this class—it increases their prestige, and often serves to increase their wealth.

Wealthy government officials usually have interlocking interests with other power bases in the country, most notably militaries and commercial/industrial interests. As Cynthia Enloe points out, in most countries, patronage is neither illegal nor even even illegitimate: "It's the glue thought necessary to hold even democratically elected regimes together."[24] Although the line separating patronage from corruption can be a very fine one, it is widely accepted that loyalty, gratitude, and "paybacks" grease the machinery of political systems. This fact is such an unremarkable feature of everyday political and business life that we speak glibly and knowingly of British "men's clubs," American "old boy networks," the Finnish "sauna society." This popular culture terminology highlights a primary characteristic of interlocked government-business relationships: they are networks of *men*. As Enloe points out, "Patronage typically serves to reinforce the attachment of *men* and the alienation of women to the political system. Most of the people in a position to hand out patronage are men; those they deem useful recipients overwhelmingly are men—for there are few women who have something politically valuable to give in return." In patriarchal societies, the power of men *as men* derives in part from their control of wealth and from the seamless web of relationships among men in government, military, and commercial elites.

The intertwined interests of men in government and men in industry serve to improve the wealth and standing of both. Many government officials make a lot of money, for themselves, their families and their friends, from the very industries and resource-extraction activities that cause environmental degradation. To the extent that this is true, one of the policy ramifications of these intertwined networks is to discourage environmental regulation or enforcement of existing environmental laws. Even the United Nations, an agency that is consistently pro government and timid in its criticism of government elites, now acknowledges the culpability of vested-interest politics in creating environmentally unsustainable pressures:

> More transparency is needed in officials' dealings with logging and plantation interests. There are many stories [in Southeast Asia] of profitable, unseen relationships between senior officials or their families, and commercial interests, that result in activities which go against governments' publically-stated social or ecological goals.[25]

The rapid deforestation of Malaysia, as a particular example, is generally attributed to the network of interlocking government and industry interests:

> Malaysian politics makes short work of good intentions. The federal government in Kuala Lumpur and the cliques that control Malaysia's two states in Borneo [where the most rapid deforestation is occurring] have a tacit understanding. The states will help generate revenues for the federal government . . . so long as they are free to run things their

own way. The men who run Sarawak are unlikely to turn even a pale shade of "green." They have too much to lose. Sarawak's chief minister hands out logging licenses at his discretion. The real owners of the licenses are usually concealed behind a veil of nominee companies. The veil was temporarily lifted during the campaigning for the most recent state election. The nephew of the chief minister at that time revealed that his uncle had granted concessions covering 1.25m hectares of forest to his eight daughters . . . friends and relatives of the nephew have licenses for 1.6m hectares.[26]

A government's environmental policies are not *always* hijacked by the vested interests of its individual members—and those interests are not always in harmony. Many democratic governments have instituted mechanisms to ensure that the private interests of men in government do not create "conflicts of interest" with the public good.

Nonetheless, environmental campaigns are often inextricably linked to broader proposals for redistributing wealth and control of resources; they are, in a broader sense, social justice campaigns, and the significance of *this* is not lost on individuals within government who have the ability to use the power of the state to protect their own wealth. In many cases, government disregard for environmentally sound policy is a response of wealthy, elite men acting to protect their vested interests *in general*, even if they may be prevented by "checks and balances" mechanisms from acting on the specifics of those interests.

Given the gendered nature of the symbiotic relationship between state and commercial power, it is important to note that when governments use the power of the state to repress environmentalists, or when they facilitate environmental abuses either by ommission or commission, they are not only acting to protect their vested interests and private wealth—they are also acting to protect their *male* privilege and the sanctity of the *male* networks that sustain their power. This being the case, then one of the explanations for the worldwide pandemic of government-level environmental subterfuge (at the worst) or ennui (at the best) is that we are witnessing the workings of male power: men closing ranks to protect one another, to protect their networks, and to protect their power.

Environmentalists as Enemies of the State

Government repression of environmental activities appears to be on the rise worldwide, and in countries where military-controlled government repression of the citizenry is the norm, environmentalists are even more likely to suffer government attack.

Government attacks on environmentalists range from relatively mild expressions of disdain to the use of deadly force. The use of state coercion to silence environmental (and other) government critics is more characteristic of militarized regimes, but even governments in "democratic bulwark" countries such as the United States, France and Britain have a long history of antagonism to environmental activists. President Nixon, who signed legislation creating the United States' Environmental Protection Agency in 1971, is quoted at that time: "[Environmentalists are] a group of people that aren't really one damned bit interested in safety or clean air. What they are interested in is destroying the system. They're enemies of the system . . . The great life is to have it like when the Indians were here. You know how the Indians lived? Dirty, filthy, horrible." [27] Similar views echo throughout the last 20 years of American administrations. Recently, a senior member of the Bush administration, the director of the Bureau of Mines, characterized environmentalists as a "bunch of nuts." [28]

The tendency might be to dismiss such views as mere rhetorical excess—except for the fact that government men who perceive environmentalists as dangerous, indeed as enemies of the state, often use the power of the state to silence environmentalists. Examples of state repression of environmentalists come from virtually every part of the globe. • In February of 1991, the Philippine military arrested, without warrant, fourteen members of an environmental group that was investigating connections between the military and illegal rainforest logging; the government representative for the jurisdiction where the arrests were made supported the military actions, telling the environmental group to "stop meddling." [29] The activists were eventually released from prison, yet remain charged with antistate subversion. • The French government, angered by Greenpeace protests against French nuclear testing in the Pacific, bombed a Greenpeace ship, the *Rainbow Warrior*, in a New Zealand harbor in 1985, killing one person and injuring several others. • In 1990, the United States' Federal Bureau of Investigation (FBI) mounted a $2 million campaign to infiltrate and disrupt the California and Arizona branches of the environmental group Earth First!; the campaign concluded with a suspicious car bomb explosion that seriously injured two activists. While the FBI denies involvement in the bombing, there is strong evidence to suggest their complicity in the campaign of disinformation that followed the bombing. [30] The FBI has a long and sordid history of infiltration and intimidation of domestic social-change groups, and the Reagan and Bush administrations have given the FBI an even freer hand in defining "security threats." [31] • The Kenyan government has long perceived environmental activists as unlawful dissidents; in a series of government crackdowns in the early months of 1992, environmentalists were targeted for attack. Wangari Maathai, a prominent leader of the Greenbelt movement, was arrested and detained several times in early 1992; on at least one occasion in March, she was beaten unconscious by government police. • In 1987, the Malaysian government cracked down on a broad range of social critics, including environmentalists, arresting more than 100 people and holding them without charges and without access to legal counsel.[32] Four prominent environmentalists were among those jailed, including three leaders of Friends of the Earth Malaysia, who were working on campaigns against a nuclear waste dump and in support of tribal communities opposing logging ventures in Borneo. One source inside Malaysia reported that:

> The government is intolerant about any public expression of dissidence toward government operations. And the government itself or its ministers are involved in many activities that environmentalists have criticized. Even an educational campaign about pesticides runs counter to government interests . . . because the government is involved in the manufacture of pesticides and in extensive palm oil and cocoa plantations that require heavy pesticide use.[33]

Rainforest Politics

Tropical rainforests have become an environmental *cause célèbre*; battles for rainforest protection make most clear the conflicting agendas of environmental reform and vested government interests.

The world's tropical rainforests are geographically located in an equatorial and subequatorial belt, a geography that overlaps with much of the "developing" world, a geography of poverty. Everywhere in the world, poverty is one of the leading causes of environmental degradation.[34] The leading cause of poverty, self-evident as it seems, is an unequal distribution of wealth, a wealth that is often derived from unequal distribution of access to natural resources. In the developing world—which is also the tropical rainforest world—centuries of colonization, which by definition is the domination of empire by a foreign elite and in conjunction with the installation of a cooperating domestic elite, followed by chaotic decolonization and, in many cases, military governance, have skewed power relations so severely that in many countries in Africa and Central and South America, the elite 1 to 3 percent of the population now controls upwards of 70 percent of domestic wealth and land.

Many of the most pitched environmental battles to protect tropical rainforests are, at root, battles over land redistribution, agrarian reform, and control of natural resources. It is this interlocked agenda of social and environmental reform that brings environmentalists into direct conflict with the agendas of governments-qua-elites, and that brings government wrath down on the heads of environmentalists. Nowhere is this more clear than in the environmental skirmishes over rainforest deforestation in Central America, an issue that has captured international attention.

Central America   Central America's rainforests contain the densest concentrations of biological diversity in the world. But today, two-thirds of Central America's forests have been destroyed, and rainforest destruction continues at a conservatively estimated rate of 1,500 square miles annually.[35] Deforestation is the direct result of the unequal distribution of land that characterizes many of the Central American economies. A small segment of wealthy landowners controls most of the arable land in most of the Central American countries; most government and military officials rise to power from the ranks of these landowners. Landowners lease land to tenant farmers, but there is no guarantee of land tenure, and as opportunities open up in the international food commodities market, landowners convert peasant-farmed land, or standing unexploited forest land, into plantations for export crops. In the 1960s, the increasing demand for cheap beef in the United States and Europe resulted in the conversion of thousands of acres of farm and forest land into cattle ranches. In the 1970s and 1980s, the demand from the rich world for "luxury" crops boomed, and vast tracts of land in countries as diverse as Kenya, Mexico, and Colombia were converted to export crop agribusiness enterprise— providing broccoli, strawberries, and other "off-season" crops for North Americans in the depths of February winter, carnations and tropical flowers all year 'round.

Uprooted peasant farmers, in search of arable land, are often pushed onto the most marginal lands—steep hillsides, where their farming escalates soil erosion, or into previously unexploited forest tracts. Rainforest soil, though, is unsuitable for agriculture (the forest's fertility is held in the foliage itself), and both peasant farmers and cattle ranchers must continually expand their forest clearings in search of fertile land. Once begun, the consequences of rainforest destruction ripple throughout the entire ecosystem. Topsoil washes away from deforested regions. Erosion of watersheds becomes a serious problem, and rivers fill with silt washed away from deforested lands. Aquatic ecosystems, such as the reefs and mangrove swamps of Central America's Caribbean coast, have become choked with silt. Increasingly heavy pesticide application is the only way to sustain the artificial plantation agriculture on tropical soils. Pesticide runoff contaminates water supplies, kills insects and birds, and people. Pesticide poisoning presents a serious health hazard in Central America. Nicaraguans and Guatemalans today have higher levels of DDT in their bloodstream than any other people on earth; the average DDT levels in Guatemalan cows' milk is 90 times higher than that allowed by US standards.[36]

Environmental reports from Haiti and Puerto Rico reflect similar patterns of destruction caused by the skewed distribution of wealth and power. An American Peace Corps volunteer wrote this description from Haiti in 1987:

> The [Duvalier] government owns all the land on La Gonave [a small island 10 miles off the mainland] and collects rent from the entire population while providing little or no services to the population. Because the government does not allow anybody to own land on La Gonave, people do not really have the motivation to take care of their land, because someone with contacts in the government can easily claim land for their own. . . . Many of the farmers have become trapped in a vicious cycle where they cut down their trees to make charcoal to sell. That in turn leads to erosion which reduces the productivity of the land which forces the farmer to cut more trees to sell charcoal to live. This cycle has led to the complete deforestation of La Gonave. Barren mountains supporting very little crops [sic]. The lack of forest cover or trees has created desert-like conditions as the soil has no protection from the tropical sun. Ecologically the island is a nightmare.[37]

Much of the world's remaining tropical forest is in countries where the military has extraordinary influence in government—countries including Burma (Myanmar), the Philippines, and several nations in Central America—where resource exploitation, and *deforestation in particular*, serves military purposes. In other countries, governments resist environmental reform not only because the elite individuals in government wish to retain ownership over their extensive landholdings and retain their privilege to take profit—at whatever environmental cost—from resource exploitation, but also because the solution to deforestation (and other environmental problems) often would entail a radical dismantling of social and political hierarchies, which are just as vociferously defended as is private wealth.

In many of the tropical-rainforest countries, deforestation is accelerated by the continuous displacement of landless tenant farmers by wealthy landowners. But in other states, displacement of the poor into forest lands has become official government policy. While governments may not see it as in their interest to alleviate poverty through radical economic reform, nonetheless, acute poverty, and the congregation of "poor masses" in concentrated pockets, can pose a threat to the viability of the ruling elite. One "solution" that a number of governments have found appealing is to move the poor. The Thai government is planning a massive relocation scheme, originally scheduled to begin in 1992.[38] The Brazilian government (in fact, through several regimes) has, for years, been aggressively promoting a campaign to move poor people away from crowded coastal cities into the forested interior; in this context, the forested interior is often spoken of as a social and political "safety valve." The catastrophic destruction of the Amazon rainforest, in which government subsidized transmigration plays a large part, is by now fairly well known. The Indonesian government's population relocation program has come under far less scrutiny, perhaps in some measure because it is supported by a number of other governments and the World Bank.

Indonesia   Starting in the early 1980s, the Indonesian military government embarked on what has been described as the "largest colonization scheme in history."[39] The government's Transmigration Program has already shifted 3.6 million people from the densely populated islands of Java and Bali to the less densely populated "outer" islands of the Indonesian archipelago, and initial plans called for 65 million more to be moved over the next 65 years. The government has recently softened its ambitious plans, but remains committed to the policy of transmigration.

It is a catastrophe both for the people and the rainforests of Indonesia. The outer islands support what is left of Indonesia's rainforests, and are home to a number of indigenous communities. The transmigration project threatens both. Internal Indonesian government reports themselves conclude that transmigration is the single most dangerous threat to the nation's forests and is likely to cause the loss of an area of forest the size of Belgium during the current five-year plan.[40] The tribal peoples of the outer islands have been dislocated, and their land expropriated.

The transmigration project was heralded as a means of alleviating poverty for the country's 165 million people, crowded onto only 7 percent of the national territory. The government lured millions of poor Indonesians into the scheme by promising land, initial relocation assistance, and a vision of unbounded agricultural riches. Others were involuntary migrants: for those whose lands had been expropriated by development schemes involving, for instance, dams and mines on Java, transmigration was their only offer of compensation. City vagrants, prostitutes, and lepers have also been rounded up and "urged" to join the program.[41]

The promised paradise never materialized. As with all agricultural schemes on cleared tropical forest land, agricultural productivity plummets after the first year or two. Migrants are forced by crop failures to move deeper into the forests to clear fresh plots. Settlers often find themselves worse off than they were before. Some transmigrants are in such dire straits that rural women have, in some cases, turned to prostitution in the islands' new towns, and there are documented cases of families selling their children.[42] In the majority of migrant families, men have sought wage labor off the farm to supplement their diminishing agricultural livelihood. In the words of one observer,

> This puts an added strain on the women who stay on the settlements while their husbands are away. Socially isolated on their individual homesteads, many struggle to feed their families. Not only do they lack government services such as schools and clinics, but they have also lost the support network of friends, neighbors and relatives that lightened the burdens of life back home.[43]

Several years now into the project, the government itself has abandoned the pretense that the transmigration effort is really alleviating poverty, and yet its commitment to transmigration has barely waned. The underlying objectives of the program are becoming more clear. The military regime sought to have total control of the far-flung empire of 3,000 islands that it inherited from the Dutch; transmigration would consolidate national control over the whole archipelago. In some critical areas, military bases have been established to ensure this control. Extermination of ethnic minorities appears to be a deliberate goal of the program, rather than an unfortunate side effect; tribal culture is seen by the authorities as an obstacle to development. There is a strong racist element in this policy, with Javanese migrants being urged to settle on the lands of the racially distinct Papuans and intermarry with them. In the words of the governor of the territory, "this will give birth to a new generation of people without curly hair, sowing the seeds for greater beauty."[44] Transmigration aids the consolidation of land ownership on the main islands, while it also helps to conceal economic inequities there.

Largely spurred by environmentalists' outcries over the destruction of indigenous cultures and the rainforest, the Indonesian government has come under domestic and international criticism for its transmigration policy, which in turn has renewed scrutiny of its other (numerous) human rights abuses. The Indonesian government has reacted to the criticism defensively, and officials within the government have complained about the impropriety of "outside interference" in internal matters.

Other governments, similarly, have decried infringements on what they see to be their sovereign right to exploit national resources on their own terms.

CONTROLLING NATURE, OWNING RESOURCES

The environmental politics of the 1990s are complex and contradictory. We have achieved the insight that many of our most pressing environmental problems are global ones: problems that include ozone depletion, global warming, deforestation, acid rain, the international traffic in "exotic" animals and plants, and the overexploitation and pollution of the world's oceans, among others. Most informed citizens as well as most policy leaders acknowledge that these are transnational issues that can only be solved by worldwide government cooperation, by forging an ethic of globally shared environmental accountability and concern. Many environmentalists have long argued that natural resources—especially large biospheres such as tropical rainforests or oceans—must be construed *not* as the property of individual sovereign states, but, rather, as a global commons. (Similarly, environmentalists often argue that natural resources *within* countries must be seen as *common* wealth, and that governments have a responsibility to temper the presumed "right" of individuals to accumulate wealth based on resource exploitation without "outside" interference. It is, obviously, this latter extension of the argument that most antagonizes government leaders and that sets the tone for government/environmentalist clashes.)

Yet, at precisely this historical moment of perceived environmental possibility, nation states continue to wrestle with each other, carving up little bits of the planet and appropriating them as their "own." There is no softening of the ideology of privatized resource ownership among the governments of rich nations—indeed some, most notably the leadership in Britain and the US, are even more aggressively pursuing resource privatization. At the same time, governments of poor nations are intensifying their commitment to protect exclusive sovereign ownership of resources. In tandem, these trends do not bode well for the development of a global environmental ethic.

Winning Over Nature

The rich nations of the world today were not always global players. Without its empire, Britain would have remained an insignificant little island on the northern edge of the map. The US, without its territorial and economic colonies, would never have achieved "superpower" status. Most of today's rich countries *became* rich through unrestricted exploitation of their *own* natural resources, but mostly by laying claim to the resources of *others*. Control over natural resources is a symbol and also a vehicle of state power.

But what does it mean to "privatize" natural resources? The ideological underpinnings of resource privatization derive from a deeper philosophical stance toward Nature—namely, the notion that Nature can be owned and controlled. A number of historians identify the historical development of this notion with the emergence of the Judeo-Christian religious ethic, an ethic based on a presumption of natural hierarchies, with God and "man" on top, "nature" on the bottom: "[In Christian belief], man shares, in great measure, God's transcendence of nature. Christianity, in absolute contrast to ancient paganism and Asia's religions (except, perhaps, Zoroastrianism), not only established a dualism of man and nature, but also insisted that it is God's will that man exploit nature for his proper ends."[45]

The Western scientific revolution of the seventeenth century and the industrial revolution of the eighteenth century institutionalized and provided the means for "man's" ownership and control of nature. The scientific revolution, shaped largely by the philosophical and scientific writings of Francis Bacon [the "father" of modern science], was predicated on the understanding that "man" could and should control the unruly forces of Nature. Lynn White writes:

> The emergence in widespread practice of the Baconian creed that scientific knowledge means technological power over nature can scarcely be dated before about 1850, save in the chemical industries, where it is anticipated in the 18th century. Its acceptance as a normal pattern of action may mark the greatest event in human history since the invention of agriculture, and perhaps in nonhuman terrestrial history as well.[46]

Feminist historians, most notably Carolyn Merchant, extend this historical argument with a gender-based analysis.[47] Merchant argues persuasively that the scientific and industrial revolutions were premised on a view of nature as female, and that the quest for control of nature was *explicitly* couched as male control of a female force:

> Sensitive to the same social transformations that had already begun to reduce women to psychic and reproductive resources, Bacon developed the power of language as political instrument in reducing female nature to a resource for economic production. The "controversy over women" and the inquisition of witches—both present in Bacon's social milieu—permeated his descriptions of nature . . . and were instrumental in his transformation of the earth as a nurturing mother and womb of life into a source of secrets to be extracted for economic advance. . . . [For] the new man of science, Nature must be "bound into service" and made a "slave," put "in constraint" and "molded" by

the mechanical arts. The "searchers and spies of nature" are to discover her plots and secrets. . . . Here in bold sexual imagery is the
key feature of the modern experimental method—constraint of nature
in the laboratory, dissection by hand and mind, and the penetration
of hidden secrets—language still used today in praising a scientist's
"hard facts," "penetrating mind," or the "thrust of his argument.". . .
The new image of nature as a female to be controlled and dissected
through experiment legitimated the exploitation of natural resources.[48]

More recent feminist inquiries into the social and ideological construction of science support the analysis that modern science and industry
serve to extend and entrench male control not only over a feminized
Nature, but throughout civic life.[49] The development of modern capitalism,
signified by the industrial revolution, drawing on ideological justifications
and technical capabilities provided by the scientific revolution, was predicated on hardening the divisions between "private" and "public" spheres
of life. Women were increasingly relegated to the "private" sphere, and
the ideology of excluding women from public and civic life still shapes
modern Western culture. In the eighteenth and nineteenth centuries,
women were excluded (in many cases by force of law) from membership
in the emerging ranks of public, civic, and industrial elites. The ownership
of nature, and the concomitant possibility of accumulating wealth through
the exploitation of privately owned natural resources, were rights that men
reserved for themselves. It is still the case that women are largely excluded
from the leadership ranks of industries, militaries, and governments, those
institutions that claim the right to own and control nature and the exploitation of natural resources. When viewed in historical context, it is clear
that the privatization of natural resources primarily serves the interests of
entrenched *male* elites—and that the maleness of those elites is not a
mere coincidence.

Private Assets in a Free Market

The philosophical stance that Nature can be owned becomes "operationalized" in government policy through twinning the practices of resource privatization with the principles of "free market" economics. It is
private control of natural resources that ultimately defines and sustains
the "free market" industrial economy in rich-world countries. These policies are the hallmarks of capitalism, and they are the cherished principles
of the governments of the most powerful global political players. The
dramatic political changes of the early 1990s suggest that the black ink of

"free marketism" will spill rapidly over the global landscape, blotting out the colors of alternative economic models: Nelson Mandela recently articulated a vision of a newly democratic South Africa with a globally-integrated free-market economy; the leaders of the new Commonwealth of Independent States are falling over themselves in the rush for free-market Western economic reform; government leaders in Eastern Europe are pursuing an ambitious privatization drive—Western newspapers are full of ads to lure potential investors: "Learn how your firm can benefit from the privatization of Eastern Germany . . . Complimentary Breakfast Seminar"; even China is tilting towards a free-market economy.[50]

There was considerable ebullience in the West about the victory of free market democracy over the "evil empire" of Marxism; in Eastern Europe, there were raised expectations about the promise of environmental relief proferred by the leap onto the bandwagon of democracy and capitalism. But the free market is no panacea for the region's environmental ills. The environmental crises in Eastern Europe are not only, or even primarily, a crisis of economic ideology, and they will not be solved by replacing Marxist economics with free market economics. Already, the bloom is off the rose; anger at the environmental neglect of the old system is giving way to powerful new worries about the expected cruelties of the new free-market regime. As one city councilman in the Russian federation wryly noted, "From the standpoint of pollution, the market will be no kinder than the old command system. Profit is the new top priority, and the environment is considered an investment with no return."[51] The environmental shortcomings of the "new" economic regime are already becoming apparent: in Czechoslovakia, a strict monetary policy has prevented allocation of many resources to environmental measures; in the Russian federation, the government has levied a lien on foreign earnings in the private sector, once earmarked for imported pollution-control equipment, now comandeered for repayment of the foreign debt; eager to attract overseas investors, the Hungarian government has promised foreign companies cheap labour and lax pollution laws.[52]

Despite its all-too-evident shortcomings, as we wend our way through the last decade of the twentieth century, free-market economics appears to be taking over the world. Government leaders in the rich industrial nations of the world are even more, not less, committed to the principles of private ownership of resources. The policies of government leadership within two of the most prominent "developed" nations, the United States and Great Britain, have, over the last 15 years, extended private control of resources. "Privatization" has been the rallying cry of the conservative governments within both countries.

In terms of recent environmental policy in the United States, the governments of the Reagan and the Bush administrations have nurtured what has come to be known as "third-wave environmentalism."[53] (The "first wave" is defined as the early 1900s conservation of forest land and wildlife, and the "second wave," the legislation and regulation of the 1970s.) "Third wave" government policy steers a course away from government regulation on environmental matters, and is predicated on the belief that economic, market-based incentives will lead to a "natural" market force for environmental protection. The executive director of the Environmental Defense Fund, Frederic Krupp, describes this as a strategy to "harness the profit motive and introduce new incentives to get business to do the right thing in the first place,"[54] rather than burdening industry with regulations. Private industry, not surprisingly, is encouraging this approach, which the US Chemical Manufacturers Association, among others, applauds as "focusing on solutions."

In 1990, the Bush administration created a "Council on Competitiveness"—a secretive agency, run by the vice-president, with a mandate to free industry from the shackles of government regulation. Within the first year of its existence, the Council undercut the regulatory authority of dozens of government agencies, including the office that oversees workplace heath and safety (OSHA) and the EPA. The Council severely weakened the Clean Air Act, making it easier for utility companies to evade pollution controls; it killed an EPA incinerator rule that required operators to recycle 25 percent of reusable materials before burning garbage; and it halted the implementation of an EPA ban on incinerating lead batteries.[55]

Similarly in Britain, in pursuit of "free market" purity, the Thatcher and Major administrations have promoted privatization at every turn. One of the most controversial privatization projects was the recent selling-off of public water authorities. The water privatization plans were drafted by the Thatcher government in 1986, in close conjunction with the interests of a powerful but very narrow industry interest group. Apart from this interest group, it was difficult to find anyone in Britain who wanted water privatized; in opinion polls, water privatization was universally unpopular.[56] Public outcry forced changes to the original document, but the final agreement to convert water provision into a private service was concluded in 1989. A century ago, in most industrial countries water supply *was* a private business. But in many countries, by the late 1800s or early 1900s, water supplies were deprivatized when it became evident that private interests could not ensure equal access to safe water: turn-of-the-century private water enterprises were typically incompetent and corrupt, providing piped or carted water to the few who could pay and safe water to no one.

There are legitimate concerns that the *re*privatization of such basic services as water supply will re-create a class-based hierarchy of access to "common good" resources; this is an even more pressing concern when ideologies (and technologies) of privatization are exported to Third World countries, where urban conditions in many ways parallel the conditions of nineteenth-century British cities. British (and French) programs for the privatization of the water-supply *are* in fact being exported to poor countries, in some cases under the aegis of the same coterie of financiers that fought for private water contracts in Britain. The first British export contract is already in the bag: a consortium is building an entirely private water system for the Malaysian city of Ipoh.[57]

"Privatization" continues to be held out to the developing world as a panacea for water systems, a philosophy that is supported through such international development agencies as the World Bank. Descriptions of water-supply projects funded by international aid nowadays contain the ominous phrase, "institutional reform," which means some degree of privatization. Two years ago, the Ghana Water and Sewerage Corporation began removing the handles from pumps in villages that would not or could not pay new tariffs imposed by international aid financiers.[58] The outcome was not more income for the Water Board, but more disease in the villages as people reverted to "free" but polluted water sources.

Privatization—the shift from community control and management of common property to state or individual ownership and control—and the economic model of the free market is responsible for much of the ecological "global wilding" of the last century. That it is a model fundamentally inequitable and exploitative of human and natural environments is clear to most people working for social justice and environmental sustainability in the poorer countries. Peggy Antrobus, the Coordinator of DAWN, a Third World feminist development and economic network, makes this point clearly:[59]

> The ecological crisis is the other side of the coin of this dominant economic model—a model that has been further reinforced by the changes in the socialist countries and the apparent triumph of the free market. I am amazed at the lack of reference to the current trends in the global economy in discussions on the environment. I want to know, is it that there is a deliberate effort to ignore these connections, to keep these things separate? Or do people not realize the implications of the unrestricted market on the environment? One of the things that is going to lead to a stalemate in Rio [the 1992 UN conference on the environment] is the failure of the chief negotiators, both from the north and the south, to recognize the contradictions between the free market and environmental protection.

And yet, privatization and the private accumulation of wealth through privatized exploitation of natural resources increasingly is the beacon of hope that rich world governments hold out to the poor. A Latin American observer comments on what is now a universal trend:

> Latin America and the Caribbean are told that privatization is a key to economic development. This is a notion that cannot be dismissed out of hand by progressive forces because the experiences of state-led development and planning have not always been positive. Nevertheless, it is not clear that people are ready to simply transfer state resources from bureaucratic to entrepreneurial enterprises.[60]

Despite compelling evidence of the negative impacts of transferring privatization ideologies to the Third World, and despite the overwhelming evidence of the ecological catastrophe wrought by the "free market," it continues to serve the interests of the governments in the rich industrial world to export this particular model of "development."

Exporting the Model: International Development

The policies and power relationships that are currently emerging under the name of "development" are surprisingly similar to those that defined the colonial system.[61] International development aid is orchestrated by large intergovernment institutions, most particularly the World Bank and the International Monetary Fund (IMF). These two institutions provide loans to poor countries for "development" projects, but in doing so they impose often coercive conditions on aid-recipient governments; governments in poor countries, once enmeshed in the aid/debt cycle, cannot extricate themselves from the demands of the World Bank and the IMF. In many Third World countries, it is now not the domestic government that controls economic policy but external rich-world governments acting through the World Bank and the IMF. Recent dramatic changes in world politics suggest that the "First World" development model is now also being exported to Eastern Europe and the former USSR. Economists, financiers, and government officials from the leading industrial democracies are now key players in the reconstruction of the Commonwealth of Independent States's political and economic system. In the eyes of one observer, "this means that the industrialized nations have started to treat the Soviet Union as a developing nation . . . in need of basic reform before new aid can be granted."[62]

Most international development aid from rich countries to poor is predicated on the assumption that for poor countries to successfully "develop" they must mimic the ideologies and economic policies of the already-developed world. The current economic development strategy for the Third World consists mostly of bringing Third World countries into the orbit of international trade by influencing them to eliminate subsistence agriculture and artisan modes of production catering to a local market, and to replace them with capital-intensive plantations and factories geared to the international market. Thus, rich-government-to-poor-government aid typically emphasizes large capital-intensive projects, large construction and infrastructure projects, such as dams and other power structures, and agricultural transformations encouraging export-oriented crop production. This model of growth, when exported to poor countries, has widely contributed to rapid exploitation of natural resources and, by increasing the gap between rich and poor within Third World countries, it has escalated poverty, which itself, in turn, has put unsustainable pressures on the environment.[63] An independent World Bank monitoring agency details the process:

> In a mad competitive rush to export to shrinking Northern markets, the nations of the South, aided by official lenders and Northern investors, have been raping their forests, polluting their rivers, poisoning their soil . . . . Bio-diversity is being lost and land ownership further skewed in the attempt to modernize and maximize agricultural export production. Little time is given to a consideration of the quality and environmental impact of these projects and investments and even less to the views of the local population.[64]

In the late 1980s, environmentalists turned up the heat on the World Bank, identifying a panoply of environmental horrors caused by Bank-funded projects. Among other projects, the World Bank has recently funded mining ventures and jungle colonization schemes in the Amazon, has encouraged huge hydroelectric dam schemes in Indonesia and India that will displace millions of people, and it has helped underwrite expanded private cattle ranching schemes in Botswana that would appropriate previously communal grazing lands.[65] In 1987, bowing to rising pressures, the World Bank announced a major policy shift, accompanied by internal structural changes, to "balance growth with environmental protection." Since then, Barber Conable, president of the World Bank, has gone on record repeatedly in support of environmental safeguards for Bank funded ventures. The efforts of the World Bank to "go green" appear to be sincere, but have received generally mixed reviews from international environmental groups.

One of the problems with "green aid" is that it is just a variation on a theme; it is not a new departure. Aid—even green aid—is provided within the context of "development," and in large measure this the core of the problem. Within the World Bank, after as before its green conversion, conventional ideologies of development remain intact. Large-scale infrastructure projects are still favored, and in fact in the 1990s a new generation of "megaprojects" are in the making, many of them water-development schemes, that are no less out of scale with the lives of the people whose land they will use than the enormous development projects of the 1960s and 1970s.[66]

International development is characterized by uneven power relationships. Since its green conversion in the late 1980s, the World Bank now imposes "green conditionalities" on many loans to the Third World. A number of Third World governments interpret these conditionalities as infringements on their sovereign rights, and as yet another attempt to ensure uninterrupted rich-world access to the resources of the poor.[67] International aid, even newly environmentally-sensitive aid, is still top-down. Governments on the receiving end of international aid complain of a new "green colonialism."

Economic relationships between the First and the Third Worlds are largely shaped by the debt crisis, a crisis that is also one of the gravest environmental threats of the late twentieth century. Governments struggling to pay off foreign debts encourage rapacious resource exploitation. In Ghana, for example, half of the farming land is used for growing cocoa for export; the government of Sudan exported food during the height of recent famines to pay debts; in Colombia, despite widespread malnutrition, flowers are grown on prime farming land to provide export revenue; UNICEF estimates that 500,000 children die each year because of the debt crisis.[68]

Moreover, "development" has been a disaster for the women of the Third World. Vandana Shiva, an Indian environmentalist, suggests that the dual victimization of women and the environment is no coincidence—rather, that it is inherent to the structure of development: "[Development is] an extension of modern Western patriarchy's economic vision based on the exploitation or exclusion of women, on the exploitation and destruction of nature, and on the exploitation and destruction of other cultures. . . . While gender subordination and patriarchy are the oldest of oppressions, through development they have taken on new and more violent forms."[69] Shiva makes clear the links between the destruction of the environment in the name of development and the subordination of women: women's economic work in the household and in subsistence agriculture are rendered invisible in the bottom line of development accounting; "unimproved" Nature, organic agriculture, and subsistence farming, all of which flourish under women's stewardship, are deemed "unproductive."

When Barbara Rogers wrote her pathbreaking book about development and women in 1980, *The Domestication of Women*, she stated unequivocally that "all those [people] who determine the formulation, design and execution of development policies, programs and projects . . . are men."[70] More than a decade later, this is still largely the case. As a small example, in February 1992, the US Agency for International Development (AID) released its annual list of "who's who" in the field; of 46 AID country directors, 3 are women; of 40 deputy directors, 7 are women.[71] The assumptions that Western men bring to development work on issues such as the appropriate or "normal" sexual division of labor, about how to measure "economic" activities, about sexuality, and about women's 'place' become imbedded in the development process—usually to the disadvantage of women.

"Development" exacerbates the hardship and oppression of women in poor countries in specific ways:[72]

• The model of economic development imposed on poor countries increases social inequities within aid-recipient countries. The introduction of capital investments and industrialization creates in Third World countries a middle class consisting largely of (male) government officials and of (male) traders and manufacturers, some of whom enjoy considerable affluence. The poor, however, get poorer. "Privatization" exacerbates pre-existing class *and gender* inequities—women, especially poor women, lose the most when human and natural resources are privatized.[73] At the same time, poor-country governments accumulate huge debts— development is financed by international loans. When repayment becomes a problem, the IMF steps in with mandated "structural adjustment programs"—mostly, programs that require the aid-recipient government to reduce their domestic expenditures (so there is more money to pay off foreign loans) by implementing "austerity" measures. These austerity measures, and other "structural adjustments," usually start with reductions in government subsidies for staples such as milk and bread. Structural adjustment exacts a terrible toll on the poorest segments of society and on families living on the most marginal resources. The means for women to provide food and shelter for their families are pushed further and further beyond their reach. The resulting increases in malnutrition and in maternal and infant mortality are predictable and ubiquitous in countries struggling under foreign debt. Everywhere in the world, women constitute the largest population in poverty. International development aid accelerates the international "feminization" of poverty.

• Industrial development within Third World countries means that Third

World governments open up their economy to foreign investment—providing international access to a cheap domestic labor force, which is most typically women's labor. Women's labor is *made* cheap by agreements between men in domestic and foreign governments:[74]

> The international political economy works the way it does..in part because of decisions which have cheapened the value of women's work. These decisions have first feminized certain home and workplace tasks—turning them into 'women's work'—and then rationalized the devaluation of that work. . . . Organizing factory jobs, designing machinery and factory rules to keep women productive and feminine—these were crucial strategies in Europe's industrial growth. Industrialized textile production and garment-making were central to Britain's global power. Both industries feminized labor in order to make it profitable and internationally competitive. Other countries have learned the British lesson in order to compete in the emerging global political economy. . . .[75]

In the current international economy, the female "global assembly worker" is the modern equivalent of the British or New England "mill girl."

• International development strategies encourage the export of raw materials (or assembly pieces) from poor countries, and the import of manufactured goods. This has the effect of putting local artisans out of business, many of whom are then forced into marginal livelihoods in urban slums. Women comprise the largest number of the small-scale artisans in poor countries, the arts and crafts workers selling to local markets who are now being displaced by export-oriented industries.

• Current international development strategies that encourage the development of export-crop agriculture are modeled on and were foreshadowed by "green revolution" schemes of the 1960s. In India, a country that invested heavily in pesticide-dependent "green revolution" agriculture, the "revolution" has displaced workers and increased worker exploitation, it has aggravated class conflict, it has disrupted traditional societies, and it has had disastrous environmental consequences. Most of these impacts have been disproportionately felt by women.[76] Current schemes for the development of large-scale export-crop agribusiness in the Third World promise no better—they are a disaster for densely populated, agrarian-based countries of the Third World. In such countries, every acre of land converted into export-crop agriculture is an acre of land made unavailable for feeding local people; thus, the export of agricultural produce is only possible at the cost of increasing malnutrition. In conditions of food scarcity, women eat less and last; when malnutrition increases, women suffer first and most.

Moreover, women are the majority of the world's farmers (though not of the world's landowners). The transfer of land from subsistence agriculture to export agriculture involves a property transfer that disenfranchises women: land is taken out of women's control and put into the control of male-dominated agribusiness. Development advisors from industrialized nations assume that men are the appropriate recipients of new seeds, tools, training, and credit, even in areas where women are the primary agriculturalists.

As Cynthia Enloe makes clear, using the example of banana agribusiness, the international economy of export agriculture is a highly gendered enterprise:

> Notions of masculinity and femininity have been used to shape the international political economy of the banana. Banana plantations were developed in Central America, the Caribbean, Africa and the Philippines as a result of alliances between men of different but complementary interests: businessmen and male officials of the importing countries on the one hand, and male landowners and government officials of the exporting countries on the other. To clear the land and harvest the bananas they needed a male workforce, sustained at a distance by women as prostitutes, mothers and wives.[77]

Vast areas of the Third World have been turned over to growing cash crops. In West Africa, for example, 70 percent of Gambia's arable land and 55 percent of Senegal's is used to grow peanuts; in Africa as a whole, the production of tobacco has increased by 60 percent since the mid 1960s.[78] Agribusiness and export-crop schemes hold the promise of great wealth, but only for a small number of entrepreneurs, usually an elite class of men with prior ties to governments.

• A recent report from Brazil carried the chilling headline "Slave Virgins Up for Auction at Amazon Mining Camps." Prostitution is omnipresent in the frontier-like settlements created by the rapid and rough expansion of industrialized mining, forestry and agribusiness schemes throughout the Third World.[79] For urban women who can no longer support themselves through arts and craft enterprise, and for rural women subsistence farmers who are displaced from their land by export-crop enterprises, prostitution is often the last resort. "Sex tourism" is one of the largest sectors of growth in the economies of many Third World countries; prostitution has become a big business prop of the international political economy.[80]

International development aid has the effect of reshuffling existing social and political hierarchies within poor countries. It reinforces *prior* inequalities between men and women, and introduces *new* market forces

BEFORE AID: Agricultural life was much simpler...

NOW WE'VE GOT AID... that's all changed....

So THEY'RE GIVING US MORE AID...!

Figure 3.1.   The debt/aid trap. (© ISIS International/Liz Mackie, by permission.)

that entrench women at the bottom of political, industrial, and food pyramids. At their apex, the organizations that control international development directives, such as the IMF, are thoroughly masculinized. Ironically though, the top-down, male-driven development strategy has mobilized women's political organizing throughout the Third World. In Mexico, women have taken the lead in fighting the austerity programs imposed on the Mexican government by the IMF; in the Phillipines, the Freedom from Debt Coalition has adopted a clearly feminist strategy. The old assumptions that with the development process the availability of goods and services would automatically increase and poverty would decrease is now under serious challenge from women's movements in the Third World, even while it continues to guide development thinking in centers of patriarchal power.

Exporting the Model: "Free" Trade

"Free trade" is the geographical arm of "free market" economics. Free trade agreements are being touted by most rich-world governments as the economic vanguard of the twenty-first century. Current government administrations in many of the world's industrial nations are enthusiastic about free-trade agreements—primarily because they facilitate the expansion of the rich-world's industrial growth, and are ideologically in synch with notions of unrestricted private-enterprise exploitation of resources. Environmentalists are worried.

In 1992 the American, Canadian and Mexican governments negotiated a North American free trade agreement. It roused considerable environmental controversy. Environmentalists are worried about the ecological implications of free trade on a continent that includes a First World superpower and a Third World country. In principle, the North American free-trade agreement *could* provide economic clout to *increase* pancontinental environmental enforcement; reality, this is unlikely to be the case. Most likely it will only further the environmental exploitation of America's poor neighbor, Mexico. The US president negotiating the agreement, Bush, pushed for "fast-track" approval of the free-trade treaty, and aides to Bush specifically said that they did not want negotiations to be slowed down by including any specific environmental assurances. Instead, Bush suggested, environmental issues could be negotiated in side agreements after free trade begins. The Bush administration has pledged to continue current levels of environmental protection along the border (hardly an encouraging promise), and to provide for future consultations between the US and Mexico on possible imposition of tougher transboundary environmental regulations. In the words of one observer, "it's the old 'trust us' mentality."[81]

The government of Canada signed a free-trade agreement with the US in the late 1980s, over vociferous domestic opposition. The agreement has crippled the Canadian economy, and Canadian observers caution their Mexican counterparts:

> We have seen in the Canada-US free trade negotiations that corporations are looking at establishing one large economic entity with common standards. Those standards are invariably the standards that best serve corporate interests and profits. We can expect that decisions will be made that will endanger the environment and people's health. . . . We are already seeing in Canada the social and human costs to the Canada-US trade deal. . . . The proposed deal with Mexico also will have heavy human costs. Though there will be a small circle of winners, there will be far more losers as workers in Canada, the US and Mexico compete with each other through lower wages, poorer working conditions, and less stringent environmental standards. . . . companies will be attracted [to Mexico] by the absence of strong health and safety as well as lax environmental regulations in Mexico.[82]

The compression of North America into a single economic zone is just the most recent in a growing number of inter-state economic conglomerations, including the European Economic Community, a proposed pan-continental Latin American economic zone, and a proposal for Asian economic zones. With the rising prominence of free-market economics, there is an accelerating worldwide trend toward lowering international trade barriers and in favor of the creation of larger economic entities with harmonized economic (and environmental) policies. There is every reason to be afraid that "harmonization" of trade regulations may favor the *lowest* environmental denominator.

The painfully slow progress toward international environmental treaties—on issues such as ocean protection and ozone depletion—may be undermined by international free-trade regulations. For example, in the Fall of 1991, a dispute-resolution panel, set up under the worldwide trade agreement known as GATT (General Agreement on Tarriffs and Trade), ruled that the US violated trade agreements when it banned Mexican tuna imports on the basis of evidence that Mexican tuna fleets kill too many dolphins. A number of government officials and environmentalists say that the GATT ruling's broad wording appears to reach beyond the specific US-Mexico dispute, holding as a general principle that countries cannot use trade sanctions to protect resources beyond their territorial limits. The precedent, environmentalists say, portends a broad challenge to a host of environmental laws and treaties.[83]

The 1991 GATT talks produced a proposed new ruling that will prevent any country from discriminating against the products of another country for their "method of production." According to GATT rules, the method in which things are produced, regardless of their ecological impact, can neither be encouraged or discouraged because such measures will act as barriers to free trade. If approved, this new ruling will force nations to open their markets to everything from unsustainably harvested timber to food grown with unacceptably high levels of pesticides. In the view of several environmental specialists, this new interpretation of GATT free-trade laws strikes at the heart of a series of recent environmental treaties negotiated to protect shared global resources, including restrictions on the trade in endangered animal species.

THE "EVIL TWINS": SOVEREIGNTY AND NATIONALISM

Sovereign Rights

Some poor countries, struggling with the burdens of "maldevelopment," have taken the lead in calling for more cooperative and equitable international systems. At the same time, many poor countries, throwing off the last shackles of colonialism, are increasingly possessive of their natural resources. Ironically, environmental fears seem to be fueling an increasingly vociferous resurgence of nationalism and sovereign claims to exclusive control of resources among governments in the Third World.

The "private" control of resources by sovereign states is the ideological mirror image of private control of resources by industrial/military/government elites within individual countries. The ideology of individual/private control of resources *within* a country mirrors and emboldens claims to sovereign/private control of resources *among* countries of the world. It is not surprising that governments in poor countries are taking on the broad lessons of "privatization" offered by the rich nation states, turning them to their own advantage in the power game of sovereign states by laying claim to sovereign/private ownership of natural resources. As an ideological stance, both the private control and ownership of natural resources and the sovereign control of those resources are conceptually rooted in the premise that "Nature" itself can be privately controlled and owned.

The governments of Brazil and China are among the most vocal proponents of privatized sovereign control of resources.

Brazil   While growing international concern about the destruction of the Amazon rainforest has raised the hopes of environmentalists and forest dwellers, it has added to the nervousness of Brazilian politicians and military leaders about foreign interference. In the early months of 1989, the government of Brazil, in close conjunction with the military, embarked on an ambitious public relations campaign to denounce the international campaign to save the forests. Dark suggestions of foreign conspiracies, and even of possible foreign invasion, were floated by military and government officials. Brazilian President Sarney made a series of high-profile press announcements, claiming sovereign rights over Brazil's resources:

> Brazil is being threatened in its sovereign right to use, exploit and administer its territory. Every day brings new forms of intervention, with veiled or explicit threats aiming to force us to take decisions that are not in our interest. . . . We are free. Brazil is ours. Nature is ours, and it is our duty to defend it. . . . We don't want the Amazon to become a green Persian Gulf."[84]

While the conspiracy theories did not seem to take hold, these complaints bolstered both the military and right-wing coalitions of landowners. The Brazilian military was given a fresh mandate, that of protecting the country from outside environmental interference. President Sarney told the armed forces high command that, "now more than ever they had the mission to defend the Amazon."[85] The slogan, "the Amazon is ours," was adopted by the right-wing Rural Democratic Union, a coalition of landowners, and by several members of Congress with ties to the construction companies that have large stakes in the continuing development of the Amazon. Seven other South American governments united behind Brazil to denounce foreign pressure to save the rainforests, saying they would not take orders from abroad on their environmental priorities.[86]

China   Four decades of headlong industrial development in China have resulted in an ecological catastrophe within the borders of that country. Outside observers claim that China's industrial growth is about to become a major global problem as well.

The most pressing environmental problem in China is raw industrial pollution. The Chinese government embarked on an ambitious program of industrial development in the 1950s, and has barely looked back since.[87] In Beijing, the air is 16 times dirtier than it is in New York, and an astonishing 35 times more contaminated than in London. Air pollution in 60 Chinese cities exceeds the "danger" levels set by the World Health

Organization. Shanghai, China's largest city, suffers from the nation's worst water pollution, and related rates of stomach, liver, and esophageal cancer are described by one health official as "epidemic."[88]According to one official estimate, 29,000 miles of China's waterways are seriously contaminated by industrial toxins, and most are polluted by raw sewage; only about 2 percent of sewage and household wastewater in China is funneled through treatment plants—the rest is dumped directly into the nearest body of water.[89] The vast Bohai Sea, outlet of the Yellow River, is "dead, or nearly so," and nearly 1,500 miles of waterways are so polluted that they no longer support fish. The acid rain problem is enormous, and close regional forest studies show a 40 to 50 percent tree kill-off rate from acid rain.[90]

The Chinese government has made a recent commitment to environmental protection. But the problems are overwhelming, and it is unclear what effect environmental regulation will really have on protecting the environment of the Chinese. Among the global community of nations, China's environmental problems would remain largely overlooked, were it not for recent worries, aired by the rich governments of the world, about the future effect of the world's most populous nation on the global environment. At present, China's impact on the global environment remains small when compared to the atmospheric and oceanic damage done by the US, Europe, and the former Soviet Union. For instance, although China is home to one-fifth of the world's population, its emissions of "greenhouse gases" make up only about a tenth of the world total. Similarly, China's release of chlorofluorocarbons (CFCs), responsible for destroying the ozone layer, is miniscule—barely one-tenth of the CFCs released by the US alone.[91] But China's industrial ambitions, outside observers claim, could prove to be a global disaster, and there is increasing pressure on the Chinese government to redirect their plans for industrial growth.

Chinese officials—like officials in most developing nations—resent Western demands that they clean up their environmental act for the "global good." "They feel that the Chinese shouldn't pay for the past ecological sins of the developed world by giving up progress." [92] "Outside interference" of this sort can also serve to aggravate internal tensions within poor countries. Environmental criticism from Western nations has increased the paranoia of Chinese government officials about the collusion of enemies of the state from within and without, and has accelerated internal repression of environmental activists. The failure of the government to halt the destruction of China's environment was one of the grievances that fueled the prodemocracy demonstrations in China in the late 1980s. A Chinese scientist known for his criticism of government environmental policy outlines the effects of his government's nervousness about environmental criticism:

> We have to tiptoe where once we walked boldly, every issue suddenly has a dangerous political dimension to it. Campaigns to protect the environment will probably become very difficult because we might be accused of 'spreading dissent,' or 'counterrevolutionary crimes' if what we say displeases the party.[93]

Eco-nationalism

Claims to exclusive sovereign rights over resources by governments in developing countries are inextricably intertwined with postcolonial nationalist movements in many of those countries. Nationalism in the guise of environmental sovereign rights is a double-edged sword. While on the one hand it may offer some degree of protection from the worst "colonizing" exploitation of resources by foreign powers, a nationalist political stance also elevates internal unity to a position of political primacy. Many of the governments most vociferous about nationalism are exactly those least able to tolerate internal dissent—including criticism from environmentalists. Environmentalists in China are labeled as enemies of the state. In Brazil, the frenzy of nationalism whipped up by President Sarney and the military in 1989 cowed environmentalists; the only "Green Party" congressional representative reflected, "These days officials are treating me like an anti-patriot, a traitor."[94] Nationalism often becomes a tool of the state for suppressing dissent from government opponents. Nationalist sentimentality about resource ownership often serves to legitimate unregulated resource exploitation within the nation-state—by local rather than foreign elites.

Whose Idea Is the Nation-State?   It will be extraordinarily difficult to arrive at ecologically sound solutions to the world's environmental problems within the context of a global system comprised of militarized, privatized sovereign states; the record to date suggests it may be absolutely impossible. As one environmentalist notes: "While military security rests firmly on the competitive strength of individual countries at the expense of other nations, *environmental* security cannot be achieved unilaterally: it both requires and nurtures more stable and cooperative relations among nations."[95] At the Rio Earth Summit in 1992, a gathering of the world's nations to assess the state of the world's environment, nationalist protectionism—vaunted by both rich and poor countries—defeated any hope for global cooperation. Nationalist chest-thumping turned the Rio meeting into a sorry and shallow exercise in futility. One observer at the conference commented, "The world is not at a state yet where we are ready to take measures to preserve the whole which override the interests of national states. It's like a pinball machine. Whenever you push the sovereignty button, all the red lights go off."[96]

The sovereign-state global system is an anachronism, one at odds with an environmentally sound system that would require cooperation and sharing for a common good. The majority of environmental problems demand regional or even global solutions that encroach upon what we now think of as the prerogatives of national governments. As Jessica Tuchman Mathews, vice-president of the World Resources Institute, points out, today's negotiating models for intergovernment agreements are inadequate.[97] Intergovernment treaties are weak tools: despite the existence about 61,000 treaties covering almost every conceivable area of international concern (only a small minority of which address environmental concerns), "these intergovernmental treaties were drafted by nation-states and designed to protect the sovereignty of nation-states, and they recognized as legitimate actors only the governments of nation-states."[98] Accords between sovereign states that do not challenge the basic assumptions of militarized, privatized sovereignty will be compromised and unwieldy tools of environmental change. Hard as it is to imagine, an effective politics of environmental globalism will require a significant seachange in the culture of sovereignty.

Despite growing awareness of the environmentally dysfunctional nature of the current global system, there is surprising reluctance to even putting the issue of sovereignty on the environmental agenda.[99] To understand this reluctance, it may be important to probe the extent to which sovereignty, nationalism, and territoriality are socially constructed forms—in which certain social groups may have entrenched vested interests.

Once the question of social/cultural origins is broached, feminist metaquestions cannot be avoided: Is it coincidental that the system of sovereign statehood is propped up and protected by world governments and militaries in which men, almost universally, prevail? Is sovereignty a *male* construct? Is it possible that the global system of territorial statehood is a product of an identifiably *masculinist* consciousness?

Feminists in political science, geography, and international relations have just started to explore these questions, and, as with other touchy issues, there is considerable debate (and some acrimony) about the conclusions.[100] Nonetheless, the notion that sovereignty may well be a masculinist project *is* suggested by a number of often-taken-for-granted features of the contemporary social structure—(or, at least, of the Western social structure; feminists in other cultures are just starting to ask similar questions): • The ethos of militarized security is underlain by a valuing of independent strength and competitiveness, attributes that are conventionally encouraged and rewarded in men, discouraged and devalued in women. • Empirical evidence suggests that states and the inter-state system have been fundamentally gendered structures of domination and interaction.[101] • If sovereignty is conceptualized in terms of power and control, then, as one international relations academic notes, "phrased thus, sovereignty seems to reflect traditionally male thinking, with its emphasis on control and its penchant for absolute and dichotomous categories."[102] • The modern European-based state system has, since its birth, had an uneasy relationship with its natural environment: natural resources and geographical spaces have been viewed primarily as resources for increasing state power and wealth.[103] The emergence of the state system was coincident with significant developments in science and industry in post-medieval Europe—developments that were most notably characterized by attempts to gain mastery and domination of the natural world. As analysts such as Carolyn Merchant, Sandra Harding, and Evelyn Fox Keller demonstrate, mastery of the natural world through the scientific and industrial revolutions was a highly gendered enterprise, rooted in male control over women and women-identified nature. The erection of boundaries has been an important part, from the beginning, of the effort to gain mastery and control over nature. The history of the emergence of modern states is thus intimately interwoven with the history of new technologies, devices, and attitudes about controlling nature *and* women.

Similarly, nationalism—the twin of sovereignty—can be understood within a gendered context. While nationalism can be a tool of the state in squelching dissent from all government opponents, more specifically, from Northern Ireland to Afghanistan, nationalism has been a tool that bolsters the power base of men, *as men,* in government. Few nationalist movements are informed by women's experiences of oppression. Nationalist governments are just as likely as colonial governments to be structured by, and to institutionalize, male privilege. Fred Halliday, an international relations expert, writes:

> If there is an assumption that national independence and national interest take precedence over the claims of any specific group within the nation, there has also been the assumption that, in general, the spread of nationalism is beneficial to women since they are part of the nation. . . . There is, however, another side to the story. Nationalist movements subordinate women in a particular definition of their role and place in society, enforce conformity to values that are often male

defined, and make it possible to delegitimize alternative politics on the grounds that these are alien. The use made of nationalist and anti-imperalist arguments to discredit and silence feminist movements in recent years is indication enough of this.[104]

Women who raise questions of their own subordination in the midst of a nationalist resurgence are likely to be accused of undermining the struggle, or, worse, of threatening the security of the state. As Cynthia Enloe suggests,

> A nationalist movement informed by masculinist pride and holding a patriarchal vision of the new nation-state is likely to produce just one more actor in the international arena. A dozen new patriarchal nation-states may make the international bargaining table a bit more crowded, but it won't change the international game being played at that table.[105]

Pious Globalism

Given the much-cherished global system of sovereign states, it is surprising to find that many government leaders, especially those of the rich countries of the world, now appear to be taking up the cause of a new globalism. There are increasingly ardent proclamations from rich world governments about the need for individual nations to manage their resources for the "global good." These pious proclamations of "globalism" have been widely received with scorn and skepticism in the Third World, and with caution by many environmental grassroots groups—and for good reason. When the governments of the rich world embrace the arguments of environmentalists, arguments that they have resisted at most every other turn, there is reason to be wary.

The new globalism often appears to be little more than a cynical sleight of hand. More than an indication of a dawning environmental sensibility, it may simply betray the nervousness of rich world leaders who are witnessing their own ideology of "privatization" running amok. When poor countries use arguments of privatization to bolster their sovereign rights, which implies limitations on rich-world access to Third World natural resources, rich governments respond with "globalism."

It is unfair to suggest that all government advocates of the new globalism are cynically motivated; many of the individuals who are emerging as spokespeople for globalism may well be swayed by environmental reason. But the green conversion of the rich-world governments is unconvincing in the absence of evidence of a concomitant wider ideological transformation. To start with, a sincere commitment to a new globalism would necessitate a willingness among rich-world leaders to reexamine their much-cherished principles of resource privatization. It is a contradiction for the governments of rich countries to promote a development ideology based on the pell-mell exploitation of privately-owned resources for themselves, (and to export this model when it serves to bring poor countries into their economic orbit), while simultaneously arguing for the "globalization" of resources, such as the Amazon rainforest, that are located within the boundaries of poor nations. It is a contradiction not lost on the governments of poor nations.

Governments of poor countries are justifiably nervous when they hear government officials in rich nations talk about "our common" Amazon . . . or "shared, global" interests in leaving large parts of poor countries unexploited. As one observer from Latin America wryly notes, "when the United States has worried about us in the past this has usually signalled bad news."[106] The timing of the new globalism is suspicious: it is just at the point when many poor countries are coming into their own that rich countries respond with rhetoric about the global good. President Sarney of Brazil is not alone in worrying that "green" globalism is perhaps a concerted effort among rich nations to hold back the economic growth of poor countries just at the moment when many of those countries are emerging as economic powers that might challenge the hegemony of the rich world.

And, in fact, the proclamations of rich governments seem to have little more than rhetorical weight. As many Brazilians point out, rich countries need to clean their own house first before pointing the finger at poor countries. Many Brazilians argue that if the Americans are so concerned about global deforestation, they should stop chopping down their own forests first. Governments in the Third World are rankled by rebukes from countries such as Britain, France, and the US—countries, in the words of one Brazilian official, with "no enchanting environmental record."[107]

DUMPING ON THE POOR

Finger-Pointing

Governments of poor countries interpret the new globalism of the rich world as a concerted effort to shift attention away from their own egregious record of environmental degradation. By any measure, the weight of evidence underscores the responsibility of rich countries for many of the global problems we now face. Most of the world's most pressing environmental problems—fossil fuel pollution (and the related threat of global warming), the use and production of ozone depleting chemicals, the resource use and pollution generated by oil dependency, acid rain emissions, production of household and industrial waste, the luxury trade in exotic animals and plant species—are clearly the result of the economic priorities and lifestyles of the rich nations of the world. Having despoiled the earth for their own gain, it would appear that rich governments now want to halt development in the poor countries of the world—in the name of the global good.

Many governments of poor and rich nations are now embroiled in acrimonious disputes about who is responsible for the world's environmental ills. Governments of developing nations accuse rich governments of avoiding responsibility for their own environmental problems by diverting attention to the environmental problems of the developing world. Many activists in the Third World echo their governments' dismay at the trend that they identify as "blaming the victim." An Indian scientist explains:

> The exercise of blaming developing countries has already begun. Until recently, it was widely accepted that that developed countries of the West consume most of the world's fossil fuels and produce most of the carbon dioxide—the main agents of global warming. In recent years, however, Western nations have been carrying out a sustained propaganda campaign alleging that deforestation in developing countries, and the generation of methane through irrigated rice farming and the raising of cattle, is also contributing to global warming. This has shifted the onus onto developing countries. . . . No effort is made to separate the "survival emissions" of the poor from the "luxury emissions" of the rich. . . . It is time for the Third World to ask the West: "Whose future generations are we seeking to protect—the Western World's or the Third World's?"[108]

Finger-pointing at the Third World also masks the ironic fact that the wealth of the industrial world is based not only on uneven access to wealth and resources within its own national borders, but on the development of uneven relations between the rich and poor nations. Rich countries have *gotten* rich by exploiting the natural resources of poor nations, and they continue to do so. Mahatma Ghandi, when asked if he hoped India to reach the same standard of living as Britain, replied, "It took Britain half the resources of the planet to achieve their prosperity. How many planets will a country like India require?" It is clear that it is in the interests of the rich world to maintain a global system that is based on *uneven* economic and environmental relations.

Rich countries continue to enhance their wealth by expanding their resource reach beyond their own borders, while frequently exporting their problems. Having stripped their own forests, they are turning to timber plundering in former colonies; choking on a glut of municipal, household, and industrial garbage, they seek to ship away their waste to small Pacific islands or debt-strapped countries in Africa.

Trashing the Poor

By conservative estimate, the United States produces at least 275 million tons of hazardous waste annually; West European nations, together, produce an annual 22 million tons.[109] Millions more tons of "household waste," not officially classified as hazardous, are generated by rich countries each year. Rich countries are choking on their own waste, and running out of space for its disposal. In addition to simply running out of room, most industrial nations have increasingly stringent regulatory standards for disposal of hazardous wastes, regulations that have sent landfill and disposal costs in industrial nations skyrocketing.

One of the simplest "solutions" to the problem of what to do with the waste has been to ship it to countries with looser environmental regulations and cheaper disposal rates—countries with cash-short governments that need foreign exchange and trade, even if that trade is in waste. The dumping of waste in poor countries has since become a major industry, and the web of the waste trade includes virtually every country in the world.[110] In the US alone, the number of notifications of intent to export waste filed with the Environmental Protection Agency rose from just 12 in 1980 to 465 in 1987.[111] Greenpeace has catalogued at least 1000 attempts to export more than 160 million tons of toxic waste from rich countries to poor since 1986.[112] Governments in countries such as the US, the Netherlands, Switzerland, and the former West Germany, which have led the way in protecting their own populations from the hazards of waste dumping, have helped in foisting the problem onto poorer countries.

The international trade in hazardous waste caught public attention in the mid-1980s, following a number of much-publicized follies of barges floating around the world carrying cargo that no one wanted. A garbage barge from New York plied the seas from port to port in 1986–1987, with nowhere to dump its garbage. The city of Philadelphia contracted with a waste-disposal firm for disposal of thousands of tons of incinerator ash. The ash was loaded on a Liberian-registered ship, the Khian Sea, in 1986. Originally destined for dumping in the Bahamas, the cargo was refused entry there after government officials were tipped off about the hazardous contents of the cargo; the Khian Sea approached, and was denied entry, in Bermuda, Honduras, the Dominican Republic, and, across the ocean, in Guinea-Bissau, and Wales. Returning to the Caribbean, the waste-disposal firm made an agreement with local military officials in Haiti, where it dumped 3,000 tonnes of ash before being ordered out of the country by the national government. Two years later, the Khian Sea turned up in Singapore under a new name with an empty hold; as one observer comments, "for all intents and purposes, the incinerator ash disappeared into the Indian Ocean."[113]

These somewhat amusing stories of the-garbage-with-nowhere-to-go helped shed light on the extensive pattern of toxic dumping on the Third World. Among dozens of other stories breaking in the late 1980s, a few examples of dumping in Africa underscore the complex government/industry networks that facilitate dumping: • Five times in 1988, ships from Italy arrived in the remote Nigerian fishing village of Koko, dumping a total of 3800 tons of highly toxic wastes; the village of Koko has since been evacuated.[114] • A Dutch firm negotiated a contract with two government officials in the Congo in 1988 for the importation of one million tons of European toxic waste; the contract was eventually canceled after the deal became public.[115] • In 1986, high-ranking officials in the government of the Central African Republic were charged with negotiating a deal to dispose of 70,000 tons of European industrial and pharmaceutical waste in return for a $500,000 fee.[116] • In March 1988, a Norwegian ship dumped 15,000 tons of material labeled "raw material for bricks" in a quarry on an island off mainland Guinea; in June, Guinea newspapers reported that vegetation on the island was dying. Investigations revealed that the "raw material for bricks" was, in fact, incinerator ash from Philadelphia. Norway's Consul, Sigmund Stromme, was arrested and charged with forging documents in order to bring in the shipment; Mr. Stromme turned out to be a principal in the waste-hauling company.[117]

Following on the heels of the outrages of the late 1980s, African nations have forged strong internal agreements among themselves prohibiting the importation of waste (most notably the 1990 "Bamako Agreement" forbidding the import of hazardous and radioactive wastes into the continent of Africa). But poor countries, the targets of "toxic terrorism," cannot stop the flow of waste on their own. Industrialized nations need to take responsibility for the toxins and waste produced by their own industries. A much-heralded international agreement on waste traffic, the UN Basel Convention of 1989, failed to outlaw the flow of toxic waste from rich nations to poor; agreements made between the leaders of industrial nations continue to leave poor countries vulnerable to ongoing waste dumping. As one of the leaders of the Organization of African Unity (OAU) said, "After Basel, Africans realized that we would have to take responsibility for protecting our own continent, as it was clear that many industrialized nations were unwilling to do so."[118] The only industrialized countries that now categorically ban waste exports to the developing world are Italy and Norway.[119]

Governments in many industrialized nations, pressured by outrage over the most publicized incidents of international dumping, have instituted waste-export regulations based on the principle of notification and "informed consent." In the US, for example, the EPA must now notify and receive consent from the government of the intended recipient before a shipment of hazardous waste can leave port; the federal agency, however, has no authority to prevent a shipment from taking place, even if it doubts the country is capable of handling it. No restrictions currently exist on the export of wastes not technically classified as hazardous, such as incinerator ash, even though many of them can be highly toxic.[120]

Intergovernment agreements based on "notification and consent" legitimate the trade rather than preventing it. In a small world, there is no "away." The only way to solve the problems of the trade in toxic wastes is to forbid export, and, within industrial countries, to implement restrictions on the *production* of toxic materials. But "free market" governments are reluctant to intervene in the regulation of private industry, a reluctance compounded by the fact that in many cases government officials have private financial interests in toxics-producing industries. The root cause of the garbage crisis is the unwillingness of rich-country governments to intervene in domestic industrial decisions. Combined with the corruptibility of government officials in many poor countries who exercise government authority for private gain, it's a dangerous situation:

> Until governments take steps to reduce the use of toxic materials, to detoxify the solid waste stream at its source, garbage will be a large and dangerous problem. And where there are large problems, there are large opportunities for self-interested business people and self-serving politicians to engage in shell games.[121]

And, so, the export of hazardous wastes continues, a practice that creates disincentives for industries and governments to reduce or manage the amount of waste they generate. Shipping toxics away gives the illusion of solving the problem.

"Informed consent" laws are weak regulatory tools: governments in poor countries often do not have the expertise to analyze the environmental impact of the contents of mixed-waste shipments; corrupt government—or military/government—officials in poor countries can often be bribed to grant shipping permission; the regulations rely on the company shipping the waste to provide information on what wastes are being exported. Government bureaucracies in Third World countries are often ill-equipped to respond to notification, and enforcement on both ends is likely to be inadequate. Many poor countries have neither the legislation nor the technology to ensure safe disposal of wastes once they arrive on their doorstep. Given the repeated failure of the United States and other industrial nations, with the most advanced technology in the world, to deal properly with hazardous waste, it is difficult to imagine that a Third World country could do a better job. The physical environment of many of the poorest tropical nations exacerbates the environmental damage of hazardous waste dumping. Frequent and heavy rains in the tropics accelerate the leaching of wastes into the soil under landfills, making water supplies prone to contamination. Health and environmental agencies within many poor countries often lack political clout and the scientific and financial resources to deal with environmental health crises.

Government officials in industrial nations resist the notion that *they* should take responsibility for evaluating the risks that export of hazardous waste poses for waste-recipient nations. Instead, government officials in rich countries invoke a cynical respect for the 'sovereign rights' of poor country recipients. In 1987, the American assistant secretary of state said, "We should not create a system that makes the US government responsible for what rightfully is a sovereign decision by others."[122] Regulatory officials shrug their shoulders at the implication of responsibility: "We only have control over what's in the country," said one EPA official, "Once it leaves, we can't do anything about it."[123] When pressed about US companies dumping in Mexico, the same EPA official repeated, "There's nothing we can do about it. It's really none of our business; it's a sovereignty issue."[124]

Waste trafficking is an international shell game. In light of African nations' strong stand against waste dumping, waste traffickers have shifted their sights elsewhere, most recently to the Pacific. Cash-strapped Pacific countries have been approached with offers of millions of dollars in return for allowing the dumping of garbage, toxic waste, and nuclear waste. • A California-based company has approached the government of Papua New Guinea with a proposal to build a "disposal facility" for hazardous wastes; they propose bringing over 600,000 tonnes of toxic waste a month for "treatment" in Oro Province. The Premier of Oro has supported their plan, primarily for the promised revenues—the provincial government would receive $45 (US) per tonne of waste imported, and the national government $15 per tonne.[125] • Another California-based company, LPT, is seeking approval to build an incinerator in American Samoa to burn US wastes and export the ashes to the Philippines.[126] • Agreements are being negotiated with the government of the Marshall Islands to dump millions of tons of American household garbage in a chain of atolls in the Marshalls. Promoters of the plan argue that the waste will provide landfill to protect the islands from the anticipated onset of flooding associated with global warming.

In addition to the Third World dumping trade, many industrialized countries exchange waste among themselves, although the flow of trade still reflects power imbalances between exporters and importers. Canada and Mexico receive the bulk of US toxic waste; Hungary, Romania, and the UK are major destinations for European wastes; China accepts waste from the USA and from Europe, much of which it dumps in Tibet.

The international waste trade is symptomatic of the larger problem of environmental "double standard" relationships between governments. The international trade in pesticides reflects similar dynamics. Under current laws, for example, US corporations are free to export pesticides that are banned in this country, unregistered, and acutely toxic, "restricted use" pesticides. And they do. At least 25 percent of the approximately 500 million pounds of pesticides exported from the US every year are not registered for use in this country.[127] The track record of many US chemical companies illustrates the importance of "unregulated" foreign markets to pesticide producers. For example, in 1988, after years of struggle by American activists, the chemical "chlordane" was withdrawn from US markets because of its carcinogenicity and persistence in the environment; two years later, overseas exports from Veliscol Chemical, the sole manufacturer of chlordane, increased tenfold, to more than 5,000 pounds a day.[128]

In 1988, the Irish government gave authorization, and a $3 million grant, to a British company for plans to manufacture soap containing mercuric iodide for export to Africa, where it is used to lighten skin color. The soap can cause fetal damage, and anemia and kidney failure in high doses absorbed through the skin, and the EEC has banned the marketing of cosmetics products containing mercury in its own member states. But production of such cosmetics for exports remains legal. An official of the Irish government, challenged on this issue, replied, "We are not in the business of taking moral decisions for other countries."[129]

Setting Our Sites on the Poor

It has not escaped the attention of industrialists that it is considerably more profitable to produce hazardous waste in poor countries in the first place, rather than produce wastes in the First World that then have to be shipped out. Locating toxic-producing industries in the Third World, an arrangement that is facilitated by intergovernment agreements, is merely a more subtle form of dumping on the poor. Increasingly, First World industries are exporting not just the waste, but the hazardous industries themselves.[130] The twin practices of dumping of wastes and locating hazardous activities in the Third World make "good" economic sense. In an internal World Bank memo in 1991, the bank's chief economist summarized the prevailing wisdom: "I think the economic logic behind dumping a load of toxic waste in the lowest wage country is impeccable and we should face up to that." He went on to argue that if pollutants cause diseases that turn up in old age, (a category that includes many cancers), it made the best economic sense to locate polluting activities in countries where people aren't likely to live long enough to fall prey to such diseases.[131] When this memo was leaked to the public, the economist claimed he was being ironic and provocative—which may be the case, but, in fact, he put his finger on the economic logic behind rich country/poor country relations.

Starting in the mid-1960s, American firms shifted many of their most noxious industrial plants, especially automobile assembly, electronics, and electroplating industries, across the border into Mexico. There are now about 1,800 American assembly plants, called "maquiladoras," in the Mexican borderlands. For US firms, Mexico offers a convenient location (the nearest Third World country) with cheap labor (mostly women workers), and lax environmental laws. For Mexico, maquiladoras mean exports, industrialization, and employment. By agreement between the Mexican and US governments, the maquiladoras import duty-free components and raw materials from the US and make them into finished products, which are then exported back to the US.

These plants also generate thousands of tons of hazardous waste annually, a product that is *not* shipped back to the US. For years, neither government paid attention to the waste accumulating in vacant lots, open landfills, and riverways inside Mexico. Under pressure from environmentalists, a 1983 agreement established regulatory guidelines, including the requirement that American factories operating inside Mexico ship their waste back to the US. These regulations are openly flaunted. In a 1988 study, of the 400 maquilladoras on the border with Texas, only 11 were returning their wastes; an Arizona-based project discovered that of 28 nearby maquiladoras some claimed to have returned their wastes to the US, but neither country could provide records of the waste crossing the border; the EPA office regional office in San Francisco reports that of the 1,000 or so companies operating below the border, it only has documentation to show that a half-dozen have returned their waste to the US.[132]

Electronics industries produce extremely toxic waste, including carcinogenic chemicals such as trichlorethylene, and a range of cyanides, degreasers, solvents, acids, and heavy metals. By all accounts, most of the wastes are being dumped randomly, including disposal into municipal sewer systems, local waterways, and irrigation canals. A Mexican study of the border industries concluded that maquiladoras are contaminating the water table and contributing to cancer, congenital defects, and nervous system damage among the borderlands Mexican population.[133]

In addition to the point-of-source wastes produced by American firms operating industrial plants in Mexico, Mexico is a favored dumping ground for waste shipments, both legal and illegal, from companies operating in the US. In 1986, the US and Mexico signed an agreement to allow American companies to ship hazardous waste to Mexico for "processing" and disposal.[134] While this "informed consent" trade is monitored under the auspices of the US Environmental Protection Agency, EPA officials admit that they have no knowledge of the quality of Mexico's waste disposal or treatment sites. The EPA has not tried to inspect those sites because it does not have the right to do so. The EPA has no authority or interest in what happens when wastes cross the border. The EPA relies on the waste-shipping company to inform them of the contents of the waste. The US Customs Service, out of concern for safety, has instructed border agents not to check the contents of hazardous waste containers against the bills of shipping provided.[135] In point of fact, the entire country of Mexico has just two legal toxic waste dumps and ten toxics recycling plants.[136] An official with the Mexican Conservation Federation says he knows of no facility in Mexico able to handle hazardous wastes properly: "Even if it is brought here legally, nobody knows what to do with it."[137]

THE DOUBLE STANDARD STARTS AT HOME

Rich-world governments send out mixed environmental messages. They export a development model based on resource privatization, yet balk when poor countries extend this claim to sovereign/private rights. When they do invoke respect for the sovereignty of poor countries, it is primarily because it serves as an excuse to evade responsibility for dubious overseas environmental activities. They argue that poor countries need to keep their shoulder to the wheel of the "global environmental good," while continuing to foul their own nest without pause. The environmental double-standard, though, is not just an international export—like virtue and violence, it starts at home.

South Africa   Modern South Africa is a state defined by the double standard. Apartheid has taken a devastating environmental toll on the Blacks of South Africa, the consequences of which will endure long after other institutional mechanisms of the racist state have been dismantled.[138]

• The bits and pieces of "homelands" in South Africa reserved by the government for Blacks, packed with people not wanted in the white economy, are among the world's most degraded lands. Whites have exclusive control over 87 percent of South Africa's land; Blacks have been relegated to the remaining 13 percent.[139] The land granted to the Blacks in the first place was the most fragile and least productive in the country; from the very beginning, Blacks were given land where topsoil is thin, rainfall scarce, and the ground rocky. Borders were carefully drawn, sometimes redrawn, to exclude anything of value.

Extreme overcrowding and poverty, enforced by government policies, have been disastrous for these regions. Soil erosion and land degradation are endemic; agricultural productivity has declined steadily. One American AID official comments, "Many of the homelands bear more resemblance to the face of the moon than to the commercial farms and game preserves that cover the rest of the country."[140] Food production has fallen so dramatically that the homelands are not self-sufficient and most food must be imported.

Most of the forests in the homelands have been stripped: two-thirds of the Black population uses wood for fuel and a study of four homelands found that women typically make treks of six to nine kilometers every other day to collect fuelwood.[141]

• The enormous costs to the government of maintaining apartheid have made the government financially dependent on mining. Profits from mining are essential to the survival of minority rule in South Africa. The government treats mining like a sacred cow, and the industry is virtually unregulated. The ecological impacts of unregulated mining are extensive. Swaths of the country have been deeply scarred by reckless mining. Air pollution over the nation's coal region ranks with the worst in the world, partly because of an energy strategy that aimed at minimizing dependence on anti apartheid oil exporters. The Chamber of Mines, a confederation of major mining corporations, keeps a tight lid on information about environmental impacts, and the government shields it from criticism. Air and water near mining and smelting operations are little monitored, and results from the modest monitoring that is done are not made public. Not just a domestic problem, South African mining and energy policies contribute to *global* environmental threats on a scale completely out of proportion to the size of the nation's population.

• Across the northern border, the South African military has collaborated in devastating herds of elephants and stands of tropical hardwoods as part of its campaign to cripple and intimidate neighboring states hostile to apartheid.

• The South African government makes money by accepting shipments of toxic waste from Europe. Most of the hazardous waste dumps and reprocessing facilities, both for imported and domestic wastes, are located within Black townships.

South Africa's homelands are not just convenient waste-dumping sites, but are the preferred location for the most hazardous waste-processing industries. The world's largest mercury-processing plant, managed by a British multinational, Thor Chemical, is located on the edge of the Kwa Zulu homeland in Natal province.[142] Mercury—one of the most noxious and hazardous chemical pollutants—saturates the soil and water surrounding the plant, in concentrations as high as 8,000 times the limit the US has defined as hazardous.[143] Black villagers who live downstream from the plant have no running water—the mercury-contaminated river is their sole source for drinking, cooking, and washing water.

The poverty and environmental destruction caused by the apartheid double-standard create multiple burdens for Black women. On a map, the "homelands" resemble an archipelago—a scattering of islands of poverty and environmental catastrophe. This is also a feminized geography. Working-age Black men are mostly absent from the homelands—either commuting hours each day to jobs in white cities, or housed semipermanently in working camps clustered around industries and mines. Women are left behind to farm and raise children. A detailed survey of homelands in one region found that 81 percent of working-age inhabitants were women.[144]

Women and their children are most exposed to the predations of pollution and environmental unsustainability: trapped in the homelands, they live with these conditions as part of their daily life. Environmental degradation makes women's work—farming, collecting firewood, preparing food, washing laundry—doubly dificult, and more hazardous. Many of the Black townships and homelands do not have even rudimentary services—many have no electricity, sewage systems, or safe water supply.

South Africa combines the pollution of a First World industrial nation, the poverty-related environmental degradation of a Third World economy, and the unregulated excesses of a militarized state. The ultimate lesson of apartheid's ecological toll is that inequitable social institutions—whether the suppression of women, tribal groups, or the poor—are not compatible with environmental sustainability. It is tempting to put aside the South African experience as an extreme, and aberrant, case study. Unfortunately, the truth is that "environmental aparthedism" colors the environmental landscape of country after country, including self-proclaimed democratic vanguard states.

The USA (1) *Without Reservations* Since 1963, the US has exploded 651 nuclear weapons or devices on the US mainland; all the nuclear bomb testing sites in the continental United States are located on Native American lands, mostly on Western Shoshone territory in Nevada.[145] The Shoshone Nation characterizes the nuclear testing program as an attack against them, but one they are powerless to repel. Now, Native Americans are facing a new assault on their health and new incursions on their territorial integrity: American disposal companies have come up with a novel scheme for handling wastes so rancid or toxic that no one wants them—"give them to the Indians."

Native American reservations in the United States are like miniature Third World islands, surrounded on all sides by a First World superpower. In many ways, they *are* nation-states—reservations retain some sovereignty, and state and local health and environmental laws generally don't apply to Indian lands. Poverty and unemployment are endemic on the reservations. Native American leaders have little political clout with the "outside world," and have limited technical expertise to evaluate the health and environmental implications of waste proposals.

These conditions are perfect for waste dumpers. The 500-plus reservations in the US offer 53 million acres of unexploited open space within "domestic" boundaries. And the fact that state environmental codes don't apply to reservations is a major advantage in holding down the costs and liability of waste handlers. Most waste-disposal contracts offered to Native Americans include explicit provisions for environmental exemptions. Negotiations currently under way, for example, with the Rosebud Sioux reservation in South Dakota for what would be one of the country's largest landfills include provisions that "in no event shall any environmental regulations or standards of the state of South Dakota be applicable to this project," and further provisos that if the tribe itself adopts new environmental standards the costs of compliance will be paid by the tribe itself.[146]

Desperate for income, many tribal leaders were receptive to the first approaches of waste dumpers; some tribal councils now are in the business of leasing portions of their land for landfills. But opposition within the Native American community is building. In many cases, it was Native American *women* who mounted the first challenges to their male tribal leaders, and women continue to be in the forefront of Native American organizing against waste dumping.[147] In 1985, a Cherokee woman, Jessie Deer-in-Water, founded a national political-action group, Native Americans for a Clean Environment; Winona LaDuke of the Indigenous Women's

Network has taken a lead in formulating proposals for a coordinated Native American moratorium on waste dumping.

But there is enormous pressure being brought to bear on the Native American population to accept the wastes of industrial America. Since 1988, according to some Native American groups, tribes have been presented with as many as 120 proposals for waste dumps, incinerator sites, and reprocessing facilities—there's barely a reservation in the country that hasn't been approached. In California, Oklahoma, South Dakota, and Washington, plans for garbage landfills, sludge dumps, and toxic waste facilities on Indian lands are under consideration. Even the federal government has jumped on the bandwagon—the federal Department of Energy has written to 565 tribes asking them to accept *radioactive* wastes.[148]

The frustration of Native Americans with the race-based politics of trash is mounting. As one activist wryly comments, "There was this white guy stumbling around in the Atlantic Ocean and when he happened on this country we had 100 percent of the land. Now we have 3 percent. Why do you have to use that for your trash?"[149]

(2) *Dumping In Dixie*[150]   In the United States there is unmistakeable evidence of an environmental double-standard so pernicious that it has been dubbed a system of "environmental apartheid."

Hazardous facilities, including waste dumps and noxious industries, are most likely to be sited in communities dominated by minorities. A 1987 study by the Commission for Racial Justice (CRJ) of the United Church of Christ concluded that "the factor of race is a more significant factor than any other in determining the location of hazardous waste dumps." Among other findings, the CRJ study concludes that: • Three out of the five largest commercial hazardous waste landfills in the US are located in mostly African American or Hispanic communities. • Three out of five African American and Hispanic Americans live in communities with uncontrolled toxic waste sites. • Sixty percent of the total black population, as well as 60 percent of the Hispanic population (three out of every five), lives in communities with one or more uncontrolled toxic waste sites. • About half of all Asian/Pacific Islanders and Native Americans live in communities with uncontrolled waste sites.[151]

A close study of waste facility siting in Houston, Texas, revealed that all five of the municipal garbage dumps and landfills were located in African American majority neighborhoods; further, seven out of the eight municipal incinerators were in Hispanic or African American neighborhoods.[152] In some instances, minority communities, forced by economics into marginal residential areas, coalesce in areas where hazardous industries already exist; but in the preponderance of cases cited in the CRJ and other studies, new toxic facilities are consistently sited in already-existing minority communities. All siting agreements require approval by local zoning boards, and, sometimes, permits from state and federal government officials.

Analysis of facilities siting suggests a geography of environmental racism. But minorities are subject to environmental discrimination in a host of other ways. Pesticide exposure among farm workers, most of whom in the US are Hispanic, causes more than 300,000 pesticide-related injuries a year. Navajo Indians are used as the primary work force for the mining of uranium ore, leading to alarming lung cancer mortality; uranium mills are predominantly located near Native American communities. Puerto Rico, America's private Third World enclave, is one of the most heavily polluted places in the world, boasting, among other dubious distinctions, the only community in North America that had to be relocated due to mercury poisoning.[153]

THE INSIDE TRACK: SETTING THE ENVIRONMENTAL AGENDA

Friends in High Places

It is not always clear who sets government environmental agendas. In principle, in democratic governance, the "public" serves a crucial oversight role. But in both the US and Britain, the "free marketeers" in power are increasingly acting to diminish citizen involvement in environmental policy making, cementing instead the closed loop network of industrial and government leaders. Industry is often the primary partner of government regulatory agencies, and, in practice, environmental issues often become matters of closed-door, private negotiation. In Britain, "free-market" regulation even has an official name—"voluntary compliance." One commentator on British politics notes that voluntary compliance goes hand-in-glove with government predilections for secrecy: "[In Britain], pollution control agencies have traditionally kept their discussions with industry confidential. . . . pollution issues have been restricted to private rather than public negotiation."[154]

Similarly, in the US, the Reagan and Bush administrations have systematically dismantled public participation in environmental policy making. Within six months of taking office, President Reagan eliminated provisions for citizen involvement in a host of government environment regulatory agencies.[155] Public involvement in the decision-making process is seen by many federal officials and agency administrators as a "disruptive influence." An American political scientist explains the shift away from citizen involvement, using the example of toxic waste regulation: "[Government regulators believe that] public involvement favors the opponents of toxic waste depositories and waste sites, confuses and alarms the public about the risks involved in siting decisions, and generally makes scientifically safe location and management of toxic waste sites extremely difficult."[156]

Industrial interests often serve to directly block the access of an informed citizenry to environmental decision making. In 1991, under pressure from consumer-goods manufacturers, the US EPA withdrew from public circulation a consumer environmental handbook produced by its own offices.[157] This handbook, intended as a general primer on consumer environmental issues, urged consumers to reduce waste, to replace chemical household cleaners with natural cleaners, and to avoid single-use and "disposable" products—recommendations that seem relatively tame in this age of green consumer consciousness. But they antagonized manufacturers such as Procter and Gamble (one of the largest producers of household cleaners), Scott Paper Company (manufacturer of disposable paper products), and the Foodservice and Packaging Institute. Proctor & Gamble went so far as to ask, "On what authority does the agency base recommendations that have the effect of intervening in the commercial sector?"; Scott Paper Company proclaimed the paramount need to "preserve personal choice in the marketplace and avoid product purchase guidance."[158] Under this sustained barrage of attack from industry, the handbook was banned within four months of its publication.

On a global scale, the stimulus behind the new drive for international free trade laissez-faire mostly comes from the large planetary corporations that have emerged in the late twentieth century, which are increasingly able to manipulate intergovernment agreements in their own favor. Corporate advisors are playing an aggressive role in setting the policies that guide international agreements on trade and corporate interests are prevailing at all levels of government environmental policy.

Industry does not always have the upper hand, and governments *do* enact legislation that does not serve industrial interests. In 1991, for example, the German legislature passed extraordinary waste-reduction laws that require manufacturers to reclaim and recycle up to 80 percent of their consumer packaging. The Japanese government, bowing to international pressure, agreed in late 1991 to ban drift-net fishing, over the strenuous objections of its domestic fishing industry. But in governments that are most committed to principles of free market economics and privatization of resources, private industry and vested commercial interests have a privileged and protected role in setting environmental policy.

And those same governments are the most inclined to restrict the role of the public in environmental decision making.

Science

More than industry, though, it is science that has the inside government track on environmental matters. Because environmental problems are characterized as physical disruptions in the biosphere, governments increasingly rely on scientists to determine whether environmental problems exist in the first place, and, if so, what to do about them. Environmental impacts are measured on a scientific yardstick, and regulatory standards are science-driven.

Exclusive reliance on scientific "rationality" is a slippery slope. In the first place, this regulatory stance presupposes that "science" *can* define acceptable environmental quality; it also presupposes that scientists are disinterested, neutral players who can arrive at 'objective' truth. Both of these presumptions are seriously flawed.

In chapter 4, I discuss at length the perils of science-led environmentalism. In the context of thinking about the role of science in government decision-making, it is important to note several main points: • Reliance on science-based environmentalism gives primacy to an expert structure that primarily consists of white men. Science is not a neutral player—like other "man-made" institutions, it is a product of culture and ideology, both of which are rooted in social constructions of race, class, and gender. • The overarching paradigm of Western science is not environmentally propitious—in its recent history, science has been an enterprise devoted to the domination and exploitation of Nature • Science has not been a politically liberating institution; it has been used primarily as a prop of the status quo. • Reliance on science does not necessarily lead to environmental clarity on issues of safety and risk; more often than not, it leaves policy-makers tossed on the horns of scientific uncertainty. In the realm of environmental assessment, there is seldom unimpeachable scientific "evidence" of anything (until it is too late). For every expert who says that global warming is imminent, there is another who says that there is no problem; for every claim that a given chemical will cause health damage, there is a counterclaim, both based equally on scientific "fact." Scientific uncertainty promotes inertia; governments that use science as their main (often, only) policy crutch can find themselves crippled. Rich-world governments that have the biggest fiscal and policy stake in science are the most likely to defer environmental action in favor of "more research." If the world ends not with a bang but a whimper, the last whimper that will echo in the void will be a weak cry for "more research. . . ."

Nations of Men

Without belaboring the point, it is important to repeat the fact that governments are men's clubs. Worldwide, the average representation of women in national-government legislative bodies hovers around 2 to 3 percent. The implications of this only occasionally intrude upon the public consciousness. When, in 1991, the US Senate held highly-publicized hearings on a sexual harassment charge leveled against a Supreme Court nominee, the angst of women watching the dismal and degrading performance of their male elected representatives was palpable. Women around the country were united by the uncomfortable insight that "*they* just don't get it!" A poll taken shortly after the hearings indicated that 69 percent of Americans agree that the country would be "better off" if more women served in Congress.[159] Although it is hard to precisely measure the discontent, it is clear that the sense of unease with men-only governments is growing. Women, especially, are fed up with their politicians, and with the inbred politics of closed-door "club" governments.[160]

But does it really matter, at a policy level, that governments are preserves of men? Skeptics point, of course, to women politicians such as Margaret Thatcher to parry claims that women's leadership can offer an alternative vision. But this is a tired example, and it is not particularly surprising that male-assimilated women who rise to power through conventional men's channels of power will continue to serve men's interests. However, one of the first political studies to specifically address the question of (American) women's participation in government concludes unequivocally that women *do* bring substantively different policy priorities and agendas to government service. In 1991, the Center for American Women and Politics at Rutgers University released the results of a two-year systematic and comprehensive analysis of the impact of women in public office—a study that concludes that women "do make a difference." The major finding from this research is that women officeholders are reshaping the public policy agenda in distinctive ways.[161] When compared with their male colleagues in office,

- Women public officials have different policy priorities. Women are more likely to give priority to women's rights policies; they are also more likely to give priority to public policies related to women's traditional roles as caregivers in the family and society.
- Women public officials are more feminist and more liberal in their attitudes on major public policy issues.

• The officeholders most active in reshaping the policy agenda are feminist, liberal, younger and African-American women; however, non-feminist, conservative, older and white women officeholders are also actively reshaping the policy agenda.

• Women who have close ties to women's organizations are more likely than other women to be reshaping the policy agenda.

The Rutgers study is especially significant in light of the fact that other polls over the last few years have shown a persistent and clear gender gap on environmental issues. Women are consistently less enthusiastic about nuclear power and arms and military spending than their male counterparts, and more in favor of tough environmental regulation and public spending on environmental protection.

In reflecting on the dynamics of the 'men's club' political machinery that have hindered a sustained commitment to environmentally sound public policy, the Rutgers study further established: • that when women move into leadership positions in governing institutions, their commitment to reshaping the policy agenda remains strong; • that women are more likely to bring citizens into the process of government; • they are more likely to opt for government in public view than government behind closed doors; and • women are more responsive to groups previously denied full access to the policy-making process.

◐

Most governments have dismal records as arbiters of the public environmental trust. None of this is to suggest that government policymakers, and individuals within governments, do not often "do their best" for the environment. There certainly *are* shining examples of successful and strong government initiatives on issues of environmental sustainability and environmental justice, both domestic and international. (Although it is also necessary to point out that, in most instances, the feet of government officials have been held to the fire by outside environmental pressure groups, and many governments have been most begrudgingly nudged into action.)

The best intentions of men and women within government are often hijacked by the bureaucratic culture and inertial structure of government itself. Military governments are the least likely to tolerate environmental reform; even among governments not under the direct rule of militaries, there is virtually none not firmly under the thumb of military priorities. Almost without exception, governments are composed of self-interested male elites, most of whom at one time or another see themselves as beleaguered and badgered by enemies of the state both within and without. The hegemony of "old boy networks," networks that presume and privilege the overlapping interests of men in industry, the military and the government, has been barely bruised by decades of feminist activism. Government officials are cushioned from the effects of environmental degradation by their class privilege, and few experience first-hand the predations of environmental racism. Nature is construed as a propriety interest: Western governments promote privatization of resource exploitation; Third World governments invoke nationalistic and sovereignty claims to resource ownership. The "new globalism" is a promising shift in paradigm, but, in practice, application of the principles of globalism are far from evenhanded. For many governments, it is often most expedient, efficient, and cheaper to export environmental problems, "out of sight, out of mind," rather than solving them.

As unpromising as it seems though, as both global and local environmental problems become more pressing, governments will become more, not less, important environmental actors. The reality is that those individuals and groups among us who strive for environmental sustainability will need to work *with* governments more, not less. But to do so effectively, we will need all our wits about us—a renewed commitment to cooperation need not be (cannot be) on the old terms. In this light, struggles against militarism, feminist struggles against the exclusivity of male power, struggles for top-down racial equity and bottom-up social justice in the fullest sense can be rightly seen as struggles for the environment. The environmental movers-and-shakers *outside* government who are shaping eco-agendas for the twenty-first century will need to embrace these struggles as their own.

A close look at the eco-establishment gives one pause.

Chapter 4
The Ecology Establishment[1]

The first photograph of the earth viewed from space was flashed around the world by the international wire services in the early 1960s. This was a defining moment in our collective environmental consciousness. The now familiar picture of a green, blue, and brown globe, wrapped in swirling white clouds, suspended in black space, started a revolution. For the first time, people around the world seemed to grasp, emotionally and intellectually, how finite and alone we are in the universe. Our seemingly vast world was reduced to a very small sphere, a warm and comfortable planet in a cold blackness, but fragile and very much confined.

This new consciousness of "Spaceship Earth," in tandem with activism of the New Left, and feminist and student movements of the 1960s, sparked the contemporary European and North American environmental movement. Earth Day was declared on April 22, 1970. It was a day to express joyous awe at the intricate ecosystems that sustain all life on the planet; a day for humans to be humble, to recognize that our fate is inextricably intertwined with the fate of rivers, trees, animals.

Those heady days of populist consciousness launched two decades of environmental legislation and regulation. The outcry for environmental action in the 1960s and 1970s catalyzed the founding of dozens of environmental action groups, from Greenpeace to the World Wildlife Foundation; it breathed new life into older, more cautious "conservation" groups, such as the Audubon Society and the Sierra Club; it also spawned sprawling governmental and legal environmental bureaucracies on both sides of the

Atlantic—a bureaucratic arrangement that conservatives and corporations are now mobilizing against, and from which liberals are now retreating. Ironically, just as the limitations of large governmental environmental bureaucracies are becoming evident, the environmental movement itself is stepping onto the establishment fast-track.

Now, twenty years after the first Earth Day, there are Greens in government throughout Europe, and environmental organizations in Washington with staffs of hundreds and spacious suites of expensive office space. From Greenpeace to the Worldwide Fund for Nature, environmental groups have started to hire professional political lobbyists and advertising agencies to package their message. Friends of the Earth (UK) hired a designer to rework its logo, for many years a "flying" turtle, to reflect a slicker, more streamlined and "modern" image. Similarly, in 1991, the American National Audubon Society replaced its flying egret, its logo for 105 years, with a blue-and-white flag. Environmental action groups are relying less on field work, and more on litigation, direct mail, lobbying, and "expert" testimony before government commissions. There is a very clear trend toward the "professionalization" of ecology pressure groups and an emergent nongovernmental "ecobureaucracy" in London, Bonn, Washington, and Toronto.

At the same time of course, there are still thousands of environmentalists who have never seen the inside of a lobbyist's office, and hundreds of community-based environmental groups that still run on volunteer labor out of someone's kitchen. The schism between the "ecology establishment" and the grassroots movement is growing, and the ecology movement, both in North America and Europe, will soon be irreversibly bifurcated, if it isn't already.

Despite the resilience and recent proliferation of grassroots environmentalism, it is the large eco-establishment groups who are in the foreground of setting the environmental agenda. This sector of the environmental movement is changing rapidly. It is becoming professionalized and bureaucratized, a trend that has wide-ranging ramifications. The professionalization of an ecology group brings changes not only in its headquarters' address, logo, and stationery; it also brings changes in its tactics, priorities, and politics. These changes are controversial; issues of changing style, substance, priorities are now being debated, somewhat reluctantly, often heatedly, in environmental circles. What is *not* being debated, what is overlooked time and time again, is the fact that these changes also alter the *gender* politics of the movement; and, in turn, that changing the politics between men and women inside the movement will change the nature of environmental activism.

THE GREENING OF MEN'S POLITICS

The European political landscape took on a new coloration when, in 1983, the West German Green party ("Die Grunen") won 27 seats in the national parliament. The Greens were shockingly unlike any previous national party. They represented a broad coalition of citizens' groups, including antinuclear, peace, feminist, and ecology movements. The Greens talked of "postpatriarchal" values, of their four guiding principles of social responsibility, grassroots democracy, nonviolence and ecology. They refused to identify themselves as either Right or Left, "just Green, and up front." And they shook up the established parties, proving that environmental consciousness was a viable electoral force. Now, almost a decade later, the conventional wisdom among poll watchers is that "value-based" divisions in Europe are replacing class divisions as predictors of voting behavior, and that environmental values in particular are rapidly rising to the top of the public agenda.

One of the most striking features of the early formulation of West German Green politics was its explicit commitment to feminist principles. Their principles of nonexploitation, nonviolence, and the vision of a future based on harmony with nature, not domination of it, drew directly from the women's movement, which had emerged as a political force during the previous decade. Feminists were prominent in the coalition that created the German Greens, and their imprint on the party is unmistakable. The Greens adopted various rules to ensure women's equal participation, including a 50 percent guideline that stipulated that at least half of those at every level of leadership should be women.

The results of this explicit commitment to promote sexual equality within the party were impressive. In the 1987 federal elections in West Germany, the Greens increased their representation to 44 seats, 25 of which were held by women. In the British elections of 1987, the UK Greens contested 133 seats; 30 percent of the candidates they put forward were women, a proportion much higher than any of the other parties chose to muster. In France, women made major inroads into political life as a result of the local elections in March 1989, a fact that is credited to the presence of "The Ecologists" for the first time in many local contests.[2] The German Greens are the first party to have a female majority in their representation to the European parliament.[3]

The green revolution in electoral politics has gained momentum. Since the early 1980s, there has been an unmistakable "greening" of mainstream politics. Green parties have won seats in national elections in West Germany, Sweden, Austria, the Netherlands, Belgium, Switzerland, Finland,

Portugal, Italy, and Luxembourg. Other parties are hastily jumping on the Green bandwagon. Former British Prime Minister Margaret Thatcher and American President George Bush, both conservatives with appalling records on ecological issues, have declared themselves "environmentalists." Other leaders and political aspirants recognize that they need an environmental platform to succeed in contemporary politics. Even in countries with no *nationally* visible Green party, there are active Green networks. In the US, Australia, Canada, and Mexico, among other places, there are national networks of local Green affiliates and nascent national Green organizations.

But this sudden inflation in political Green currency has come at a high cost to the Greens themselves. While the Greens have introduced green values into conventional political arenas, the relationship has cut both ways. Mainstream politics have simultaneously transformed the Green agenda, even the internal structure of Green parties. Many of the established Green parties have become fractured by internal divisions and fragmented by hot disputes over policies and strategies. Green values have not always fared well in translation. The Green commitment to *feminist* values seems to have especially suffered in the course of the Green transformation into a mainstream political force.

The feminist/Green alliance is not necessarily a natural or comfortable one. It is an alliance that needs constant tending, constant support. The ecological *values* of feminists and Greens certainly overlap to a great extent, but the realities of operating in a formal political context always threaten to overwhelm this philosophical compatibility. The Greens in Germany and elsewhere combined the philosophies of the "New Left" and the values of feminism in a structured political forum, but never resolved the deep tensions between these two—the differences may have been momentarily smoothed over, but they were not reconciled. As one observer reminds us, ". . . as all of us knew who were part of it, the new Left had problems—sexism, anti-intellectualism, a dearth of political experience, and a lack of critical historical perspective."[4] These problems often divided men and women in the New Left, and these unresolved gender differences surfaced in the new Green parties.

Ynestra King, a prominent American peace and ecofeminist activist, points out the dangers of seeking to achieve change through the conventional political channels, if those channels are not themselves changed:

> in drawing on the Western democratic tradition, the greens are still working with a political legacy that is founded on the repudiation of the organic, the female, the tribal. . . . Even as greens are resucitating

(and reinventing) the democratic tradition, I am mindful that the original citizen in that tradition is male, propertied, and xenophobic.[5]

This political heritage means that a viable alliance between feminists and men of the Left, new or old, requires constant vigilance—vigilance against slipping into traditional roles, conventional power hierarchies, and petty sexism. The vigilance, even the *intent* to be vigilant, of the German Greens has weakened as they have become more accustomed to operating in circles of conventional male power.

Gender-based leadership quotas, while clearly necessary, are often mistaken by men within Green parties as an adequate response to sexism. But Green feminists are not so sanguine. Women within the German Greens, for example, often complain that in order to be respected and to be awarded political jobs by the male majority, they must "work in the way of men"—that is, they say, they must make aggressive arguments and allow no emotions or feelings into discussions.[6] Many women are not interested in learning how to operate in this style. A recent interview with some German Green women describes the internal dynamics of the party:

> The patriarchal structures here and the hectic schedules mean that all interpersonal contacts are superficial and everyone functions on the outer level. The men like that because its safe. Only the women are conflicted, and each woman must try to figure out how conflicted each of the others is. . . . some men too are dissatisfied with the patriarchal style of politics but put up with it 'because of their greater need for a public identity.' Many thoughtful men told us that they know there is something wrong with *trying to make new politics by using the old style* [emphasis added], but could not suggest any changes or concluded that the standard, "efficient" ways are necessary since the Green party is small and faces such huge tasks[7]

This last is a common claim, one that cuts across political affiliations. Men often justify their unwillingness to question patriarchal ways of conducting business by claiming that a more urgent agenda takes precedence. Women are often told that "their" issues will be dealt with later, "after the revolution" as it were. Women are becoming wary of this tactic. The weight of history suggests that if women's issues are not taken up in the midst of social change, indeed as central *to* the agenda for social change, then they will not receive a hearing in the flush of whatever new social order replaces the old. And more importantly, this delaying tactic proposed by men is predicated on the belief that the most important and pressing problems are ones that can be explained and solved without reference to the power relations between men and women.[8] That is, the Green men who hide behind this procrastination are implicitly presuming that relations between men and women don't have bearing on changing the socioecological order that they see as their first priority—that ecological issues take priority, and that gender issues can somehow be dealt with later, both in a vacuum.

The German Greens are now struggling with a number of fractures in their party line. One of the most serious divisions has opened between the Green "Realos" ("realists") and the "Fundis" ("fundamentalists"). The Fundis oppose any compromise in the radical Green agenda; the Realos argue that compromise and coalition are often a pragmatic necessity to accomplish real gains in the current political system. In the short run, the politics of compromise and "reasonableness" won the day. Just a few short years after securing their place in conventional political arenas, Green spokesmen were promoting a more mainstream image. Helmut Lippelt, one of their deputies in the Bonn parliament, says "the Greens have outgrown their utopian notions and are moving toward the center."[9] In relinquishing their radical politics, the Greens may have squandered their political base; in the 1990 elections, the West German Greens lost all of their 48 parliamentary seats.

Early on in the struggle, women's issues fell into the chasm between Fundis and Realos. Jutta Ditfurth, a Fundi spokeswoman, described the Green debate on a recent "women's" issue:

> Another issue that was raised that has been troubling the [German] Green movement enormously over the past months is the right of married men to rape their wives. As leftists inside the party, we have adamantly opposed rape under the veneer of marriage and have supported women in their attempts to prevent their husbands from taking possession of their bodies against their will. A recent party congress called for a minimum two-year sentence for men who commit marital rape. But the realos hedged on this question and asked only for a one-year maximum sentence. As you know Germany has a very strong patriarchal history behind it, and the women's movement has had the greatest difficulty trying to counteract it. It disturbed me enormously to find that the realos are not prepared to confront this patriarchalism directly. . . .[10]

The Green party also divided on opposition to West Germany's restrictive abortion law. Petra Kelly, a founder of the German Greens, described the scene of one debate:

> We had a big discussion and I presented the position that we must get rid of this law because it is unjust to women and . . . there

should never be a situation where a board of men—some doctors and judges—tells a woman what she must do. . . . Of the 10 women in the *Fraktion*, eight of us voted for my position, but the majority of the men agreed [with the witness, a Catholic nurse] that much of the law should remain.[11]

To the extent that a feminist agenda still exists in the German Green party, it is an increasingly segregated one. Social issues *are* very much analyzed through feminist lenses. But in the main Green policy document, their "Federal Program," the feminist perspective is nearly absent on issues of militarism, economics, healthcare, education.

Throughout her life, Petra Kelly, recently deceased, spoke out against the structural sexism that permeated the Green party:

In the Green party there are still many men who are able to be in politics because their women are taking care of their children. We had one case where a woman on the Executive Board had to resign because she couldn't cope with her child alone and continue to be in politics. If Green men stand there and say, well, that's her problem, then they have not understood what the Green party is about. . . . Men will claim that they are talking about "World Politics" and that [issues of violence against women] are "women's issues." They don't connect to it, and those connections are probably the most important. That is the problem with the Green party: many men will speak about women's equality in public, but often say, "my girlfriend is typing my speech."[12]

Petra Kelly's outspokenness and her international visibility made her a controversial figure in the Green party. She was ridiculed and condemned by some of her peers. Like many progressive women of prominence around the world, she was the subject of pornographic lampooning, most recently in *Penthouse* (the men's magazine). The price of being a public woman in a man's political world is often excruciatingly high, and the Green promise of forging a new political contract, one based on feminist values and methods, one that would include, not alienate, women, has not withstood the test of political reality. Women appear to have lost the most in the German Green ascendency to the world of "real" politics.

By placing women on the policy back burner, the German Greens are squandering their political base. In most Western countries, women comprise the majority of the electorate, and almost all observers agree that the strongest support for Greens, from Italy to Australia, comes from women. Other parties are starting to recognize the political advantage of having women on their side, and in West Germany, strong appeals to feminists are coming from the Social Democratic party.[13] If the Greens turn away from their feminist agenda, they will lose a major claim to women's support.

Once in the corridors of power, the Greens seem unable to sustain commitment to the values that draw women to the party in the first place. Jean Lambert, a co-chair of the British Green party, explains why she thinks the Green party campaigns in the UK in 1987 had a special appeal to women:

> It's the way in which the party works. It's not a hierarchical party: it doesn't have the patriarchal leader figure which a lot of parties have, which I think puts women off. It's also a party that has very strong feminist roots and therefore recognizes women's needs—at most things, for example, there's a creche [child-care facility] for children so that women can still be involved.[14]

She also points out that the nonjudgmental attitude toward people's contributions encourages women to come forward: "No one's going to expect highly professional, very slick skills, so therefore you can actually offer what you have and people will accept that gratefully." Judging by the West German experience, it is not clear that the British Greens will continue to "be grateful" for the skills of all, and especially all women, if the Greens become more powerful players in the political hierarchy.

Elsewhere, emerging Green parties are already proving less hospitable to women than might be expected, even at the very early stages of their political development. In profiling the Mexican Ecology Movement, one journalist points out that "Unlike the European Green movement, the influence of feminism and pacifism is not great."[15] Feminism is "only latent" in the French Greens, and, according to one observer of the 1989 local elections in France, feminist issues are "thus far invisible to the [Green] voters."[16] At the 1988 annual general meetings of the Canadian Green group in British Columbia, twin proposals to include feminism as a Green value in the constitution and to continue to use consensus decision-making were defeated.[17]

In the face of this defeat, the British Columbian women are doing what women alienated from male political movements have always done—they are forming a separate feminist Green caucus. Separate organizing is increasingly proposed as a viable option for German Green women too. As one German woman commented:

> They [the Greens] do have a good radical programme on women. But it's difficult for feminists inside the party to get the programme

accepted. . . . Many feminists in it are pretty frustrated, and there have
been calls to leave it. Some who've been in it say it's not worth it: it
takes all our strength just to get things through the party.[18]

If Green women are driven to organize autonomously from the Green
party, not only will the party suffer, but the rare opportunity for women to
be integrated into political party machinery, to acquire and exercise formal
political skills, will slip away. As the West German commentator says, "To
withdraw from parliament back to autonomous women's centres would
mean giving up all instruments of power."[19] Petra Kelly also expressed
dismay at the prospect of women withdrawing from the Greens: "If we
have our own Green women's program, I want it integrated into the full
political program. I don't want a separate program because then the men
will end up saying, "That's the women's part."[20]

Everyone will pay a price for the alienation of women from Green
parties. Women will end up being once again marginalized in political
arenas. The Greens will jeopardize their base of support. The Green agenda
will lose its analytical acuity and saliency if its feminist commitment is
reduced to symbolic gestures of equality. And Green parties will have a
hard time maintaining their credibility as ushers of radical and genuine
change.

THE FRIENDLY ECO-ESTABLISHMENT:
AN ENDANGERED SPECIES?

The story of "the feminists and the Greens" bears on the debates today
around the professionalization of environmental pressure groups. Environmental organizations, on the whole, have so far proven to be *relatively*
hospitable workplaces for women. Many women who work in the environmental movement report that there is less overt sexism, more respect for
them, and less pressure on them to conform to male work constructs than
in more conventional businesses or organizations. Kathryn Fuller, who
was appointed president of the World Wildlife Fund in 1989, the first (and
only) woman to head a major international environmental membership
organization, remarked that, "I have never felt limited by gender—which
is one of the enormous luxuries of working in Washington DC for the kinds
of organizations I have."[21] Many of the men who work in environmental
organizations, women say, tend to be sensitive and gentle men; as many
women point out, men with big egos tend not to be attracted to the small
salaries that are (or, were) standard in the ecology business.

In addition to these considerations, women are drawn to environmental work because, in the words of one insider, "these are the issues that women feel strongly about anyway. . . . the holistic principles behind it appeal to women." Francesca Lyman, a long-time activist in Washington environmental circles, points out that women also have special skills that bring them into public-service work: "Fundraising. Women are good at fundraising. All our training in social skills, the 'lady' training we get as girls, translates well into the organizational skills needed to coordinate fundraising drives, and the social skills needed to cajole, entertain, and persuade potential funders."[22] Others point out that women feel especially comfortable in environmental work because of the decentralized structure of many environmental organizations, the emphasis on consensus decision-making, and the absence of rigid hierarchies.

These women-friendly, nonhierarchical, consensus-seeking workplaces may soon be on the "endangered" list. The recent trend toward professionalization in the ecology establishment is bringing rapid changes in environmental tactics, organizational ethics, and workplace structures. Environmental organizations need to start a dialogue about what these changes mean for working relations between men and women in the environmental business—and about what *those* changes mean for the nature of environmental activism. But, judging by responses to a 1989 survey sent to over 30 of the largest environmental organizations, this dialogue has not begun, and few leaders inside the environmental movement recognize the value of even asking basic questions about the gendered nature of their business.[23]

Sexism in the Movement

The existing power structure of the environmental establishment in North America and Europe is ubiquitously male, and mostly white. While grassroots groups everywhere in the world are primarily run by women, virtually all of the large international and major national environmental organizations are run, and have always been run, by men. While most of these environmental groups were started by radical amateurs, the creation of single men and women driven by vision and zeal, many have now become big businesses, with boards of directors, corporate officers, large staffs, and big budgets.

The size of environmental organizations today is staggering. As interest in environmentalism surges in the popular consciousness, these groups are growing at a breakneck pace. According to early 1990 figures, Greenpeace recorded an American membership of 2 million; the Sierra Club has doubled in size in the last ten years, to reach a current membership of more than 500,000 and a budget of almost $35 million; the World

Wildlife Fund counts 3.7 million members in 28 countries; the National Wildlife Federation has the largest membership—5.8 million, a doubling since 1970—and a budget of over $85 million.[24] Canadian environmental groups, though smaller than their American counterparts, report similar growth: the Canadian Wildlife Federation reported a 1989 membership of over 550,000, and a budget of $7 million; Greenpeace Canada counts 60,000 members, and operates with a budget of $1.7 million.[25] Friends of the Earth UK and Greenpeace UK both report that their membership has doubled in the past decade.

In contrast with the overall robustness of environmental organizations, the representation of women within these organizations is appallingly low and growing at a snail's pace:

• Greenpeace, founded by David McTaggart in 1972, has a current operating budget of $27.5 million, with main offices in 20 countries; the executive director of Greenpeace International and the executive directors of all 20 of the national offices are men; as of 1989, no woman had ever held a national Executive Director post. • Friends of the Earth, (FoE) founded by David Brower, has never had a woman President of either the UK or the US office. (Both Greenpeace and FoE consider themselves to be the most progressive of the large environmental organizations, and Greenpeace especially prides itself on being outside the eco-establishment "inner circle.") • In Washington, there is an environmental elite that calls itself the "Group of Ten," which includes relatively conservative

Table 4.1. The Growth of the Eco-Establishment in Britain[a]

	Membership				Income (£000s)			
	1980	1985	1989	1991	1980	1985	1989	1991
Greenpeace	10,000	50,000	320,000	411,000	175	600	4,500	8,000
Friends of the Earth	12,000	27,000	120,000	240,000	200	414	2,902	5,700
World Wide Fund For Nature	51,000	91,000	202,000	230,000	1,646	4,601	20,760	22,300
National Trust	950,000	1,323,000	1,750,000	2,300,000	24,560	37,328	55,800	77,600

[a] Membership and income of selected British environmental groups.

Sources: John McCormick. *British Politics and the Environment.* London: Earthscan Pub., 1991; *The Economist,* June 6, 1992.

organizations such as the Sierra Club and the National Wildlife Federation, and relatively progressive organizations such as the the Natural Resources Defense Council and the Wilderness Society.[26] The executive directors of all ten are men, and no woman has **ever** held the top post in any of these organizations. The • Earth Island Institute, founded by eco-iconoclast David Brower in 1982, is governed by a 16-member board of directors, 5 of whom are women (31%); the 10-member Board of Directors of the Worldwatch Institute includes three women (30%); the relatively staid National Audubon Society is run by a Board of Directors of 36 members, 11 of whom are women (30%).[27] • In all of the environmental groups surveyed, women predominate in traditionally female occupational slots—they work in "Membership Services" or in "Administration and Support." One woman who works for an American environmental group observed that men seem to be more comfortable with women in support roles, and very uncomfortable with women in leadership roles.

The view from the top obviously does not tell the complete story. There *are* a number of highly visible and influential women in middle management positions (and even a few in "upper management") in many environmental organizations, certainly more than in most conventional organizations, and perhaps even more than in other social-change organizations. But the uniformity of the male top-heavy hierarchy, and its universality among groups that are otherwise deeply divided by political and philosophical differences is striking. The 30 percent barrier is also striking—this invisible barrier seems to represent the "glass ceiling" limit for women's representation in environmental management.

While some women are bemused by the sameness of sexism within and without the movement, other women are not so good-natured about men's power plays within environmental organizations. Many women who were in the forefront of the recycling movement in Europe and the US in the 1970s, for example, are angry that now that recycling is becoming part of the social mainstream, it is also becoming big business—and it is men who are reaping the benefits. Florence Thompson, a private recycling consultant in Connecticut, comments that, "When it's volunteer work it's left to the women, but when the bucks come, the men take over. . . . There are many women who are in recycling but they never seem to be in the position of decision-making."[28] Another community recycling administrator, Suzanna Benson, adds that, "There is a feeling among some women that 'This [creating the recycling movement] was our doing' and now we've got all these blue suits and red ties coming in."[29] Recycling and rubbish disposal used to be community-based enterprises; they are now among the strongest growth industries in the US (Waste Management Inc., for example, the largest American firm in this business, reports annual revenues of over $6 billion).[30] As national and multinational firms have transformed local problems into global businesses, community women have been edged out by business men—a classic example of the gender-based process of environmental "professionalization."

It is paradoxical that a movement that has been fueled by women's concerns and largely sustained by women's labor, one that offers the potential to provide a more hospitable working environment for women than most, should be so mired in conventional male power structures. Janet Marinelli, a former editor at *Environmental Action* suggests that this may be because the organizational model that the environmental movement measured itself against, perhaps unconsciously, during its formative years was the very male, very traditional example provided by the older, mainstream conservation groups. Groups such as the National Wildlife Federation were successful, well-established, and well-funded enterprises at the time that some of the newer groups, such as Greenpeace, were struggling to establish themselves in the environmental world. These early conservation groups were (and still are) organizationally patterned as men's clubs. Even though the newer, progressive environmental groups joining the scene in the 1960s and 1970s differed considerably in environmental philosophy from the old-line conservation organizations, it was groups such as the Audubon Society, the National Wildlife Federation, and the Izak Walton League that laid the groundwork and set the tone for the emerging environmental bureaucracy. And today, when the executive directors of the Washington "Group of Ten" (all men) get together for their occasional informal breakfast meetings, it is clear that the men's club ethos has been essentially unchanged by the entry of supposedly more radical environmental groups into the ecobureaucracy; the culture of eco-establishment politics remains essentially male.

Sexism is measured not only in organizational representation, but in the setting of priorities and agendas. Despite their apparent philosophical differences, the agendas of the "new" environmental groups share surprising affinities with the agendas of the older conservation groups. The early conservation groups directed their efforts to saving wildlife primarily in order to maintain "stock" for recreational hunting and fishing; for many conservation groups, this continues to be a primary *raison d'etre*. The prominence of wildlife/wilderness issues on the agendas of the "newer" environmental groups may be largely attributed to the fact that these groups essentially grew out of the same male-oriented fishing/hunting tradition as the older conservation groups. A recent comprehensive survey of American environmental groups, the first of its kind, makes clear the extent to which wildlife-nature concerns dominate the environmental agenda:[31]

Table 4.2. Priorities on the Agenda of the Eco-Establishment[a]

Fish and wildlife management and protection	19%
National forest, parks, and public lands management	12%
Private land preservation and stewardship	11%
Toxic, hazardous, and solid waste management	8%
Protection of waterways (rivers, lakes, coasts)	7%
Water quality	6%
Urban and rural land-use planning	4%
Wilderness	4%
Agriculture	4%
Air quality	3%
Economic, sustainable development	3%
Marine conservation	3%
Energy conservation and facility regulation	2%
Zoological or botanical gardens	1%
Mining law and regulation	1%
Nuclear power or weapons	1%
Other	11%

[a] Average percentage of organizational resources (staff time and money) spent on various issues and programs among major US environmental groups, 1990.

Source: Donald Snow, ed. *Inside the Environmental Movement.* Washington DC: Island Press, 1992.

This skew in priorities has interesting gender implications. A recent report suggests an intriguing gender dimension in American attitudes toward wildlife.[32] Two wildlife experts conducted a survey of a broad spectrum of Americans in the late 1980s; their conclusion is bold:

> Male vs. female differences in attitudes toward animals were dramatic. The strength and consistency of male vs. female differences were so pronounced as to suggest gender is among the most important demographic influences on attitudes toward animals in our society.[33]

Specifically, they found that men, by a significantly higher margin than women, endorsed "dominionistic" and "utilitarian" attitudes toward animals—that is, men showed a much greater willingness to exploit animals, to usurp wildlife habitat for increased human gains, and to derive personal satisfaction from the mastery and control of animals; further, men viewed wildlife in terms of whether animal populations could sustain particular levels of "harvesting." Eighty-five percent of the hunting population in the survey was male; 29 percent of all men reported having hunted during the previous 2 years, compared to 4 percent of the women. Women in the study valued animals differently—while generally less knowledgeable about wildlife and about "harvestability/sustainability" issues, women displayed significantly higher "scores" in terms of moralistic and humanistic concerns about the treatment of wildlife—women were far more bothered than men about the possible infliction of pain and suffering on animals. The men predominantly belonged to sportsmen's organizations, while the women were much more likely to be members of humane and animal welfare organizations. The authors of this study conclude that the wildlife management/conservation industry (of which environmental groups are arguably a part) evaluates wildlife in male-rooted terms: the success of wildlife and wilderness management is typically measured in terms of sustainability, yield, and acceptable rates of harvesting. Other aspects of the importance of these differences become more clear when we consider, at the end of this chapter, the nature of specific environmental campaigns on "animal rights" issues.

It is clear that women *are* concerned about conserving wildlife, and women are no less "nature-lovers" than men—although for significantly different reasons. But more than this, women's environmental interests often start with concern for human health and habitat, issues that the large environmental groups have been slow to take on.

While wildlife and nature preservation predominate, human health issues remain a low priority within most mainstream environmental groups, and these organizations have yet to take seriously environmental issues that may bear more directly on women's lives. For example:

- In 1960, one American woman in 20 contracted breast cancer; by 1990, the rate had escalated to a near-epidemic 1 in 9, an increase that experts are increasingly attributing to environmental factors.[34]

- In industrial societies, the home is the most polluted environment.[35] While there is some indication that "indoor pollution" is emerging as a nascent concern on the environmental agenda, environmental organizations still commit negligible money or staff time to this issue.

Racism in the Movement

The exclusion of minorities from the environmental movement seems, too, to be deeply entrenched. At the beginning of 1990, the Audubon Society in the US had 3 minority staff members on staff, out of 315 (.9%); the Sierra Club could find only 1 minority staff person, out of 250 (.4%);

the Wilderness Society had no minorities on its board of directors, and, in a work force of 130, only 4 in professional positions (3%); the Natural Resources Defense Council counted 5 minority staff members out of 140 (3.6%), while Friends of the Earth US had 5 out of 40 (12%) minority staffers, including secretaries.[36] The record of minority representation in environmental groups in Britain, Canada, and throughout Europe is equally dismal.

The whiteness of the green movement is not just a staffing problem. Minorities are underrepresented in the membership base of these organizations. The primary membership support for eco-establishment environmental organizations still comes largely from the white middle-class. This is not a function of the wider sociolology of "joining"—in fact, repeated studies show that African Americans have higher rates of affiliation with voluntary associations of a social, political or religious nature than do their white counterparts.[37]

When pressed, leaders of environmental groups acknowledge their poor record of hiring and promoting minority employees, and their equally poor record of including minorities in the membership, but they deny racism. The standard litany of explanations put forward by environmental groups runs something like this: We can't find anybody, or, we can't find anybody qualified . . . We asked them, but they didn't come . . . or, We have members who are people of color, but they never come to meetings . . . People of color aren't concerned about environmental issues . . . or, race is not our issue. This debate should sound familiar to anyone active in the (predominantly white, middle-class) women's movement in Europe and North America of the 1970s and 1980s. Those of us who have struggled with racism within other social justice movements realize that it takes concerted and conscientious effort to reverse the white tide. It also requires introspection, a willingness to deal with discomfort, and a close scrutiny of the presumptions embedded in organizational structures.

Environmental groups need to be willing to take on this challenge, and to look to the struggles of race and class in other movements to learn how to do so most effectively; there are certainly lessons to be learned from women's movements. For example, environmental groups won't find people of color to work for them if they continue to look for mirror images— people who look exactly like they do, individuals whom they define as their social or professional counterparts. In the first place, environmental groups shouldn't try to find people who look like themselves—a broad-based movement benefits from having many different perspectives. If people of color don't come to meetings, or don't stay, where are the meetings being held? Are they in places comfortable to people of color?

Do people of color have a chance at participating as equals, or are they added as afterthoughts? Is the group's agenda already set? The "agenda" issue is perhaps the most crucial. Commitment to broadspectrum environmentalism requires a rethinking of environmental priorities, the most urgent of which may be expanding the focus from narrowly defined conservation-related goals to a consideration of environmental issues as *social justice* and public health issues. If the eco-establishment continues to commit most of its money and staff expertise to issues of conservation of wildlife and preservation of "nature," this will continue to situate environmentalism outside the circle of communities whose main concern may be *urban* public health and safety issues: "The environmental movement has been dominated by 'wildlife people' in the public and media, and not by people who believe that the environment starts where you are—dog-mess, graffiti, litter, and all."[38] Racism has a particular geography: residential segregation concentrates African American and Hispanic populations in the US (Caribbean and East Asian in the UK) in areas where risks from industrial and urban pollution are often extreme. Racial minorities are disproportionately victimized by environmental health hazards: blood lead levels in minority children are dramatically higher than in the population as a whole, as are certain cancers and respiratory diseases. The "containment" of minorities into urban neighborhoods, and the concomitant whiteness of "the countryside," locates minorities outside the sphere of traditional environmental wildlife and nature concerns.

Urban environmental issues are not the same as wilderness environmental issues. The lessons learned and the expertise required to mount campaigns to save the whales or pandas are not directly transferable to urban environmental campaigns. Urban environmental issues are public health issues (such as lead paint poisoning), public policy issues (such as the provision of efficient and accessible public mass transportation), and social equity issues (such as access to adequate housing). As one critic of Greenpeace remarks: "Its eco-warriors cut bold figures when getting in the way of whalers on the open seas. Making Greenpeace make a difference in places like [a] Brazilian slum requires a lower profile— and messier politics."[39]

Environmental groups need to develop new paradigms, and more concretely, new expertise to address concerns of urban environmental viability. This is not just a "minorities" issue. By the turn of this century, half the world's population will live in cities. As cities grow in absolute size and proportional importance, they play a larger part in shaping the global environment. Urbanization puts enormous stress on natural habitats, and as cities grow, their internal ecology changes: the world's urban centers are increasingly hazardous environments in which to live. Environmental campaigns for the next century must focus on efforts to make cities more humane and safe places to live—and more environmentally benign.

Beyond the immediate relevancy of urban issues, there is a host of other environmental issues of particular importance to people of color in the US and the UK.[40] For example:

- In the US workforce, African American and other minority employees continue to be disproportionately concentrated in high-risk, blue-collar jobs that tend to have health-threatening environments.

- Race is one of the most significant variables in determining the location of commercial and industrial hazardous-waste sites. In the US, the three sites accounting for more than 40 percent of the nation's total disposal capacity are located in predominantly African American or Hispanic communities; the nation's largest hazardous-waste landfill, in Sumter County, Alabama, is in the heart of the state's "Black belt"; the predominantly African American and Hispanic southeast side of Chicago has the greatest concentration of hazardous-waste sites in the nation.

- Environmental racism extends beyond national borders. The practice of dumping toxic wastes from Europe and the US in Third World countries perpetuates old patterns of imperialism, colonialism, and racism.

Social justice and environmental protection are inseparable. This is certainly true in the Amazon, where a wealthy landowning elite is trying to suppress indigenous movements for sustainable use of rainforests; this is certainly true in Central America, where a wealthy landowning elite has drenched the land in pesticides and herbicides, trying to squeeze every drop of cash-crop profitability out of a sometimes marginal land base; it is no less true in Brixton or Chicago, where people of color confront environmental degradation on a daily and domestic level. If environmental alliances are to reach across classes and cultures, the movement must be broad-based and connected with day-to-day survival and social justice issues.

This is not to suggest that people of color and minority communities are only concerned with day-to-day survival. Nor is this to suggest that all people of color and minority members live in the inner city and don't care about wildlife; in fact, in one recent American survey, 80 percent of African American respondents said they had an interest in wildlife, wanted contact with wildlife, and felt that Blacks should concern themselves with wildlife issues—but only 38 percent said they would join an environmental organization.[41] Multiracial awareness means acknowledging that there may be different priorities in different communities, and that the exclusion of certain groups from the environmental movement is as often as not the result of priorities and practices that are skewed in the interests of the dominant class. While a number of environmental groups, most notably Greenpeace, have made recent efforts to broaden their agenda to include coalition-building, outreach, and community-based programs, the pace of change is slow, leading many critics to doubt the extent of their commitment. The trend to "professionalization" within the ecobureaucracy is a growing structural impediment to such change.

PROFESSIONALIZATION

The trend toward a more "professional" environmental style is contro versial, resisted by many old hands in the environmental movement, seen by others to be pragmatic, inevitable and productive. Many people both inside and outside ecology circles worry about the future of environmental politics: Will a professional movement lose touch with its grassroots base? Will it lose the ability to generate waves of popular support, which in the past derived from direct action, outrage, and emotional appeal? Will this make it more or less effective in promoting social change?

These are all urgent and important questions, but the current "crisis of growth" in the environmental movement is not unique. Most modern Western social movements, many rooted in the antiestablishment politics of the 1960s, have developed a two-tiered structure as they have matured: an insider, "reformist" wing working for legislative reform within the established system, and a grassroots-based "outsider" radical core. Most Western feminist movements, for example, now have "femocrat" structures coexisting, and occasionally conflicting, with community-based women's activism. The perpetual debate about whether social change is best mediated from within or without the existing system remains unresolved in most of these social movements.

To date, debates between reformists and radicals within the environmental movement have fueled a robust introspection. Environmentalists wrangle with each other over tactics, style, issues, campaigns, leadership, finances, publicity—everything seems grist for the mill in the environmental debate over growing pains in the movement. Everything, with one notable exception: what environmentalists are *not* asking are questions of *gender relations* within the movement. Does it matter that the leadership staff and structure of the environmental establishment in Europe and North America is increasingly male, and white? Does it matter that this leadership structure replicates the structure of the corporations, militaries, and governments that are often their environmental adversaries? Does it matter that, as it "matures," this progressive movement apparently cannot sustain a progressive vision of gender and power relations? Does it matter that the schism in the environmental movement is increasingly between a mostly *male* -led professional elite and a mostly *female* -led grassroots movement?

These issues certainly matter to feminists who think about the ways that power and male privilege operate in this world. But, more importantly, they also matter to women and men who work, or who want to work, in the environmental field. And, ultimately, they matter for the environment. As environmentalists forge new directions and new leadership strategies for the coming decades, they need to be doing so with a feminist sensibility.

In the recent social history of Europe and North America, the process of "professionalization" of any social group or discipline has, without fail, resulted in the exclusion of women, the undervaluing of women's contributions, and the trivialization of women's concerns. Professionalization is a process whereby men have wrested control of activities away from women (who often conducted them as extensions of their work in the private sphere) and reconstituted them as exclusively male activities within the public sphere.[42] The relationship between maleness and "professionalization" is so strong that in the literature of radical labor sociology and labor history, the extent of male control of any activity or enterprise is often used as a measuring stick of the extent of professionalization; complete male control indicates full professionalization.[43]

Women have learned hard lessons about the costs to them of professionalization. In an essay reflecting on the role of women in British party politics, one woman observes:

> The important story for women about political parties is the growth of professionalism. . . . It is the pressure for professionalism which shapes the form and functioning of today's parties. . . . For women in general, who would like to see politics as a force which they can use to improve the quality of their lives, the professionalization of politics is an obstacle. It makes the party more immune to local and grassroots campaigns.[44]

Professionalization may boost individual career women, just as it also marginalizes some iconclastic men. But in general, professionalization shifts the balance of power to men, it entrenches male prerogative, and privileges masculine values and priorities at the top of the hierarchy. The "professionalization" of computer programing (which started out as a secretarial function), the professionalization of librarians, health care workers, teachers, and flight attendants, and the professionalization of most academic disciplines has created an internal pyramid of power, with men at the top and women on the bottom.[45] Similarly, the professionalization of other social movements, such as the civil rights movement in the US and peace movements on both sides of the Atlantic, has tipped the balance of power in men's favor. There is no reason to think that the professionalization of the ecology movement will build a different social architecture.

The characteristics of professionalization are not difficult to identify, and most of the people who are shaping the new environmental profile are specifically aware of the direction in which they are taking the movement.

Low Voices, No Tears   In a recent interview, Jonathon Porritt, the former UK director of Friends of the Earth, articulated some of the changes being wrought in a newly professionalizing Friends of the Earth:

> Part of the image of the Greens has always emphasized an emotional response, and it has attracted those kind of people who are not ashamed of an emotional response *[i.e., women]*. And that is very much the spirit of the movement. We cannot afford to lose it. . . . BUT that image, that appeal to emotions has blunted our impact on the people who should, ultimately, be affected by our arguments—the decision-makers.[46]

Porritt goes on to say that Friends of the Earth is intensifying its research effort, to "lend us an element of authority." Without scientific and research backup, Porritt says, "we look like cranks . . . and that's what alienates those decision-makers." The former chairman of the US Sierra Club, Michael McCloskey, echoes Porritt's comments: "We don't need emotional rhetoric. We need sustained response to insistent problems."[47]

The newly professionalized environmental movement is one in which "pragmatics" and "credibility" are given privilege over "emotionalism," which is equated with "amateurism." Increasing primacy is given to slick communication skills, pragmatic politics, and a professional appearance—as measured by the most conventional yardsticks. The "reasonable man" is replacing the "emotional woman" as a green archetype, a presumptive dualism that diminishes both men and women.

Changing the Expert Structure   Professionalization enhances male privilege. Not because women can't be "professional," competent or pragmatic: indeed, as some environmentalists are quick to point out, some of the best and most prominent experts in the field are women. But as one critic notes, professionalization "places an emphasis on three key players: the lobbyist, the litigant, and the expert—a kind of holy triumvirate in setting the goals of the environmental movement."[48] Lobbyists, litigants, and experts are mostly men. Women are still a minority in most scientific professions, in environmental law, and among political power brokers— the very fields that are being given primacy in professional environmentalism. As male-dominated fields become more central to the environmental movement, women will become more marginalized.

The new leadership of environmental organizations is increasingly drawn from the ranks of management experts rather than coming from the ranks of local protest groups and environmental activists. Leaders are increasingly selected from outside the movement, not from within it— among other things this means that women who start off in the movement as volunteers, door-to-door canvassers, and stamp-lickers are more likely to remain in those positions, while management "wonder-boys" from outside increasingly pull the strings. In looking for his successor, Michael McCloskey of the Sierra Club said, "We're looking for a person who's very strong in finance and budgets, who has a track record in management, who can offer entrepeneurial leadership, who is alert to changes in the market place."[49] A representative of the Wilderness Society echoed Mc-Closkey's remarks: "The question is whether the organization will be run by well-paid, skilled professionals or whether we will cling to the bleeding heart concept. If we continue with the latter, I believe we are doomed."[50]

Because women's social roles are still defined by domestic responsibilities, career costs to women in leadership roles are greater than those for men. As one woman activist notes, "Because cause-related work is not a nine-to-five job, women are in a position of constantly having to choose between jobs and home, dancing between male and female roles and courting chronic fatigue."[51]

The prominence of men as experts—as the "credentialed" public spokespeople for the environmental movement—was reinforced by the wave of green publishing in the late 1980s. While women have been writing and publishing prolifically on environmental issues, their writings—and their feminist analysis—remains incidental to the larger body of current environmental thought. "Old-boy" social and professional networks are evidenced in the selection of authors for collaborative works, and it is clear that women have yet to break into those networks. Most environmental anthologies include little writing by women, and many include none. In 1989, *Scientific American* published a prestigious special issue devoted to "managing planet earth"; with the exception of one article co-authored by a woman with two male colleagues, the eleven articles were all authored by men, including those essays on agricultural sustainability and on population—topics about which women know a great deal. While one might expect little else from the conservative *Scientific American*, progressive environmental publications have much the same gender profile. A 1988 reader on ecological philosophy included only one article by a woman out of a total of twenty-three authors; a 1991 "Earth First!" reader included only a handful of articles written by women.[52] A joint American-British anthology published in 1991, billed as a comprehensive popular "green reader" consists of 58 short essays and reprints, only 10 of which were identifiably authored by women, many of whom are co-authors with men.[53] A 1988 British ecopolitical book, heralded as a green manifesto for the next century, ignores the contributions of women to environmental understanding and activism, as a reviewer noted: "Although the writers implicitly acknowledge them from time to time, there is no specific mention of women. Nor is there mention of the important role they will probably have to play in the future. Yet women already see many of their intuitions, values and attitudes echoed in Green politics."[54] Most environmental writing, whether conservative or progressive, assumes that the active human subjects in environmentalism are male.[55]

Changing the Reference Group (Going to Lunch with the Boys)   For the environmental movement, if "becoming more professional" means making changes to become more credible in the eyes of the existing political and corporate elite, as Jonathon Porritt suggests, then this means that environmentalists must clothe themselves in the same garb of those elites (sometimes literally). If it is assumed that the movement needs to be made more credible to "men in grey suits," it follows that environmentalists must then mirror those men's appearance, language, and ethos. Thus professionalization favors men first, and the "assimilated woman" second—the woman who does not challenge male orthodoxy, the woman who camouflages herself in male language and business clothing, even if she dons this camouflage somewhat cynically and reluctantly. Women who want to succeed in these organizations without becoming surrogate men (and there are many) fare less well. The professionalization-driven tendency to homogeneity will further exacerbate the middle-class and white profile of the European and North American environmental movement.

The reference group of the eco-establishment leaders is shifting. Simply put, men in power within the environmental movement increasingly want to be taken seriously by other men in power. At the beginning of 1990, Greenpeace magazine ran a series of "short-take" interviews with women and men within the environmental movement on the topic of "where we go from here." The difference in response between the male director of Greenpeace USA and the woman leader of a grassroots coalition is startling:[56]

> *Peter Bahouth:* "For the first time in history, activists are beginning to confront powerful institutions—the bomb makers, the petrochemical industry, the planet's bureaucratic managers—as equals."
>
> *Lois Gibbs:* "What must we do? First we must not compromise our children's futures by cutting deals with polluters and regulators. Environmental justice cannot be sold or traded. Some environmental groups might feel comfortable to sit with polluters and talk about how to minimize pollution, but no one has the right to legalize pollution."

Lois Gibbs's words were echoed by Peggy Antrobus at the 1991 Global Assembly of Women for a Healthy Planet:

> Certainly we need good legal, scientific and technical advice . . . but we also need to act outside those structures. And we must give special support to those women who are at the forefront in their action and advocacy outside those structures.We must never lose sight of the fact that the women's movement and the environmental movement are primarily *revolutionary* movements. If we give up that political challenge to the dominant paradigm, there is no hope for change.

As environmental leaders forge ever-cosier relationships with politicians and industrialists, they run the risk of losing their grassroots support. In 1991, the leaders of the Sierra Club ran into a storm of opposition from their membership over the issue of a compromise on commercial logging of old-growth forests in the US. A Sierra Club member from Texas quipped, "They are losing sight of the fact that their mission is to save ecosystems, not to have lunch with Senators."[57] Nonetheless, it is increasingly clear that the men in charge of the eco-establishment do not like to be left out of the "loop"—and that going to lunch with the right person is increasingly valued. As one environmentalist enthused in the early 1980s, "Before, we filed lawsuits and held press conferences. Now we have lunch with the assistant secretary to discuss a program."[58] The loyalties of men who make it to the top are inevitably tilted toward other men of their ilk, not to the people who are relying on them to represent their interests.

Men have a much higher stake than do most women in being taken seriously by and treated as equals by other men, even when those other men may well be their ideological adversaries. Women are not *inherently* less susceptible to the "cooptation" of power—but, more accustomed to being outsiders in the halls of power, women are in a position to be more critical of the rules of the game in ways that men on the inside can't or won't be. Male eco-establishment environmental leaders don't want to be "outsiders" any more. They want to be invited to lunch.

The Business of Ecobusiness   As men in the environmental leadership increasingly look to businessmen as their peers and exemplars, the lines between the two blur. Leaders of the eco-establishment have wormed their way into the heart of the interlocking network of power that binds the interests of industry, the military, and government—and, now, of environmentalists. Networks of powerful men grease the wheels of politics and the economy, and these networks are essential props of male power in our society.

Interlocking directorates are the building blocks of the social architecture of power in modern industrial economies. As environmental leaders insinuate themselves into the conventional networks of male power, over-lapping bureaucratic arrangements are becoming more common. The boundaries between environmental groups and their adversaries, once drawn in sharp distinction, are becoming permeable. Dean Buntrock, the chair of Waste Management Inc., sits on the board of directors of the National Wildlife Federation; Waste Management, the largest American landfill operator, has a dubious environmental track record, including criminal convictions for pricefixing, EPA fines, and polluted disposal sites.[59] Browning Ferris, the second largest American waste disposal firm, and one with a similarly troubled environmental history, hired a former director of the Environmental Protection Agency, William Ruckelhaus, to head the company in 1988. In 1989, in an attempt to recoup public favor after the Valdez oil-spill disaster, the president of the Exxon Corporation set aside a position on the Exxon board of directors for an environmentalist.[60]

Building bridges between the environmental and corporate communities—helping people "get to know each other"—has, on occasion, led to demonstrably good results. But it is fraught with dangers of cooptation and compromise. Lorna Salzman, a disaffected former staff member at Friends of the Earth, expresses the growing dismay of those outside the circle of the eco-establishment:

> Just look at the boards of directors of these environmental groups, most of them anyway. They're the power elite. . . . There's no way, looking at the boards of those organizations, you're going to get a populist, decentralist, bioregionalist or radically ecological or socially ecological view. There's no way—ever.[61]

Large bureaucracies are expensive to maintain. While most environmental organizations are membership based, small donations collected one by one from individuals are a shaky foundation for sustaining large organizations. Large environmental organizations are increasingly soliciting corporate financial support, a strategy that inevitably leads to questions of propriety and conflicts of interest.

The World Wildlife Fund (WWF) has come under especially heavy scrutiny for its close relationships with industry; corporate sponsorship accounts for an unprecedented 20 percent of WWF income.[62] The roster of corporate donors to the Conservation Foundation (now merged with the World Wildlife Fund) includes Dow Chemical, Monsanto, DuPont, Exxon, General Electric, Union Carbide, Ciba-Geigy, Chevron, and Shell Oil.[63] Further, the World Wildlife Fund is largely sustained by an elite corps of funders, dubbed the "1001 Club." The roster of wealthy individuals in the 1001 Club includes British and European royals (many of whom are avid hunters), oilmen, mining and chemical magnates, arms manufacturers, and property developers. It includes individuals such as Daniel Ludwig, a prominent tanker owner who has devastated large chunks of the Brazilian rainforest for his wood-pulp enterprises; Sir John Eastwood, whose money comes from beef and who still owns huge ranches in South America; President Mobutu of Zaire, who auctioned off large portions of Zaire to private consortia and who has presided over one of the greatest animal massacres in Africa, especially of elephants.[64] Outside observers suggest that the social and economic complexion of the 1001 Club goes a long way in explaining the conservatism of the World Wildlife Fund. The WWF prides itself on its "moderate" and "responsible" positions and its close working relationship with the captains of industry. Many grassroots environmentalists consider the WWF little more than an industry front, "claiming neutrality and objectivity while advancing industry efforts to delay or oppose environmental regulation."[65]

While a few environmental groups, such as Greenpeace, refuse corporate alliances, most other environmental organizations are treading the same path as the WWF. In 1991, grassroots environmental groups raised a furor over the inclusion of Chevron Corporation in the Environmental Grantmakers Association, an American association of foundations that fund environmental activities; Chevron is a major oil company with, among other things, a record as the largest air polluter in the San Francisco Bay area.[66] One woman who opposed Chevron's membership put the problem succinctly: "There's a lot of discomfort here. Some of us feel troubled to be supporting a group fighting Chevron and at the same time sitting next to Chevron, talking strategy." The National Aubudon Society relies on funding from General Electric to sponsor its television nature shows;

General Electric released more cancer-causing chemicals into the environment in 1988 than any other US company, and it ranks number one in the "Superfund" list of national hazardous sites.[67] In 1991, the Hong Kong branch of Friends of the Earth split from the international organization over the issue of business sponsorship; the Hong Kong office receives substantial financial support from Shell Oil, and rather than withdrawing from that sponsorship, the branch withdrew from FoE. North American organizers of 1991 Earth Day celebrations came under considerable criticism for soliciting corporate support; across Canada and the US, Earth Day sponsors included chemical companies, oil companies, energy corporations, and high-profile corporations such as McDonald's.

Critics of big business environmentalism argue that environmental leaders are losing touch with their constituency and are trading in environmental principles for a self-aggrandizing "respectability." A scientist with the Environmental Defense Fund, Michael Oppenheimer, worries that environmental groups will become mesmerized by the possibilities of expanding their influence in Washington "and forget where their strength comes from."[68] Barry Commoner sees the large environmental groups developing "a congenital arrogance that they are in charge."[69]

Free-Enterprise Environmentalism   Professionalization is also ushering "third-wave environmentalism"[70] into the vanguard of the environmental movement. Third wavers believe in free enterprise. The premise is that economic, market-based incentives for environmental protection are the next step after the first wave, the early 1900s conservation of forest land and wildlife, and the second wave, the legislation and regulation of the 1970s. The executive director of the Environmental Defense Fund, Frederic Krupp, describes this as a strategy to "harness the profit motive and introduce new incentives to get business to do the right thing in the first place,"[71] rather than burdening industry with regulations. For many grassroots environmentalists, "third wavism" offers compelling evidence that at least some of the ecoelite have become enmeshed in an abstract world of appeasement just when the struggle is reaching critical mass.

Without environmentalists in the courtrooms and congressional halls, the environment might well be in worse shape than it is. There *are* strategic advantages to a more cooperative relationship with industry, but the premise that environmental hazards can be assessed, mediated, and negotiated between environmentalists and industry requires meeting on common ground, and using a common framework of reference. Environmentalists do not control this common ground. Entry into establishment political arenas has clearly changed the environmental movement; it is not clear that the establishment has been changed. As environmentalists don the attire, reasoning, and behavior of the political establishment, they lose sight of their distinctive mandate to promote visionary, novel, or unconventional ideas. Rather than insisting, for example, that industry acknowledge the importance of aesthetic, cultural, and ethical responsibility in environmental decision making, the environmentalists are on the brink of yielding the higher ground to science and economics.

The use of "rational" scientific and economic yardsticks to measure environmental values is fraught with danger. The underlying premise is that public environmental policy can be (and should be) based on 'neutral, scientific, and objective evidence' about the state of the environment— not based on "intangibles" such as polemics, emotions, wishes, morality, aesthetics, or inherently "immeasureable" quality-of-life considerations. The environmentalists, in an effort to appear more "professional" are increasingly reliant on an economics-based approach that leaves the structure and system of industrial production in place, unchallenged, and casts environmental hazards as "market failures" that can be addressed by negotiations with industry.

There is always a gap between polemical demands of environmental purity (the ideal of perfectly clean air, water, and food), and the "real-world" constraints of costs and expertise that would be needed to achieve these ideals.[72] The implementation of environmental policy and regulation *is* obviously based, in part, on a compromise forged between an environmental ideal and economic considerations and technical constraints. But it should not be the role of the environmental movement to *advance* these constraining factors; compromise is often necessary, but it should not originate with environmentalists. The environmental movement will lose its momentum if the men who lead it, afraid of appearing emotional and less professional than the corporate men they are measuring themselves against, give up their place as defenders of an "irrational" and utopian ideal, and become, instead, cogs in a mediative process.

Where the mediation between environmentalists and corporate representatives takes place—literally—makes a big difference. The center of gravity of eco-establishment activism is shifting from the kitchen table to the boardroom conference room. The plush, mahogany-paneled corporate boardrooms where negotiations between environmentalists and industrialists are increasingly likely to take place are settings that reinforce compromise and cooptation, and that, not incidentally, reek of cosseted male privilege. At best, they are not settings that encourage innovation.

The Science Imperative   In tandem with an economics-driven policy, the definition of a "professional" environmental movement is increasingly one that relies primarily on research-based "science." Jonathon Porritt is proud that Friends of the Earth relies less on polemics ("that alienates the decision-makers") and more on research; Greenpeace has been making "genuine efforts to beef up its scientific credibility."[73] Science may be useful—indeed essential—to modern environmentalism, and there are clear instances in which science has saved the environmental day. But there is also good reason to be wary of an uncritical embrace of the science establishment.

There are a number of problems with the tilt toward science, not the least of which is that it increases the reliance of environmental organizations on often narrowly-trained, male scientific experts. The expert structure of environmentalism is shifting as science becomes the arbiter of environmental concerns. "Science" takes environmental assessment further and further away from the realm of lived experience—which is, not coincidentally, the realm in which most women are expert. Reliance on "scientific facts" (and on the experts who collect such facts) pushes amateurs to the fringes, and undercuts the valuable environmental knowledge that amateurs have accumulated over the years. In point of fact, it is the work of "amateurs" as much as scientists that has led to some of the great environmental exposes of our time—such as the poisoning of Minamata Bay in Japan, and the industrial and military catastrophes at Love Canal and Fernald in the United States. Since a higher proportion of environmental "amateurs" are women, and women are still very much a minority in the ranks of scientists, an integral byproduct of the shift towards science-based environmentalism is the increasing marginalization of women within the movement.[74]

Science does not take place in a vacuum. Feminist theorists have made clear the extent to which we must understand science as a culturally rooted product:

> Science is a socially produced body of knowledge and a cultural institution. Our culture is deeply and fundamentally structured socially, politically, ideologically, and conceptually by gender as well as by race, class and sexuality. It then follows that the dominant categories of cultural experience (white, male, upper/middle class, and heterosexual) will be reflected within the cultural institution of science itself; in its structure, theories, concepts, values, ideologies, and practices.[75]

A growing number of Third World scholars add their voices to the rising chorus of critiques of Western science to underscore the fact that science is a labor-based product, and that, inevitably, the products of science reflect the gender, class, and race profile of its practitioners.[76]

The ideological underpinnings of modern Western science are not environmentally propitious. A critique of the historical development of science underscores the extent to which modern science is based on a paradigm of the domination and exploitation of nature. Feminist analysis has added to this critique by exposing the extent to which the mechanistic scientific approach to nature is rooted in a masculinist worldview that conflated the domination of nature with the oppression of women.[77] As a practical enterprise, the resources of science are disproportionately directed to developing technologies of destruction—by some estimates, upwards of 70 percent of Western scientists are engaged in developing weaponry and other military-support programs.[78] The imposition of Western science on the periphery of the industrializing world was integral to the history of empire-building and colonial expansion of Europe; it continues unabated today. The result is the stifling of rich, alternative scientific traditions and a delegitimation of locally based, indigenous knowledge systems throughout Asia and Africa. The fact that science and technology evolved in directions that took control of natural resources away from people who use them for sustenance and survival, puting them into the hands of people who use them for profit, has caused tremendous environmental dislocation and deterioration throughout the Third World.[79]

In its recent history, science has not been a politically liberating institution. In fact, it has been used primarily as a prop for conservative (and environmentally regressive) agendas. As one Third World critic of Western science reminds us, "modern science is an instrument that has been used to serve the functions to which its controllers assigned it."[80] In environmental struggles, "scientific facts" are often thrown up in opposition to community-based and "amateur-collected" evidence against environmental violators: thus, to use a familiar example, women's hand-drawn maps of carefully collected data on neighborhood cancer clusters are routinely trounced by corporate "experts" (mostly men) who, with a flourish, "scientifically prove" that the cluster is an insignificant statistical burp. It appears as though it soon may not only be environmental violators who reject hand-drawn and home-collected information; based on the current noises out of the environmental movement, it may well be men from the movement itself who rise to the fore in rejecting "amateur" environmental assessments. If environmental leaders use science to create a hierarchy of reliability, if it is used to distance the newly professional men from the neighboorhood and "amateur" women, then the environmental movement will lose what has been to date its most valuable resource base.

Equally important, reliance on science throws environmentalists into the pit of scientific uncertainty. It is almost impossible to "scientifically prove" that the chemicals dumped in an unlined settling pond *caused* the brain cancer of the child who lived next door to the dump; it is impossible to scientifically prove that global warming is upon us. The inherent uncertainty in science always plays in the favor of environmental *offenders*, not environmental defenders. Evidence that must pass through the science sieve almost always favors environmental uncertainty. Indeed, some of the most egregious corporate and military polluters routinely depend on scientific uncertainty to "prove" that there is no "scientific" evidence of ozone depletion, of acid rain effects, of Agent Orange damage. Scientific uncertainty justifies delaying action on a host of issues. The call for "more scientific evidence" or "further research" is the rallying cry of polluters and poisoners; the environmental movement should be extremely leery of taking it up.

The reliance on science is moving ecologists away from humanistic values in assessing environmental efficacy. The "human rights" environmental initiative now under consideration in the EEC, for example, is insupportable within a scientific paradigm—"landscape aesthetics," "environmental values," or quality-of-life issues are not measurable on the yardstick of science. Science threatens to undercut some of the most innovative environmental work now under way.

A further danger lurks along this particular path toward professionalism. Science may quickly expand from being but one tool in the environmentalist's quiver, to being *the* leading paradigm for environmental action. In short order, environmental campaigns may have to be chosen on the basis of which issue will generate the most reliable "scientific" results. What's the point, after all, of relying on science if the scientific results turn out to be ambiguous and uncertain? The eco-establishment may choose not to take up an issue if their own scientific evidence does not solidly support their case. In the end, environmental policy and campaign choices may thus be limited to a narrow range of issues that can be supported by "scientific evidence."

Environmentalists are jumping on the scientific bandwagon without taking stock of existing critiques of science, among which feminist critiques are especially important.[81] There appears to be little introspection within the environmental movement about the uses of science in society, and about the extent to which science is a prop for conservative political and social values. The use of science privileges the existing social power structure and provides the rationale for maintaining the environmental status quo. Reliance on science, without a well-formed critique in hand, leaves existing social structures and structures of knowledge unchallenged. In this culture, the existing social and knowledge structure is one that privileges men over women, and positivist proof over other ways of knowing.

Even environmental "quality of life" issues are being converted into scientific undertakings—as evidenced by the tremendous surge of interest in the social science of "environmental impact assessment." In practice, environmental impact assessment has a number of problems, not the least of which is the striking unwillingness of its proponents to acknowledge that "impacts" of environmental degradation cannot be generalized for an undifferentiated population. The standard-setting studies in the EIA field consistently disregard the fact that the "impacts"of any given environmental phenomenon will vary tremendously with age, class, race, social situation, and gender. While our knowledge of the ramifications of all these realms of "difference" is still rudimentary, we do know that women and men, for a start, have different perceptions of environmental issues, that they often suffer distinct health effects from environmental degradation, and that they have different relationships to the institutions that are primary environmental players. As in other fields of scientific inquiry, EIA too often serves to keep the realities of women's lives invisible, and to maintain the tradition that "women's issues" are peripheral to EIA, or can be subsumed under the category of environmental impacts on "men" or "people."[82] Feminist critiques of the structure of research identify a number of areas where sexism intrudes in research: through androcentric perspective, through the use of sexist language, concepts, and methodology, and through a sexist interpretation of results.[83] It is impossible to escape the conclusion that the theoretical development of EIA overlooks "difference" precisely because the people who are movers-and-shakers in the field of EIA are white men who are accustomed to universalizing their own experiences.

We are clearly heading toward a science-led environmentalism, but environmental leaders will lose much of the ground they have gained over the past decades by giving up experience-based observation in favor of science-based "fact." They will certainly alienate a large segment of the population that has, so far, been the backbone of environmental action, and they will drive a wedge between men and women in the movement. Environmental leaders, eager to jump on the science/research bandwagon, remain willfully incognizant of the gendered implications of science and research; yet an environmental analysis that chooses to ignore the ideological questions posed by its own framework of inquiry will ultimately discredit itself.

Hierarchies   In the course of writing this book, I talked with dozens of women working within environmental organizations. Almost without exception, these women emphasized that they were initially drawn to work in the movement because power was decentralized, because the management structure was not rigidly hierarchical, and because diversity and unorthodoxy were valued. Professionalization, in the particular form being introduced to the movement, very clearly undercuts these features. Women *can* work, and excel, in a professional, "grey suit" ecobureaucracy: but it is not clear that they will be welcome into that "men's club" world, *or that they will want to be.*

Despite the often-heard protestations of the men in these organizations that "it doesn't make any difference" that men are the ones in the top positions, many women in these organizations have a different view. One woman who worked for a decentralized, radical environmental organization joked that "about every 6 months, at staff meetings, men would raise 'division of labor' issues—which was always just a euphemism for them wanting to be given bigger jobs with formal titles!" The complaint that men are uncomfortable with decentralized, nonhierarchical work systems was echoed by a number of women. One woman staffer with an animal rights organization said that men are always trying to impose (or re-impose) patriarchal process in the environmental workplace. By "patriarchal process" she meant a system that encourages inter- and intra-organization competition, not cooperation; discourages leaders in different wings of the environmental movement from forming bonds with one another; isolates women from each other; and places a premium on external credentials (which more men have) over experience-based credentials (which more women have).

Glimpses of an Alternative

Feminist analysts assert that it *always* makes a difference when there is a gender-skewed distribution of power in any structure or institution, despite the protestations of men that it "doesn't matter" that they are in charge. But it is often not easy to identify the consequences of power imbalances in terms of specific, real-world examples. Given this, and given the urgency of the tasks facing the environmental movement, critics might suggest that it is dilettantish to be worried about whether it is men or women in the environmental corner office. But, the fact that the environmental movement is increasingly male-led and male-defined *does* have specific, identifiable consequences in terms of environmental priorities, policies, and practices—which, in turn, have specific and identifiable consequences for the environment itself.

One of the reasons that it is difficult to discern the consequences of having a male-dominated environmental movement is that there is not a parallel women-led ecobureaucracy available for comparison. (The women-led grassroots movement offers other lessons, but it does not offer a comparable example). When all of the major groups are male-led, and have always been so, it is difficult to imagine how things might be different if the gender of power was different. The elucidation of difference is further complicated by the fact that some of the women who do make it into the upper echelons of power are the most likely to be the most "male-assimilated" women—and thus their ascendancy may make no difference whatsoever to the nature of organizational politics. It is entirely possible to have women in the movement, whose presence challenges women's inequality within the system, without challenging the system itself. To answer the question "what difference does it make?" we must thus not merely look for differences between individual men and individual women as they behave as power brokers. Rather, we must look for signs of differences between organizations and policies informed by patriarchal principles and those rooted in feminism: a tall order to fill, given the state of the world. But with a recent change in power in the Italian environmental movement we may be able to discern the outlines of difference.[84]

The Italian environmental movement, like most of its counterparts around the world, depends largely on women's initiative and women's labor, and, in an also familiar pattern, the leadership has been largely appropriated by men. In 1988, Renate Ingrao, a feminist, was elected as the first woman director of the Environmental League, one of the largest environmental organizations in Italy. Observers say that Ingrao is slowly redefining the priorities and principles of the Italian environmental movement in ways that are new, distinct, and clearly informed by her feminist principles. She is deconstructing the internal hierarchy of the Environmental League. In terms of external relationships, a central principle of her leadership is to break down the barriers and the hierarchies among the various Italian environmental groups. She is building alliances *across* the environmental movement, and is trying to redefine the "empire-building," organization-specific, competitive approach to environmental organizational relations that previously prevailed. Some of the men worry that this will cost the Environmental League its distinct identity and its primacy in the movement; Ingrao answers that organizational identity takes second place to the needs of the movement as a whole. Striking at more fundamental issues, Ingrao is trying to redefine the priorities of the the movement, steering away from the seduction of cost/benefit analyses and economics-based priorities toward a human-needs approach, and relocating the arena of environmental action from national to community-based interests. Whether she will succeed, and whether she will survive the challenges to her authority, remains to be seen. But the policies and priorities that she is proposing could have enduring consequences, both for the Italian environmental movement and for Italy's environment itself.

Ingrao's attempt to reorient environmental priorities away from economics-based analyses toward policies based on human needs represents a significant shift. This is certainly the direction in which the most progressive leading edge of the movement is moving. The EEC, for example, which has been leading the way on many environmental issues, has commissioned a study to address the question of whether environmental quality should be defined as a "human rights" issue. The "human rights" perspective in environmental politics is an innovative angle from which to argue environmental efficacy. It suggests that the environmental movement might do well to develop more humanistic measures of environmental safety to complement the conventional measurements of, for example, atmospheric or water pollutant levels. The human rights initiative, for the time being, is moving slowly in Europe, (and there is no equivalent initiative in North America), but it does represent a distinctive perspective that environmentalists should not let slip away.

As it happens though, "professionalization," which is sweeping the movement in Europe and more efficiently in the US, is moving environmental policy in the opposite direction, away from human-needs toward economics-based analysis. The male-dominated professionalization trend is propelling the movement even further away from intangible, "emotional" assessments of human-based needs and quality- of-life issues, moving swiftly into the camp of econometrics and cost/benefit analysis.

◑

Many commentators have warned of the loss of passion in the eco-establishment. But passion is not being "lost"—it is being consciously and specifically traded in for a cloak of professional respectability. It is a specific choice made by men in environmental leadership positions, trying to match a standard of officious and corporate masculinity, who are aiming to replace a values-based, emotional environmentalism with a rational, scientific, economic, "reasonable" paradigm for environmental action. It is not clear that the public will follow them in this direction: they are already losing some of their constituency, and particularly women, to the more militant, disorderly, and "emotional" grassroots coalition of community-based environmental groups, animal rights groups, and single-issue pressure groups.

Feminists within mainstream environmental movements are increasingly aware of the shortcomings of male-dominated environmentalism. One of the most eloquent discourses on this topic is Pam Simmons' introduction to the 1992 special feminist issue of *The Ecologist*; it is worth quoting at length:[85]

A shadow lies over the environmental movement: patriarchy. Like so many movements before it—socialist, conservationist, civil rights, national liberation—the environmental movement is failing to acknowledge and criticize its own attachment to male power and privilege. . . . By shying away from the challenge of feminism, men (and many women) in the movement are blocking out opportunities and perspectives that will be indispensable for reaching the solutions they are anxious to find. Environmentalists cannot credibly discuss the effects and future of development, equality and justice, conflict-resolution, the preservation of diverse cultures, the industrial and military complex, the reconstruction or preservation of economic self-sufficiency or the dynamics of people's movements without discussing feminism. . . . If the movement does not face up to its own patriarchal base, it is excluding potential allies, while creating a hierarchy ripe for a sell-out.

TAKING ISSUE

Having asserted at the beginning that "it matters" that men are in power, the discussion so far has suggested some of the ways in which the environmental movement, environmental policy, and environmental action are shaped by men who are espousing traditional masculinist values—and the extent to which these values are increasing, not diminishing, in importance in the eco-establishment. A close analysis of three environmental issues further illuminates the masculinist underpinnings of the environmental agenda.

Fur, Meat, and Misogyny

One of the most striking signifiers of sexist and misogynist tendencies within the environmental movement is the infamous Greenpeace/Lynx anti-fur advertisement that was released in Britain in 1985. The ad, which was plastered on billboards throughout the UK, and which ran as a filler in movie houses, shows a woman in high heels and a slit skirt, dragging behind her a fur coat that is trailing blood. Women were shocked by the extent to which this ad played on imagery common in pornographic violence against women. But adding a direct insult to the implicitly threatening tone, the caption to the ad states, "It takes up to 40 dumb animals to make a fur coat . . . *But only one to wear it.*"

This advertisement is a stunning example of the reluctance of environmental organizations to take on the broad lessons of feminism in their approach to constructing environmental campaigns. The ad caused an uproar among feminists, but Lynx leaders dismissed criticisms that the ad was offensive to women. They were, instead, carried away by reports of the ad's "effectiveness," and they refused to stop its distribution. Nor was the ad treated as a one-time error in judgment, and allowed to fade away gracefully. Indeed, the ad is still very much in circulation, and five years later, full-page notices appeared in American animal rights journals heralding the "arrival" of this ad, and of Lynx, on the American animal rights scene.

In 1990, Lynx produced a "sequel" ad to the "dumb animals" campaign: a white, privileged woman draped in a fur coat is curled up coquettishly on her side, staring seductively and arrogantly at the audience; to her left, stark against a white background, is a dead fox, curled in a similar pose, dripping blood. The bold lettering beneath the woman says, " Rich bitch;" beneath the fox, "Poor bitch." The implicit threat of violence against women in this ad is not lost on most women observers. The intentional parallel between the woman and the dead fox implies that as one "bitch" dies, so could (should?) another.

In anticipation of the "rich bitch" ad campaign's release, Lynx director, Mark Glover, said: "We are all very excited about this dramatic new Lynx poster. It is designed to carry the campaign forward into the 1990s and will continue the momentum started so brilliantly by the 'Dumb Animals' poster."[86] This rationalist stance in the face of humanistic objections— the notion that the ends justify the means—is a familiar masculinist stance. Women encounter objectification of their bodies and the threat of implicit violence in the media and advertising industry on a daily basis, always justified by claims of "effectiveness." In a sexist and violent culture, it is not surprising that the objectification and denigration of women in mass communications makes for "effective" advertising.

Blaming and ridiculing women for the horrors of the fur trade is an easy out. There are more creative and more sophisticated possibilities for creating anti-fur advertising—possibilities that male-dominated groups such as Lynx overlook, but that women-led groups, among them the Friends of Animals (USA), have tapped effectively. A good example of a non–woman-blaming, nonsexist, anti-fur campaign is the advertisement created by Friends of Animals that shows, in three still sequences, a (male) trapper approaching a coyote trapped in a leg-hold, the trapper tormenting the trapped animal, and finally, stomping on its throat to kill it (one of the few ways of killing trapped animals without damaging their fur). The caption reads, simply, "Take a stand on fur." This advertisement is arguably at least as "effective" as the woman-hating ads of Lynx. And it manages to convey an anti-fur message without blaming women, threatening them, or holding them responsible for the entire fur industry.

Rich bitch.

If you don't want millions of animals tortured and killed in leg-hold traps don't buy a fur coat.

"It takes up to 40 dumb animals to make a fur coat.

But only one to wear it."

Poor bitch.

Visit the Lynx Shop at 79 Long Acre, London WC2
Or write to: PO Box 509, Dunmow, Essex.

Figure 4.1.   Blaming women for the fur trade. Advertisment of a British animal rights group.

The threatening and demeaning ads used by Lynx are not only anti-women, but they perpetuate the enormous deceit that women are the main culprits in the fur industry. Nothing could be further from the truth. The fur industry is a preeminently male enterprise. The hunting, trapping, killing, skinning, auctioning, sewing, selling, and buying of furs are all male activities. Even the *demand* for fur is largely a male creation, a status symbol of conspicuous consumption that men bestow on women; the fur coat, bought *by* a man for "his" woman is a potent symbol of male wealth

TAKE A STAND ON FUR.

The fur industry depends upon the suffering and death of millions of wild fur-bearing animals each year. It also depends upon the ignorance of fur-wearing humans.

Terror-stricken animals are trapped and left to struggle, sometimes for days, before being strangled or bludgeoned or stomped to death. How can the wearing of fur garments possibly

justify this suffering?

The fur industry is counting on your looking the other way. They know that the practice of peeling the skins from tormented animals to feed human vanity is something most people would not tolerate.

If the only thing standing between you and the decision not to wear fur are the facts, take a good look. Once

you do, we believe the stand you'll take on fur will be *against* it. And that's a step in the right direction.

Friends of Animals

Figure 4.2.   An anti-fur advertisment that does not attack women, developed for an American (women-led) animal rights group. (Courtesy Friends of Animals.)

and of male benevolence. Even though women may wear most of the consumer furs produced in the US and the UK, fur-wearing by women is largely the vehicle of men's fantasies and men's sense of self-importance. The fur trade is a big business, run *by* men, *for* men, with enormous profits reaped by men.

Lynx and other animal rights groups are in a position to expose the gender-specific nature of the fur industry. It is not difficult to chart the gender politics of the fur industry. The fact that women are easy targets of blame and abuse in our sexist society blunts this curiosity. It is easier for animal-rights groups to blame women for the fur trade than it is to take a critical and realistic look at the complicity of men and male values in the annual slaughter of millions of animals.

Animal Rights   Beyond the anti-fur stance of a few environmental groups, most environmental organizations have been leery of entering into the broader realm of animal rights. The eco-establishment typically becomes involved with animals only at the point at which a species is in danger of extinction. The predominant concern with "nature in the wild" and the conservationist underpinnings of environmentalism in the US and the UK steer environmentalists away from larger issues such as the treatment of animals in factory farming food production systems or their abuse in experimental military and scientific laboratories. The animal rights movement, conversely, has not extended its reach to broader environmental issues. The schism between the two movements can be characterized, in part, as a schism between men and women.

Women are considerably more active in the animal rights movement than are men. Best estimates suggest that 70–80 percent of animal rights activists are women. Not surprisingly, then, feminist theory overlaps with animal rights issues to a greater extent than on any other environmental topic.[89] Feminists have constructed a broad animal rights agenda that includes: consideration of the parallels between the oppression of animals and the oppression of women; a rejection of the prevailing rationalist, mechanistic world views that treat animals, nature, and women as instruments that can be manipulated to further scientific and economic ends; a close examination of ideologies and cultural assumptions that associate meat-eating with masculinity; and, an analysis of the environmental costs of sustaining meat-eating, animal-product based economies.

Women have not steered away from the "emotionalism" of the animal rights issue, and, in fact, many women posit emotional connection with animals as the paradigmatic opposite of the learned desensitization and detachment that allows people in laboratories and factory farms to inflict horrible pain and deprivation on animals. For at least the last 150 years,

Table 4.3. The Gender Politics of the Fur Industry[87]

- The first step in the fur industry is either the **trapping** or the **"farming"** of fur animals. Approximately 60–70% of the fur animals in the US and the UK are "farmed." Almost 100% of *commercial* trappers are men; most fur ranches in the US, Canada, and the UK are owned by *men*.

- The **killing** of animals caught in traps, often by stomping, or by gassing or electrocution on the "farm" is *men's work*.

- **Skinning** and **"fleshing"** (removing fat and gristle from the skins) of the dead animal is *men's work*. • Manual **scraping** of the skins is *women's* work.

- Most pelts are sold at auction houses in Leningrad, Copenhagen, Montreal, New Jersey. The **trading** and **auctioning** of fur pelts is a male-dominated occupation, a high-status position of power and privilege. Most of the **buyers** of furs at these auctions also are *men*, because it is men who own the fur factories that produce consumer fur.

- The **tanning** and **dressing** of pelts purchased at auction is *men's work*.

- The **designing, cutting and sewing** of fur into coats and other wardrobe items is a highly skilled craft, and like many other skilled-craft, closed-door industries, it is an *entirely male* subculture. As the Fur Retail Association says, "Fur working is a skilled craft, and fur workers are a lifetime in the making . . . skills are passed on from *father to son*." The factories where designing, cutting, and sewing take place are owned by men, and in many cases have been owned by men in the family for several generations. • Sewing linings into fur coats is *women's work*.

- **Retailing** of furs is controlled by *men*; most fur retail stores are owned by men, and most of the salespeople are men.

- **Fashion designers** are predominantly *men*, and most of the big-name fashion houses are owned and controlled by men.

- The **consumer purchasers** of furs are also mostly *men*. Recent figures from the American Fur Industry provide this glimpse of the structure of fur consumption:[88]

1970	Percentage of furs bought by men for women:	85%
	Percentage of furs bought by women for women:	12%
	Percentage of furs bought by men for men:	3%
1990	Percentage of furs bought by men for women:	60%
	Percentage of furs bought by women for women:	30%
	Percentage of furs bought by men for men:	10%

animal rights activists in the US and the UK have made links between women's struggles for autonomy and freedom from oppression and the rights of animals. The earliest antivivisection activists in modern Europe were women: "Every cat or dog strapped down for the vivisectors knife reminded them of their own condition."[90] Many women saw the imagery of vivisection as an image of dominance, one that particularly echoed the imagery of the sexual dominance of women by male gynecologists and pornographers:

> When observers like Frances Cobbe and Louise Lind-af-Hageby witnessed these demonstrations [public demonstrations of vivisection], they felt as though they were in the presence of a new pornography, and all the animals stretched out writhing on boards were like women on the gynecologists table or bound to some chair devised by the riding master. Why did the language of gynecology with its stirrups, saddles, and straps, resemble that of pornography?[91]

A number of women have drawn parallels between the treatment of animals in factory farming, and especially in experimentation, and the treatment of women. One of the frightening aspects of this overlap is the use of male violence as a tool of domination. Tennyson's poetical ditty, "Man is the hunter; woman is the game. The sleek and shining creatures of the chase, Who hunt them for the beauty of their skins; They love us for it and we ride them down," strikes a chord of realism in women. Women who have been sexually assaulted by men often speak of the "dehumanizing" and "animalistic" quality of their attack and their humiliation by men. Rape, especially gang rape, can strikingly resemble hunting: both feature the stalking of prey, the cruelty and thrill of the capture, the degradation of the victim, and the enjoyment of the victim's terror and defeat.[92]

Violence against animals often evokes a sexualized violence: the repeated forced pregnancies of dairy cows, the "rape racks" used by breeders, animals bound and splayed on dissection tables in laboratories. Both women and animals are the "other" to a distant, male scientific research establishment that directs considerable energy to manipulating the reproductive processes of both animals and women. Similarly, violence against women is often "animalized." *Hustler*, a male pornography magazine, shocked even its regular readers in 1985 by running a cover illustration of a woman being fed into a meat grinder, and coming out hamburger at the other end. A photograph in an earlier issue, captioned "Beaver Hunters" shows a nude woman tied with ropes to the hood of a jeep in which two men dressed as hunters sit with rifles. The text under the photo says,

"These two hunters stuffed and mounted their trophy as soon as they got her home." Women are commonly referred to as "meat," "pussy," "chicks." Illustrations hang in bars and poolhalls showing a naked woman, blue-penciled and identified by body part, like a carcass of meat. There is considerable evidence, in both popular culture and scientific enterprise, of the intertwined ideologies of the domination of women and the domination of animals.

Feminists have long argued that the oppression of animals serves as the model and training ground for all other forms of oppression.[93] Nowhere is this connection more clear than in hunting. Hunting animals for "recreation" is a male ritual, cherished as a rite of passage into manhood, one that critics say trains boys into calculated callousness and cruelty. Hunting primarily teaches men to feel victorious, not compassionate, when they kill or maim a creature. Hunting "normalizes" male violence, and it is both women and animals who suffer the ripple effects of the validation of ritual violence; an inferential indicator of this may be that battered women's shelters throughout the US report a sharp upturn in calls to crisis hotlines during hunting seasons.[94]

The historical record provides ample evidence of a privileged connection between manhood and meat eating. Anthropological research suggests a correlation between plant-based economies and women's power, and animal-based economies and male power.[95] In nineteenth-century Europe and America, working-class families usually could not afford sufficient meat to feed the entire family; in these cases, meat was reserved for the men of the family. In times of economic abundance, sex-role assumptions about meat are not so blatantly expressed, and the diets of men and women become more similar. However, it is still true today that where poverty forces a conscious distribution of meat, meat is reserved for men. During the past two World Wars, when government rationing restricted meat eating in both Britain and the US, meat was reserved for soldiers. One commentator observed that World War II began a "beef madness . . . when richly fatted beef was force-fed into every putative American warrior."[96] Meat eating was an assumed prerequisite for virility.

Even under conditions when both men and women eat meat, meat-eating is still particularly associated with masculinity, an association that is most clearly suggested in the Anglo-American "popular culture" of food. In the 1980s, a whole genre of humor developed around the self-ironic phrase, "real men don't eat quiche." A "meat 'n potatoes" man is a hard-working he-man—a "real" man. Most butchers are men; women are assumed to have less expertise and knowledge about the selection, preparation, and retailing of consumer meats. Often the only household cooking that men do—reserved as their exclusive privilege—is the ritual backyard barbecuing. Jean Mayer, an American nutritionist, reflects on masculinity and meat: "the more men sit at their desks all day, the more they want to be reassured about their maleness in eating those large slabs of bleeding meat which are the last symbol of machismo."[97]

Non-feminists may scoff at an analysis of hunting as a vehicle of male violence and power. The machismo of meat-eating may be dismissed as little more than an amusing cultural artifact. Skeptical readers might question what this has to do with a discussion of the *environment*. And why should *environmental* groups be concerned with a feminist-centered analysis of animal rights?

The fact of the matter is that meat-eating and hunting are fueling global environmental catastrophes. The massive diversion of resources into meat production is environmentally untenable. Millions of acres of Central and South American rainforests have been cleared to make way for cattle-ranching for the meat-export industry; between 1960 and 1985, 40 percent of all Central American rainforests were cleared to create pasture for beef cattle. In Mexico, the amount of cropland devoted to feed and fodder for the livestock industry has climbed from 5 percent in 1960 to 23 percent in 1980, even though an estimated one-quarter of the Mexican population is chronically malnourished.[98] Whole ecosystems in the Third World have been decimated to provide hamburgers and steak for North American and European consumers. Cattle ranching is responsible for an estimated 85 percent of topsoil erosion in the US, and similar devastation in Australia and Canada. Within the US, half of all water consumed is used to grow crops that are fed to livestock; meat production requires at least 10 times more water than grain production. More than 50 percent of water pollution in the US can be linked to wastes from the livestock industry, including manure, eroded soil and synthetic pesticides. More than two-thirds of all US cropland is devoted to livestock. Meat production places enormous demands on energy: the 500 calories of food energy from one pound of steak requires 20,000 calories of fossil fuel. Sixty percent of American imported oil requirements would be cut if the US population switched to a vegetarian diet.[99]

While loss of habitat, urbanization, and pollution take their toll on animal populations, hunting, whether for big-game profit, for "recreation," or to service the trade in wild animals and wild animal parts, is a significant cause of animal extinctions and near extinctions. Uncounted millions of animals are killed by hunting each year, pushing entire ecosystems to the brink of collapse. Recreational hunting has caused the extinction of the Arabian oryx, the Mexican jaguar, and the Labrador duck, among dozens of other species. Hundreds of millions of birds are killed annually as they migrate north from Africa into Europe; hunters in countries of the

Mediterranean rim slaughter doves, falcons, songbirds, egrets—virtually anything that flies—in the name of (male) recreational ritual.[100] Animal extinctions are escalating at a staggering pace; by the end of this century, 100 animal species a day may be forced into extinction. In the late twentieth century, relatively few communities rely on the hunting of wild animals for food and clothing. Most hunting in the world today is for either "recreation" or for profit, and the lines between the two are often blurry.

Much of the hunting in the world is conducted to supply the international trade in wildlife. The wild-animal trade is a big business, worth over $5 billion a year, that has decimated animal populations in Africa, Central and South America, and East Asia.[101] It's a shadowy business, and hard facts and figures about the animal trade are elusive. But the documentary record that does exist paints a dramatic picture of massive animal slaughter. Every year, for example, dealers in Argentina export an average of 1.3 million reptile skins and almost 200,000 live birds to the luxury markets in North America and Europe; 170,000 live parrots are exported from Africa annually, and another 130,000 from Asia—and this is only the reported trade.[102] African rhinos are being hunted to the brink of extinction for their horns—which end up as male aphrodisiacs in Asia or as handles for men's ceremonial daggers in the Middle East. The slaughter of Arctic walruses, which have elephant-like ivory tusks, is increasing at an alarming rate. In the last twenty years, the Asiatic black bear was hunted into extinction—killed for organ meats that are turned into male aphrodisiacs.

The biggest global importers of wild animals and wild animal products are rich Americans, West Europeans, and Japanese—or, more accurately, rich *men* in those countries. Like the fur trade, the wild trade in animals is almost entirely a men's industry. Almost all of the hunters, poachers, and trophy seekers are men, as are the middlemen, international brokers and dealers, and the end consumers.

In a pathbreaking book on the international traffic in women, Kathleen Barry writes:

> considering the numbers of men who are pimps, procurers, members
> of slavery gangs, corrupt officials participating in this traffic, owners,
> operators, employees of brothels and lodging and entertainment facil-
> ities, pornography purveyors, wife beaters, child molesters, incest
> perpetrators, johns (tricks) and rapists, one cannot but be momen-
> tarily stunned by the enormous male population engaging in female
> sexual slavery. The huge number of men engaged in these practices
> should be cause for declaration of an international emergency. . . .
> But what should be cause for alarm is instead accepted as normal
> sexual violence.[103]

The parallel with the treatment of animals is striking. The feminist environmentalist version might well be this: considering all the men who are hunters, poachers, fur "farmers," hunting suppliers, hunting guides, skinners, trappers, furriers, fur designers, consumers buying furs for women, importers of exotic animals, hunting-lodge owners, safari leaders, trophy collectors, ivory carvers, aphrodisiac users and ivory-dagger toters, one cannot but be momentarily stunned by the enormous male population engaged in the destruction of wildlife. The huge numbers of men engaged in these practices should be cause for declaration of an international emergency . . . But what should be cause for alarm is instead accepted as normal male behavior.

Wildlife preservation and the protection of animal species are high on the agenda of most mainstream environmental groups. But while environmental groups take up the issue of animal extinctions in their save-the-elephant, save-the-panda, or save-the-tiger campaigns, they seldom take on the issue of hunting *per se*. This is a curious myopia. Perhaps this is male passivity: the male-dominated environmental movement may be reluctant to challenge the male-dominated hunting industry. Or perhaps it is because hunters and fishermen are the traditional constituency and financial supporters of many environmental groups, especially older conservation organizations such as the Audubon Society, the Sierra Club, and the National Wildlife Federation. Or perhaps it is because the hunting and "harvesting" of wildlife is so normalized as a male activity that men in the environmental movement don't see it as the ecological catastrophe that it is. Whatever the underlying cause, it is instructive to note the disjuncture between the enthusiasm for environmental campaigns that identify women as the culprits in the fur trade, and the absence of environmental analysis that would make clear the complicity of men in the slaughter of the world's animal populations.

Population

One of the major tourist attractions in Baltimore, Maryland, is the aquarium, an imposing and sophisticated new complex on the harbor. In 1989, one of the most ambitious exhibits in the aquarium was a re-creation of a tropical rainforest. As visitors entered the exhibit, they were presented with two digital counters displaying rapidly changing numbers on a screen. The first counter was a continuous tally of global rainforest loss, a mesmerizing and disturbing display. On the second digital display, the numbers flicking, almost as rapidly, kept tally of the world's population growth. There was no explanation for the juxtaposition of the two—visitors were left to assume a correlation between rainforest loss and population growth. In the 1990s, assumptions such as these are coming to shape popular environmental attitudes as well as official environmental policy.

Population is a thorny and divisive issue in the environmental movement. The "liberal" (and, now, most common) position is that uncontrolled population growth is putting intolerable strains on the world's ecosystems. Population growth, in this view, is identified as a major force of global environmental degradation and, thus, the argument follows, population control must be high on the environmental agenda. The more "radical" view is that population is not the root problem—rather, that maldistribution of resources in the world is the root cause of environmental degradation, and that a high rate of population growth is a symptom of social inequity, driven by poverty, insecurity, and scarce resources. Even mainstream proponents of this radical analysis, however, often add that population control is an environmental imperative. Environmental groups, all along the spectrum from radical to conservative, are increasingly taking positions of population control advocacy: the Sierra Club has an official position paper on population, calling for population control; the Audubon Society is mounting a major campaign in favor of population planning; proponents of "deep ecology" call for population *reductions;* the World Bank characterizes population growth as a major environmental threat. The agendas of many environmental groups are converging with the traditional mandate of population control agencies around a shared concern about overpopulation.

But the relationship between population and the environment is complex. While it is easy to slip into the assumption that there is a direct correlation between the world's growing population and the world's growing environmental problems, the evidence does not necessarily bear this out. Whereas population growth is now a Third World phenomenon, by anyone's reckoning the large-scale global environmental crises we face are largely the product of the voracious resource appetites of "First World" industrial economies and lifestyles. In terms of human burdens on the atmosphere, for example, one economist estimates that the "cost" of one US citizen is 16 times that of a Third World person; a West European is 5 times the burden. The average American consumes 300 times as much energy as the average Bangladeshi, and in an average lifetime, an American will consume 26 million tons of water, 21,000 gallons of gasoline, and 11,000 pounds of meat.[104]

If we tick off a checklist of the world's most pressing environmental problems—fossil fuel pollution (and the related threat of global warming), the use and production of ozone depleting chemicals, the resource use and pollution generated by oil dependency, acid rain emissions, the glut of household waste, municipal waste, and toxic industrial waste, the luxury trade in exotic animal and bird species, and the stripping of the world's forests—it becomes clear that the biggest contributing cause of these problems is not the "excessive" number of people in the poor countries of the world, but the excessive pressures that the richer countries exert on the world's biosphere. Global rainforest destruction, to return to the Baltimore aquarium example, is primarily the responsibility of a few rich landowners converting forest to cattle-lands, a few rich multinational corporations that are stripping forests in the Third World for the international timber trade in tropical hardwoods, and government-forced relocation schemes.

The script about overpopulation also glosses over historical causes of environmental destruction: the rapacious resource depletion of colonies by imperial powers; the forced introduction of monoculture and plantation agriculture in the tropics, displacing subsistence and indigenous agriculture—both an historical and a contemporary process; the distortion of household structure by colonial wage systems that gave colonized men considerably more economic clout than their female counterparts.

This is not to suggest that growing numbers of people do not burden natural ecosystems; they do. Destruction of wildlife habitats, localized cutting of forests for subsistence agriculture and fuelwood, and urban water-quality deterioration, to mention a few pressing environmental problems, stem in some measure from the sheer pressure of human need. Ultimately, though, most of these are problems of poverty and misdirected public priorities more than problems of absolute numbers of people. Military priorities often are the biggest obstacle to environmental problem-solving: for the price of one British Aerospace Hawk aircraft, 1.5 million people in the Third World could have clean water for life; in 1989, the annual cost of a proposed anti-desertification program for Ethiopia was the equivalent of two months of Ethiopian military spending. At the same time, the activities of militaries, whether in war or peace, are a far greater cause of environmental degradation than is population growth. A number of environmental economists are increasingly concluding that the decisive factor determining environmental quality is the nature of the prevailing technology characteristic of a population, not the size of the population.[105]

Ironically, environmental problems themselves promote high population rates. In countries where infants die regularly from gastrointestinal disease brought on by lack of clean water, for example, women have to bear higher numbers of children simply to meet population replacement levels. For poor families, without access to either good health care or economic security, large families are a necessary survival strategy—high rates of reproduction ensure that some children can live to the age when they can contribute to the family's welfare. In the absence of adequate social services, children become the "social security" of the parents. In poor families, children are an economic resource—they are agricultural workers, domestic helpers, and, often at an early age, wage earners. Thus the solution to environmental problems is not simply, or even primarily, population controls; indeed, the reverse may well be true.

A complex set of economic, political, and social relationships shape a family's reproductive decisions—a complexity that most population control advocates overlook. Poverty is one of the major driving forces of high population growth rates. Sexism is another. In most cultures, men are valued over women, and boy children over girl children. Families and male partners often exert extreme pressure on women to keep on bearing children until a son is born—and until a son survives through early childhood. The widely presumed "prerogative" of men to unrestricted sexual access to women, and the exercise of male violence to ensure this access, also fuels population rates. Where adequate contraceptives are not available, women may wish to choose abstinence as a birth control method—a choice that is often not respected by male partners. Further, in many societies, perhaps in most, women are most valued for their capacity to bear children; children are often the only "social capital" of women.

Advocates of population control often ignore the history and ideological underpinnings of the movement, especially the association of the movement with repressive regimes and genocidal governments. Throughout the history of population control advocacy, it is always the rich, the elite, and the powerful who view the growing numbers of poor as a threat to their political and economic power. This dynamic operates today on a global scale: rich countries, and rich elites within poor countries, see the poor "masses" as destabilizing threats. The inflated language of population "explosions" and population "bombs" makes the fertility of people from the Third World appear to be an existential threat to all people on the planet.[106]

Specifically, *women's* fertility is implicitly (sometimes explicitly) blamed for the global environmental crisis. Population control is a euphemism for the control of women.[107] Intervening in reproduction always means, above all, intervening in women's lives, in female reproductive organs, and in the exercise of individual reproductive freedom.[108] Population control always implies the exercise of centralized authority—a government, often aided by international development agencies (all of which are dominated by male actors)—in imposing restrictions on women's reproductive activities. All too often, little regard is shown for the human dignity of women, and little concern for their health. Women are international guinea pigs for birth control wonders produced by pharmaceutical conglomerates in the rich world; Third World women are the dumping grounds for medications that industry can no longer sell at home. Women have been subjected to mass sterilizations, without consent—in Puerto Rico, in India, in China. Unbridled racism and rampant sexism are intertwined with the politics of international population control. If nothing else, this recent history of flagrant human rights abuses in the name of population control makes its advocacy by predominantly white, male, First World environmental groups particularly dubious.

Many women in the world, and especially in the Third World, *do* regulate their fertility and *do* want to further restrict the number of children they have; as an Indian woman comments, "Women realize better than anyone else what an accelerated population growth rate means."[109] The extremely high numbers of abortions in many countries—an estimated 5 million in Brazil every year, 6 million in India—attest to the efforts of women to exercise reproductive choice. It is clear that access to birth control expands women's personal choices and improves their access to a more equitable share of social resources. But many women's groups in the Third World reject contraceptive knowledge and dubious chemical fixes imported from the West, imposed on them through centralized male systems of governments and aid agencies. Women want access to reproductive planning services; they do not want to be instruments of population control.

Global population growth may well be a "problem," but not in the ways that most of us are encouraged to believe. The fact that there is a debate within the environmental movement on the population issue is not in itself problematic. There is good reason to be concerned about the environmental pressures exerted by population growth, just as there are good humanitarian and feminist reasons to promote programs that provide women everywhere with access to reproductive planning services. What is disturbing, though, is the manner in which men have framed the public environmental debate on population, and the extent to which women's lives are, once again, made invisible within the debate.

In 1991, the Boston Women's Healthbook Collective convened a meeting of women environmentalists, feminists, and health care activists to consider the issue of population growth and the environment. Many of us from the Boston meeting took up the issue again a month later in Miami at the first international conference of the Global Assembly of Women for a Healthy Planet. The draft statement of the ad hoc Feminist Task Force on Population and the Environment charts a path for recasting the population/environment debate in women-centered terms. The preamble to this draft statement sets the context for a feminist rethinking of population/environment issues:

We are troubled by the resurgence of attitudes and statements that single out population size and growth as a major cause of environmental degradation. Unchallenged, this analysis lays the groundwork for the re-emergence of top-down, demographically driven population policies and programs which are deeply disrespectful of women.

The major causes of environmental degradation lie elsewhere: • In economic systems that exploit and misuse nature and people in the drive for short-term and short-sighted goals • In war, arms production, and military appropriation of national resources • In the disproportionate consumption patterns of the affluent the world over. Currently, 17% of the industrial nations population consume 82% of the world's resources • In the displacement of small farmers and indigenous peoples by agribusinesses, timber, mining, and energy corporations • In technologies designed to exploit but not restore natural resources.

Following from this, it is clear that people concerned about environmental degradation should work in the first place not for population control, but for global demilitarization, redistribution of resources between and within nations, reduction of rates of consumption among affluent classes and by industrial systems, and the development of socially just and environmentally responsible technologies. Blaming population growth for the world's environmental ills is seductively simple. It makes the solution to environmental degradation also seem simple: it is easier to envision a global program to hand out contraceptives to women around the world than it is to imagine dismantling social and economic structures of inequality.

At the same time, feminists *do* argue for improved reproductive planning services, and encourage international aid that makes these services available to women everywhere. Support for programs for family planning and reproductive choice, though, should not be confused with support for population control. A women-centered reproductive planning approach to population issues puts women's needs in the forefront. Women need access to safe contraceptives, the safest of which are barrier-method contraceptives, not chemical fixes. Women must be able to operate in a social and political climate that respects their reproductive choices, including access to safe abortion services. Women need consistent health-care services, not just a one-time sterilization or contraceptives-dumping run by health technicians who are never seen again.

Perhaps most important, it is an established fact that the most effective means to achieve declining birth rates is to improve women's political, social, and economic status. Historically, population growth rates have always fallen as women achieve measures of social, economic, and political equality with men. The solution to controlling population growth is not through population control, but through measures to improve women's autonomy, and especially measures that increase women's control over their own reproductive capacities and that provide protection from male violence and sexual coercion. Bureaucratically mandated population control does just the opposite.

This Earth Is Not Your Mother

As the control of "motherhood" moves up on the environmental agenda, the reification of "Mother Earth" stands in ironic contrast. The most ubiquitous icon of modern environmentalism is the image of the earth, floating in black space, with the caption "Love Your Mother." The conceptualization of the earth as a mother has a long and honorable history: Earth as Mother, as a sacred and honored female life force, is a powerful icon in non-Christian, non-EuroAmerican, mostly agricultural, cosmography;[110] it rejuvenates a contemporary women-centered spirituality movement; it inspired a generation of Earth Day activists. But it is disingenuous for a spiritually hollow, urban, technical, male, ecobureaucracy (and one that is consciously becoming *more* invested in these characteristics) to adopt the mother imagery. Not only is this a terrible irony, but "Earth as Mother" is a deceptive paradigm for environmental politics.

The earth is *not* our mother. There is no warm, nurturing, anthropomorphized earth that will take care of us if only we treat her nicely. The complex, emotion-laden, conflict-laden, quasi-sexualized, quasi-dependent mother relationship (and especially the relationships between *men* and their mothers) is not an effective metaphor for environmental action. It suggests a benign distribution of power and responsibility, one that establishes an erroneous and dangerous assumption of the relations between us and the environment. It obfuscates the power relations that are really involved when we try to sort out who's controlling what, and who's responsible for what, in the environmental crisis. It is not an effective political organizing tool: if the earth is really our mother, then we are children, and cannot be held fully accountable for our actions.

Beyond this, in a patriarchal culture in which female status is cast as subservient status, there are inherent pitfalls in sex-typing an inherently gender-free entity. A number of alternative environmental groups, deep ecologists and ecofeminists among them, are reclaiming the sex-typing of the planet as part of a radical environmental agenda. But sex-typing of a non-gendered entity invokes a male/female, greater than/lesser than cultural dualism; the limitations of female identification in a male-dominated culture undercut the claim that sex-typing the planet can be "radical."

Further, the sex-typing of the planet, in imposing on the earth a human

Figure 4.3.   The ubiquitous icon, Mother Earth.

imagery, also reinscribes an odd anthropomorphism. To describe the earth in human terms in order to understand it implies that, without this human veneer, we and it are separate and *other*.[111] Ascribing human archetypes to nature also suggests design: that we are nature's favored progeny. The term "mother" is a human invention, and evokes uniquely human characteristics. To propose that this forms the essence of our relationship with nature exalts our place within nature, and reiterates claims that "man is the measure of all things."[112] To base this dualism on male and female identities only further reinforces cultural hierarchies— it certainly will do nothing to subvert the patriarchy, and will do little to further environmentalism.

Many people might argue that the "Mother Earth" metaphor is a harmless device, not worth dwelling on. And yet, the use of this metaphor is not always benign—"hiding behind Mom's skirts" is a convenient device to deflect accountability. In 1989, a vice-president of Exxon invoked Mother Earth imagery in defending his company's cleanup operations in Alaska after the Valdez oil spill. His words suggest the cynical and facile use of the Mother Earth "defense":

> I want to point out that water in the [Prince William] Sound replaces itself every 20 days. The Sound flushes itself out every 20 days. Mother Nature cleans up and does *quite* a cleaning job.
> —Charles Sitter, senior vice-president of Exxon, May 19, 1989[113]

There may be a broader subtext to the "Mom-will-pick-up-after-us" school of environmental philosophy. In 1989, a Boston journalist gave voice to the suspicions many women increasingly harbor:

> Sometimes I think the problem boils down to this: . . . most men have had women to clean up after them. In fact, it wouldn't surprise me one bit to find out that science has been covertly operating on the Mom-Will-Pick-Up-After-Me Assumption. . . . Men are the ones who imagine that clean laundry gets into their drawers as if by magic, that muddy footprints evaporate into thin air, that toilet bowls are self-cleaning. It's these over-indulged and over-aged boys who operate on the assumption that disorder—spilled oil, radioactive wastes, plastic debris—is someone else's worry, whether that someone else is their mother, their wife, or Mother Earth herself.
> —Linda Weltner, *Boston Globe* columnist, April 28, 1989[114]

Chapter 5

The Eco-Fringe:
Deep Ecology and Ecofeminism

FRINGE ELEMENTS

The eco-establishment is the big kid on the environmental block, and to the extent that environmentalists will be participants in setting the environmental agenda for the closing years of the twentieth century, it will be the eco-establishment that has the biggest say. But these groups do not have the environmental stage all to themselves. Community-based grassroots environmental groups, ecofeminists, and deep ecologists are all important, and in some cases becoming *more* important, players on the environmental scene. In the US alone, there are hundreds of small groups that represent grassroots environmentalism; worldwide, there are literally thousands. We'll meet the grassroots activists in the next chapter; here, I want to take a look at the two ideological movements, ecofeminism and deep ecology, that represent the sharpest challenge to mainstream environmentalism.

Grassroots activism, ecofeminism and deep ecology suggest alternative paradigms for environmental action—they expand not only the sense of the "possible" in environmentalism, but the sense of the "necessary." Deep ecology and ecofeminism both emerged, in part, out of dissatisfaction with mainstream environmental politics. Many of the current leaders of the deep ecology movement are activists purged from the ranks of the eco-establishment; many ecofeminists are women alienated by the masculinist establishment, and frustrated by the limited opportunities within the environmental arena for an expression of women's values and feminist politics; others locate their environmental activism within traditions of communities steeped in a sense of appreciation for place, while still others are rooted in the women's spirituality movement.

But deep ecologists and ecofeminists have more in common than cynicism about mainstream politics. Both movements posit the need for affirmation of a "deeper" human relationship with the earth—a relationship that at its best comprises elements of mysticism, awe, and an appreciation of the "sacred" in nature. Both movements couch their environmentalism in "woman-identified" terms, and deep ecologists are the only environmentalists other than ecofeminists to explicitly assert an affinity with women's culture and feminist politics. Feminists have not always welcomed the overtures from deep ecologists, for reasons that will become apparent, and the result has been a not-always-cordial dialogue between the two.[1] Nonetheless, the two movements *are* linked by their overlapping philosophical and spiritual sympathies.

DEEP ECOLOGY

For a hopeful moment in the mid-1980s, an environmental wave sweeping Europe and North America seemed to offer a new vision and a *counterculture*, in the fullest sense of that word: "deep ecology" was an appealing, puzzling, and exotic environmental movement. Deep ecology was the pin set to burst the bubble of environmental hubris on which we build our human privilege. Its philosophers demanded that we ask probing questions of ourselves, of the nature of "being." Deep ecology represents an environmental *philosophy*, but at the same time it was a philosophy that actually spawned an activist wing with a distinct identity—deep ecology was not just an ideology, it was also a practice. The principles of deep ecology seemed to offer a challenge to patriarchal attitudes toward nature; its practice suggested a potential challenge to patriarchal methods of environmental organizing. Deep ecology offered hope and a refreshing vision to people who were concerned about the environment but who were disillusioned with the bureaucratic, reformist, and presumably co-opted mainstream environmental groups.

The term "deep ecology" was coined in the early 1970s by a Norwegian philosopher, Arne Naess. Naess articulated an ecological approach that posed "deeper" questions about life on earth than mainstream environmentalists allowed themselves to ask.[2] The deep ecology he articulated was rooted in recasting the religious and philosophical interpretation of human relations with the natural world, starting with the necessity of shifting from human-centrism into biocentrism, a commitment to revaluing humanity's oneness with nature, and an appreciation of the intrinsic worth of all life forms.

Working from Naess's starting point, American deep ecologists in the 1980s elaborated an eight-point manifesto of "basic principles," among the most important of which are these: a reification of "biocentrism," which is a philosophy that nonhuman Life on Earth (capitals in original) has intrinsic value *in itself* independent of its usefulness to humans; that humans are too numerous and that a "substantial decrease" in human populations is required for the well-being of the earth; that humans must change their basic economic, technological, and ideological structures; and that everyone who subscribes to deep ecology has an obligation to try to implement these necessary changes.[3] The imperative to take direct action in defense of the earth is central to the philosophy of deep ecology. In the US, a loosely structured national group (with international affiliates) called "Earth First!" emerged as the organizational vehicle for translating the broad philosophical principles of deep ecology into an operational environmentalism. Although Earth First! disbanded in 1990, several EF! splinter groups remain in place as the organizational foci for the deep ecology philosophy.

Deep ecology is a big tent, under which many environmentalists gather—many of whom may disagree with one another on specific tactics or campaigns, but all of whom would broadly ascribe to the basic principles outlined above. As a Western environmental movement, deep ecology is also characterized by a distinctive tone and a particular set of associations: • Deep ecology environmentalism is suffused with a ritualized vision of the Earth as Mother, and of the Earth as a independent, self-regulating female organism.[4] • Many deep ecologists celebrate an earth-centered "paganism"—the editions of the American deep ecology newspaper, *Earth First!*, for example, were designated by the names of Pagan holidays ("Brigid" for February, "Eostar" for March, etc.). • Many Earth Firsters consider themselves a "tribe," and many of the American deep ecologists posit an affinity with indigenous, Native American ecological sensibilities. • At heart, deep ecology is concerned with the preservation and protection of wilderness. Deep ecologists revere undisturbed wilderness as the pinnacle ecological state. Their militant defense of "Mother Earth" is rooted in an unflinching opposition to human attacks on wilderness. To the committed Earth Firster, the preservation of wilderness takes precedence over all human need.

Earth Firsters quickly established a high profile among environmentalists as guerilla-theater, lay-your-body-on-the-line afficionados. In Australia, Earth First! protesters buried themselves up to their necks in the sand in the middle of logging roads to stop lumbering operations; in the American Southwest, Earth Firsters handcuffed themselves to trees and bulldozers to prevent logging; in California, they dressed in dolphin and mermaid costumes to picket the stockholders' meeting of a tuna-fishing company. Earth First! actions were often choreographed with a beguiling sense of humor, and carried out with daring and panache—their most endearing and environmentally useful characteristics.

Deep ecology is not confined by a single script. It is meant to be defined through its actions as well as its philosophy.[5] If, then, the measure of deep ecology is to be taken in the actions of Earth First!-style environmentalists, the conclusions are troubling. Deep ecology, in practice, has been transformed into a paramilitary, direct action ecology force. Some of the tactics employed by Canadian, Australian, and American Earth First! contingents are questionable: for example, pouring graphite (or sand or sugar) into the fuel tanks of bulldozers and road-clearing equipment involved in logging and mining operations to seize the engines; or, more controversially, tree-spiking, a practice of hammering long nails into trees to "booby-trap" them (spiked trees are dangerous to cut, and most loggers won't work a forest that has been spiked.) While many Earth First! groups renounced the use of destructive tactics, others embraced a "no compromise" environmental fundamentalism.

The *practice* of deep ecology, as it defined itself over the decade of the 1980s, suggests a peculiar mix of influences and archetypes—almost equal parts of survivalist paranoia, frontiersman bravado, anarchist politics, and New Age spiritual sensibility. With this profile, many observers have been tempted to dismiss deep ecology as the "lunatic fringe" of modern environmentalism—but it is a movement not to be easily dismissed. Deep ecology has become an important player on the American (and to a lesser extent, the European) environmentalist scene—because deep ecologists present themselves as the only viable alternative to mainstream environmentalism, because their direct action tactics have attracted considerable media attention, and because the controversy they engender has opened fractures in the American ecology community. They have achieved some measure of success. They have spun off 50 mostly autonomous Earth First! groups across the US and overseas, created an international network of rainforest activists, and have successfully halted or forestalled a number of ecologically irresponsible projects. As one observer notes, "The mainstream environmental groups have been caught running in place trying to regain the publicity and the place in the public imagination that Earth First! has seized from them. More respectable

wilderness activists and ecologists . . . have been able to take much
stronger positions than before as a result of Earth First!'s uncompromising
presence . . . "[6]

Further, deep ecology, at first blush, appears to offer a philosophy that
speaks to feminist values. The call by deep ecologists for a major overhaul
of the political, economic, and ideological system is a necessity that
feminists have been arguing for years. Deep ecologists speak of the inter-
connectedness of life, a reverence for nature, a nonexploitative relationship with wilderness, a valuing of intuition over rational, "anthropocentric"
linear thought—all essentially women-identified ideas.

Deep ecology is virtually the only wing of the environmental movement
to make specific overtures to women in general and to feminists in particular. In the abstract, if not in the practice, deep ecologists are astute enough
to recognize that it is not possible in the 1990s to make credible claims
of a "radical" agenda without some nod to feminist analysis. A number of
male deep ecologists argue that the philosophy and theory of deep ecology
is in sympathetic harmony with feminism. Others, too, assume an affinity:
Kirkpatrick Sale, a noted bioregionalist, for example, referred to deep
ecology as "that form of environmentalism that comes closest to embody-
ing a feminist sensibility," continuing by saying "I don't see anything in
the formulation of deep ecology that contravenes the values of feminism
or puts forward the values of patriarchy."[7]

Deep Machismo   One commentator, musing on deep ecology, noted,
"Freely mixing pseudo-scholarly tomes and spit-in-the-can barroom philosophy, there is something in Earth First! to offend just about anyone."[8]
To this assessment, I would only add that women have been especially
offended. Despite its surface overtures to feminists, the transformation of
deep ecology into an environmental force has been characterized by
deeply misogynistic proclivities. Macho rhetoric of the most conventional
and offensive sort riddles the written record of deep ecologists. Their
common practice of using women-identified terms as taunts, such as
calling their critics "wimps," "sops," or "effetes," panders to a blatant
sexist and homophobic bias—as though the worst thing in the world is
to be womanly! Edward Abbey, the recently-deceased guru of the American
Earth First! movement, nurtured a reputation as a crusty, take-no-guff
curmudgeon—and as a mysogynist. For instance, in a rebuff to Murray
Bookchin, a critic of deep ecology, Abbey reveled in a well-worn combination of sexist stereotypes:

> And he [Bookchin] needn't worry that I will attack him . . . Fat old
> women like Murray Bookchin have nothing to fear from me. (What?

'Fat old women?" Did I say that? Am I not only a fascist, a racist, a cultural chauvinist, but—God forbid—a male sexist pig as well?)[9]

Deep ecology is saturated with male bravado and macho posturing. The American Earth First! movement is particularly symptomatic of the masculinist ethos that suffused representations of deep ecology's philosophy. With very few exceptions, the self-styled leaders and spokespeople of Earth First! were all men, as was a considerable proportion of its membership (in contrast with all other environmental groups). Nonetheless, the sex ratio of Earth First! membership varies considerably among local affiliates and some local groups even have higher numbers of women than men. It is clear that Earth First! is attractive to women who want to participate in environmental change; it is not clear how feminist women within Earth First! reconcile their involvement with the deeply misogynistic face of the national and international branches of the movement. One feminist Earth Firster! recently mused about the struggle to "feminize" Earth First!:

> I see no contradiction between deep ecology and ecofeminism, but Earth First! was founded by five men, and its principle spokespeople have all been male. As in all such groups, there have always been competent women behind the scenes. But they have been virtually invisible behind the public Earth First! persona of "big man goes into big wilderness to save big trees." I certainly objected to this. Yet despite the image, the structure of Earth First! was decentralized and non-hierarchical, so we could develop any way we wanted.[10]

Moreover, Earth Firsters are not just men; they are "men's men." Dave Foreman, one of the founders of the American Earth First! movement, represented the tone and tenor of the group when he said that, "I see Earth First! as a warrior society."[11] The leaders of Earth First reveled in an image of themselves as beer-swilling, ass-kicking, "dumb-cowboy rednecks" coming to the rescue of a helpless female—in this case, Mother Earth. Throughout most of the 1980s, the logo of the US organization was a clenched fist encircled by the motto "No Compromise in Defense of Mother Earth"—a bewildering mixed metaphor if ever there was one. (See Figure 5.1).

American Earth Firsters! often relied on antifeminist women as spokespersons for "the woman's" point of view (a tactic widely employed by other mostly male organizations, such as militaries and multinational corporations), to undercut feminist criticism. A 1989 article in *Earth First!* by a prominent Earth First! woman, for example, was a lengthy piece

THE COMPLEET
RADICAL ENVIRONMENTALIST

SIZE 11 - MOUNTAIN - CLIMBIN', WOODS - HIKIN', DESERT - WALKIN',
BUTT - KICKIN', ROCK - N - ROLLIN' WAFFLE STOMPERS !!!

Letters to the editor are welcomed. Lengthy letters may be edited for space requirement. Letters should be typed or carefully printed and double-spaced, using only one side of a sheet of paper. Be sure to indicate if you want your name and location to appear or if you wish to remain anonymous. Send letters to POB 5871, Tucson, AZ 85703.

Figure 5.1. Earth First! makes a statement. From top left: a self-ironic portrait, *EF!* August 1, 1988; the motto of EF!; the masthead for the "letters to the editor" column, *EF!* newsletter.

entitled, "No, I'm Not an Eco-Feminist: A Few Words in Defense of Men," in which the author launched an attack on women, including such muddled nonsequiturs as, "It's all over if men become feminists. Mothers and their children, alone, do not make a human society," and an even more bizarre rationalization for excluding women from decision-making roles:[12]

> In many tribes, a woman of childbearing age is excluded from the decision-making body; but when she has passed child bearing, she becomes the wise old woman. . . . There is good reason for this age differentiation. Many women are paranoid for two years after the birth of each child. . . . Such a person cannot make valid decisions for a group; but once this child-making period is over, such a person is a truly wise decision-maker. This [exclusion] is age-specific, not strictly sex-specific.

While deep ecologists represent themselves as forging a radical new relationship with nature, they give no credit to the women who broke this path for them. Deep ecologists speak reverently of rediscovering intuition in relating to nature; of breaking through the barriers of hierarchical thinking; of the necessity of viewing life on earth as an unbroken continuum; of celebrating the interwoven connectedness of us all—sensibilities that have been scorned by men for years as female-identified traits. But deep ecologists (mostly men) never say that. They exalt as though they are the first to discover this cooperative, noninvasive, and holistic life philosophy. Sharon Doubiago is one of a number of commentators who has taken the deep ecologists to task for this "oversight," in an uncompromising essay, "Mama Coyote Talks to the Boys:"

> The deep ecology movement is shockingly sexist. Shocking because deep ecology consciousness is feminist consciousness. . . . But nowhere is this acknowledgment made. Instead, papers, books, and repeated efforts are made to establish a tradition to show the similarity of deep ecology consciousness with "intuitionists, mystics, and transcendentalists," with "the New Physicists," with Buddhism and traditional American Indian philosophies toward nature—all fields of study which are exotic, removed and masculine.[13]

Moreover, the invocation of "native ways of being," which *is* acknowledged as a source of inspiration, is distressingly shallow, often coming down to little more substance than a sentimentalized mythologizing of the "noble savage"—a view that can, and does, easily slip over into racist assumptions and simplistic misrepresentations of a complex culture to which few white Americans can really claim access.

Disregarding Difference   Despite their putative tilt toward feminism, deep ecologists are unwilling to include gender analysis in their analytical tool kit. Deconstructing and then *re*-constructing the human relationship to nature is absolutely central to deep ecology environmentalism. Yet, most deep ecologists are not interested in the social construction of attitudes toward nature, nor are they curious about the divergence in Western history (if not universally) of male and female attitudes toward wilderness and nature. Thus while there is an explicit criticism of destructive cultural attitudes toward nature, there is no apparent curiosity about the extent to which those "cultural" attitudes may be gender, race, or class specific.

And yet there is, now, a rich literature that explores the cultural and gender differences in human relationships to wilderness. Highly respected (and widely available) research by Annette Kolodny, for example, blazes a trail for gender-specific landscape/wilderness study.[14] Kolodny rewrites the history of the American frontier, establishing that images of "conquering the wilderness," "taming nature," "mastering the wild," and the like—images that North Americans take to be the standard fare of the European encounter with new lands, and images that continue to be the standard fare of popular culture—were, in fact, *male* fantasies and *male* imagery, and were not shared by their women counterparts in the wilderness. A recent study of women's attitudes toward nature and landscape in the American Southwest reinforces the argument made by Kolodny that, whereas European men saw the American West as a virgin land, ready to be raped and exploited, women typically regarded the landscape as "masterless." Rather than seeking conflict with the wilderness, women sought accommodation and reciprocity.[15] The conventional wisdom about the American wilderness experience is that "for most of their history, Americans [of European extraction] regarded the wilderness as a moral and physical wasteland, suitable only for conquest";[16] Kolodny's persuasive reply is that massive exploitation and alteration of the continent do not seem to have been part of women's fantasies.

Deep ecologists are attempting to reconfigure our contemporary ecological sensibilities by refuting this exploitative ethic of wilderness. But *whose* ethic needs reconstruction? It surely is important to understand that this exploitative ethic is neither universal nor shared across the cultural spectrum—if it can be generalized as the predominant ethic of any group at all (and this seems unlikely), it is only of white Euro-American men. But deep ecologists are not curious about the social construction of our contemporary environmental ethic, and beyond offering simplistic paeans about "women being closer to nature," they are certainly not interested in its gender construction. Men and women are painted with the same broad stroke, and are equally indicted by deep ecologists for "anthropocentric" attitudes to nature. This is one of the major points of departure between feminists and deep ecologists.[17]

The generalizations of deep ecologists blur distinctions not only of gender, but of race, class, and nationality too. Many deep ecologists portray the human race, as one species, as a sort of "cancer" on the earth that has devoured its resources, destroyed its wildlife, and endangered the biosphere. This sweeping misanthropy lacks social perspective—it is analytically unsound to make no distinction among peoples, nations, or cultures in assigning accountability for ecological destruction. Humanity is not an undifferentiated whole, and it is not credible to lay equal "blame" for environmental degradation on elites and minorities, women and men, the Third World and the First, the poor and the rich, the colonized and the colonizers.[18] Nonetheless, many American deep ecologists are insistent on this point. Dave Foreman, the US Earth First! founder, explains the disinterest in social analysis: "We are not opposed to campaigns for social and economic justice. We are generally supportive of such causes. But Earth First! has from the beginning been a wilderness preservation group, not a class-struggle group."[19] Paul Watson, a well-known American ecoiconoclast, casts Foreman's sentiments in stronger language: "My heart does not bleed for the third world. My energies point toward saving one world, the planet Earth, which is being plundered by one species, the human primate. . . . All human political systems developed to date, be they right or left, are anthropocentric in philosophy and support the exploitation of the Earth."[20] In his germinal article, Arne Naess, the "father" of deep ecology, expressed concerns about inequalities within and between nations. But his concern with social cleavages and their impact on resource utilization patterns and ecological destruction appears to have gotten lost in the translation, because it is all but invisible in the later writings of deep ecologists.[21]

Even taking the deep ecology agenda as given, a number of critics point out that "class struggle" is inextricable from the struggle to preserve wilderness. An Indian environmentalist points out, for example, that the emphasis on wilderness preservation can be positively harmful when applied to the Third World: "Because India is a long settled and densely populated country in which agrarian populations have a finely balanced relationship with nature, the setting aside of wilderness areas has resulted in a direct transfer of resources from the poor to the rich."[22] The dilemma of wilderness vs. people is not an issue solely in the Third World (although it is in poor countries that the competing demands for land are often in sharpest relief). But wilderness advocates must at least come to terms with the analysis that suggests that efforts to protect wilderness in the Third World, particularly as they are orchestrated by Western environmentalists, may be received in those countries as just another example of imperialism, the same imperialism that pushes the poor and others into the wilderness in the first place.[23]

The "Population Problem"   It is their stance on population, "overpopulation," and overpopulation "solutions" that has most alienated women and short-circuited an intellectual alliance between feminists and deep ecologists. A deep belief of deep ecology is that there are "too many people" on the planet. "Substantial" and fast *reductions* in the human population (not just a stabilization of population growth rates), deep ecologists say, are essential for the survival of the earth.[24] Some deep ecologists have tried to put actual figures to these reductions—Arne Naess, for example, proposed that for the health of the planet, "we should have no more than 100 million people";[25] an Earth First! writer using the pseudonym "Miss Ann Thropy" suggested that the US population would have to decline to 50 million.[26] (Given that the current population of the earth is just over 5 billion, and of the US almost 300 million, it is clear that the reductions called for by deep ecologists are drastic and would require catastrophic action to implement). Other deep ecologists have proposed a 90-percent reduction in human populations to allow a restoration of pristine environments, while still others have argued forcefully that a large portion of the globe must be immediately cordoned off from human beings.[27]

The logistics of achieving such population reductions don't daunt deep ecologists. Deep ecology spokespeople have proposed a number of solutions to the "population problem"—"solutions" that range along a short spectrum from tame, vague, and muddled at one end to racist, sexist, and brutish at the other end. At the tame end, some deep ecologists have issued vague calls for widespread birth control programs—which is neither new, nor radical, nor much different from the population policies of most mainstream environmental groups.[28] There is no evidence of the much-vaunted feminist sensibility in the discussion of deep ecology population programs: women are once again rendered invisible, there is no linkage made between "population policies" and the daily lives of women around the world, nor is there any discussion or even acknowledgment of the fact that birth control policies inevitably bear disproportionately on Third World women. There is also no consciousness that, in issuing this call for population control, deep ecologists are preaching the same gospel as other men before them: that controlling female reproduction by technical means will solve the problems of "nature." Despite these "oversights" in the ideology of deep ecology population control, the call for birth control is not what has roused so much controversy.

Many of the other population solutions proposed by deep ecologists are not so benign. David Foreman, the Earth First! founder, proposes reducing the human population by halting life-saving medical interventions and aid for famine and disaster victims around the world. Speaking of the famines in Ethiopia, Foreman said, "The worst thing we could do in Ethiopia is to give aid. The best thing would be to just let nature seek its own balance, to let people there just starve."[29] This is an unconscionable and arrogant argument for a well-fed American to make, and is also based on the false premise that the famines in Ethiopia are somehow "natural." Mass starvation in Ethiopia derives not from a natural proclivity to famine, but from years of internal warfare, military spending bloated at the expense of social and environmental programs, corrupt governance, and regional environmental degradation.[30]

Edward Abbey, the guiding light of the American Earth First! movement,

set off another firestorm of controversy by linking environmental population pressures with immigration policy. In the mid 1980s, Abbey started
off by advocating that the US close its borders to Central American and
Latin American immigrants, and went on from there with an escalating
racism that was only thinly wrapped in a concern for the environment. In
a 1986 letter to *The Bloomsbury Review*, Abbey wrote:

> In fact, the immigration matter really *is* a matter of "we" versus "they"
> or "us" versus "them." What else can it be? There are many good
> reasons, any one sufficient, to call a halt to further immigration
> (whether legal or illegal) into the United States. One seldom men-
> tioned, however, is cultural: the US that we live in today, with its
> traditions and ideals, however imperfectly realized, is a product of
> northern European civilization. If we allow our country—*our* coun-
> try—to become Latinized, in whole or in part, we shall see it tend
> toward a culture more like that of Mexico. In other words, we will be
> forced to accept a more rigid class system, a patron style of politics,
> less democracy and more oligarchy, a fear and hatred of the natural
> world, a densely overpopulated land base, a less efficient and far
> more corrupt economy, and a greater reliance on crime and violence
> as normal instruments of social change.[31]

Abbey's racist remarks were never repudiated by Earth First!, and were
celebrated and repeated by many.[32] Some Earth Firsters argued that
allowing Central Americans to use the "overflow valve" of fleeing to the
US has two consequences: a) it increases the US population, and b) it
allows Central American governments to continue their irresponsible
ways. Once again, there is no discussion of social context or historical
reality: if people are fleeing Central America in increasing numbers, it is
in some measure because large parts of the region have been rendered
all but uninhabitable by US-backed large landowners mismanaging the
land and by US-backed military destruction. Nor is there discussion of the
century of US intervention in and manipulation of the governments and
economies of virtually every Central American country.

Racism lurks just beneath the surface of most discussions of "the
population problem" in the deep ecology literature, and sometimes
doesn't even lurk. Unfortunately, most deep ecologists have not taken on
the issue of racism seriously, and instead of considering the issue anew,
the anti-immigrant, and in particular anti-Hispanic, sentiment has been
given considerable visibility in Earth First! literature.

By far the most controversial issue forwarded by a few deep ecologists
was the suggestion that AIDS may be an environmental blessing in dis-
guise. Because it has the potential to wipe out vast numbers of people

worldwide, some deep ecologists suggest that AIDS will benefit the planet in the long run, and further, that it might be a disease "intentionally" introduced to lessen the human strain on the Earth's ecosystem. In a number of articles in 1987, writers in *Earth First!* put forward this amazing proposal, rousing an acrid cloud of controversy that has yet to disperse:[33]

> Barring a cure, the possible benefits of this [the AIDS epidemic] to the environment are staggering. If, like the Black Death in Europe, AIDS affected one-third of the world's population, it would cause an immediate respite for endangered wildlife on every continent.[34]

The proponents of this view base their argument on linked assumptions, starting with a consideration of the disease itself. AIDS, they say, has particular and unusual characteristics that distinguish it from other infectious diseases, such as its long incubation period during which it cannot be detected, its ability to mutate, and its durability. The authors of the AIDS argument then link this with a discussion of the Gaia hypothesis.[35] If the earth is a self-regulating organism, they argue, then Gaia (the earth) might act like other living organisms that are capable of defending themselves against biological invasions. If humans have stressed the earth's biosystems to the breaking point, then Gaia might strike back with a deadly disease that reduces the human burden "she" carries. AIDS, they say, has the characteristics of a "designer" disease, and Gaia herself may be its designer. In the original article, Miss Ann Thropy expressed only glancing concern for those stricken with the disease: "None of this is intended to disregard or discount the suffering of AIDS victims. But one way or another, there will be victims of overpopulation. . . . To paraphrase Voltaire: if the AIDS epidemic didn't exist, radical environmentalists would have to invent one."

On Deep Being and Nothingness There is a further profound philosophical rift between deep ecology and feminism. Deep ecology is premised on the necessity of breaking down barriers between human and nonhuman species. Deep ecologists exhort us to cultivate an ecological consciousness of total connectedness—to subsume our sense of self to a greater sense of oneness with all things, both sentient and insentient, around us.

For centuries women have been told that they have no singular identity. Women have always been subsumed by culture and by men, and denied independent existence. Selflessness, unbounded oneness, total connectedness, and denial of independent identity have been central to women's oppression. So, while it sounds ecologically ennobling to "think like a mountain,"[36] the male deep ecologist theorists who are promoting this notion refuse to acknowledge this historical and social context within which they are telling women to merge identities with the earth: they ask women, in effect, to once again embrace nonbeing.

It seems unpleasantly coincidental that at this historical moment, just as women and other oppressed groups are coming into self-subjectivity, deep ecologists are telling us that such self-presence and self-awareness are to be condemned.[37] The sublimation of identity urged by deep ecologists parallels in many ways the "relativism" of postmodernism—a trendy intellectual stance that has come under considerable criticism from feminist philosophers. Sandra Harding, for example, notes that "relativism appears as an intellectual possibility . . . for the dominating group at the point where the universality of their views is being challenged. Relativism is fundamentally a sexist response. . . . "[38] The revolutionary heart of the feminist movement is rescuing women from obscurity, acknowledging women as fully-embodied and well-defined beings in historical and contemporary culture, empowering women to develop a strong sense of self, to embrace their selfhood, to develop a self-conscious subjectivity. If anyone needs practice at selflessness, it is not women.[39]

Lost Promises   While deep ecologists have galvanized a renewed sense of environmental urgency and agency, the movement falls short of its claims as a radical challenge to the environmental status quo. With its macho-redneck style, its woman-hating rhetoric, its bordering-on-racism political stances, and its misanthropic bias, deep ecology is at the same time tediously old hat, and frighteningly new. Ultimately, deep ecology represents an unfulfilled promise and a lost opportunity for women and men who were genuinely interested in transforming their relationship with the external world. Once again, it is male posturing that has poisoned an apparently promising coalition of feminist analysis and environmental philosophy.

THE ECOLOGY QUESTION IN FEMINISM[40]

The combination of male resistance within the environmental movement and the reluctance of feminists to cast their analytical net into environmental waters has meant that the points of entry for women into the arena of organized environmental activity are few and far between. Where they have entered, they have slipped in mostly through the back door.

Grassroots organizing, discussed in the next chapter, is the entry point for many women into environmental activism—most of the community-based environmentalists around the world who are leading the fight in local battles against pollution and exploitation are women. But, especially in Europe and North America, it has been the peace movement that has opened a route into environmentalism for many women. As the urgency of "peace work" currently seems to be diminishing, many women peace activists are shifting their focus to environmental issues. However, even though grassroots organizing and peace activism bring women into the environmental arena, it cannot be assumed that they automatically bring a *feminist perspective* to environmentalism.

Women who take the lead in community organizing are not necessarily feminists, nor are they necessarily aware of or interested in feminist analyses of power, culture, sexuality, structure. In fact, many women who are in the midst of a struggle against a daily-life threat express the view that feminist questioning is diversionary.

In peace politics, the most distinctive *women's* contribution has come from a vantage point characterized as "maternal feminism"—an ideological stance that has received a mixed reception among peace activists and feminists. Many women coming into environmentalism from the peace movement in the late 1980s bring with them a tradition and philosophy of women's activism based on maternalism. Not surprisingly, there is thus a convergence of "maternal thinking" with "ecofeminism." The emergence of ecofeminism cannot be attributed solely, or even largely, to the crossing over of peace-working, "maternalist" women into environmentalism; however, in many ways, "maternalism" is to the peace movement as "ecofeminism" is to the environmental movement—and for reasons that will become clear, both are marginalized within their respective movements, and both are controversial among feminists.

Peace on Earth

The institutional culture of the organized peace movement in the US and Europe in the 1960s and 1970s would be familiar to anyone working within environmental organizations today. Drawing on a tradition of progressive activism and the constituency of the political Left, peace organizations were (and are) fueled by woman power, but the leadership was mostly male. The agenda for peace activism, forged by this male leadership, often revolved around issues of comparative armaments, military strength, war tactics, and technical expertise; the language of peace was often reduced to throw-weights, missile size, and kill ratios. This emphasis privileged men as experts and left many women feeling alienated and marginalized within the movement.

Women struggled to redefine priorities within the peace movement, and to carve out space within the movement to express more humanistic, quality-of-life, and more broadly antimilitaristic concerns. They created outlets for a distinctively women-identified peace politics, such as the women-only peace camp at Greenham Common in the UK, the Women's Strike for Peace, and the Women's Pentagon Action in the US.[41] The most vivid visual images from the peace movement are those of women's activism—perimeter fences at nuclear bases decorated with ribbons and bangles, women dancing on missile silos, and quilts for peace. Women's activism, such as this, has been a powerful force in shaping peace politics. But because much of the women's activism has been driven by a "motherist" ideology, it has also been controversial, as much among feminists as among other peace activists.

Sara Ruddick, whose recent book, *Maternal Thinking*, situates her in the middle of feminist debates on maternalism, identifies three characteristics of a women's politics of resistance: its participants are women, they explicitly invoke their culture's symbols of femininity, and their purpose is to resist certain practices or policies of their governors.[42] It is the invocation of conventional "symbols of femininity" that is so controversial.

Throughout this century, women have worked for peace on the basis of traditional and idealized visions of womanhood and femininity.[43] Claiming their right as mothers (the givers and preservers of life) to speak for life, women have spoken for peace. The ideal of "moral motherhood" has provided the rationale for middle class women's peace groups in all the major wars of the twentieth century. Like many of the grassroots environmental activists today who fight for environmental safety for the sake of protecting their children's health, women have fought for peace in the name of their children. Indeed, the two have been so closely intertwined that feminists have not found it easy to articulate a *women's* perspective on peace that does *not* rely on the imagery and symbolism of women's traditional roles as mothers and wives.[44]

The motherist political strategy is problematic. On the one hand, the "moral mother" argument draws women's political strength from the very arena to which they have been relegated to assumed passivity. Sara Ruddick points out that "Peace, like mothering, is sentimentally honored and often secretly despised. Like mothers, peacemakers are scorned as powerless appeasers."[45] This "reclaiming" of strength from presumed weakness is a hallmark of feminist organizing and a brilliant strategy used by many oppressed peoples; it leaves women culturally located, but redraws the boundaries of what is possible from *in situ*. The contradiction between violence and maternal work is readily apparent, a contradiction that can be tapped to loosen the grip of militarism. Maternalism gives women the right to speak out from a position that their adversaries, who are often conservative promoters of "family strength," find unassailable.

However, a peace movement that draws its strength from invoking stereotypes and traditional sex roles can backfire by reinforcing both sex typing and militaristic assumptions. It leaves in place, in fact it celebrates, the constricting presumptions about women's—and men's—appropriate place in society. In so doing, motherist politics can undermine the cause of peace itself by reinforcing a gender system that encourages male violence in the state (and family) in the first place.[46] The idea that women should assume nurturing and caring roles fits distressingly well into military ideologies about the role and purpose of women in relation to military objectives. Critics of "maternal thinking" argue that if women build a peace politics based on an affirmation of conventional sex roles, there will be no forward movement in breaking down those roles that may themselves perpetuate militarism. When peace activists leave standing the assumption that it is women's place to be nurturing and peaceful, and men's place to be aggressive and warlike, then there is little impetus to challenge those stereotypes themselves.

While no one challenges that women, whether feminist or not, should pursue peace, should they do so simply as progressive people concerned with issues of social justice, as women claiming a special voice on the basis of their life-giving associations, or as feminists equipped with theories about the distribution of power in society at large and in the peace movement itself in particular? The central question, then, is whether a "motherist" politics *is* a feminist politics. A number of prominent feminist commentators, including Simone de Beauvoir, have warned against mother-based politics: "Women should desire peace as human beings, not as women. And if they are being encouraged to be pacifists in the name of motherhood, that's just a ruse by men who are trying to lead women back to the womb."[47]

The problem that many feminists find with maternalism is not that it's "not feminist" in a doctrinaire sense, but rather that it appears to undermine some of the basic understandings of Western feminism, and it comes uncomfortably close to countering the efforts of feminists who seek to reorder private and public gender assumptions and conventionally defined sex roles.[48] Further, a politics based on maternalism leaves little ideological space for those women who are not mothers or for those who have little inclination to be maternally nurturing. In this regard, mother-based peace organizing can and does open rifts between women who are not mothers (of whom many are lesbians) and women who are, and between women for whom mothering is *not* salient in self definition and women who construct a personal and political identity based on mothering.

The dilemma posed by maternal peace activism has sparked an active debate in feminist circles.[49] The controversy has often swirled around particular symbols and expressions of women's peace activism, such as the all-women's peace camp at Greenham Common. The underlying ideological debate is about whether or not a *women's* peace movement based on maternal politics is a *feminist* movement—or whether it might not be ultimately counterproductive and self-defeating for women.[50]

While the debate continues, the peace issue itself seems to have moved to the back burner. In the late 1980s, peace was overtaken by other issues; most peace organizations recorded dwindling memberships, slackening media interest, and lower public visibility. As worries about the eternal chill of nuclear winter give way to worries about the endless summer of a greenhouse globe, many peace activists have shifted their primary focus to environmentalism.

Many women in the peace movement find that their activism for peace is readily transfered to activism for the environment. The institutionalized violence that militarism represents is connected to and paralleled by institutionalized destruction of the environment; the connections between feminism, ecology, and militarism are strong. Much of the language, the iconography, and the structural analysis constructed by women in the peace movement is salient to an analysis of environmental issues:

> We see the devastation of the earth and her beings by the corporate warriors and the threat of nuclear annihilation by the military warriors, as feminist concerns. It is the same masculinist mentality which would deny us our right to our own bodies and our own sexuality, and which depends on multiple systems of dominance and state power to have its way.[51]

However, as in the peace movement, if women want to carve out a distinctive niche in the environmental movement—if they want not just to participate in the shadows of men's environmentalism—they need to be able to articulate a distinctive analysis of environmental issues. Once again, so far, the most visible expression of women's special contribution to environmental issues, at least in the West, has been a "maternalist," spirituality-based movement known as "ecofeminism."

The Nature-Woman Continuum

"Women are to nature as men are to culture." Behind this familiar adage lurks a complex ideology that has shaped scientific and social presumptions and the development of Western institutional structures over the last several hundred years. Feminist historians of science trace the institutionalization of this idea to the European "scientific revolution" of the sixteenth and seventeenth centuries, when the dominant imagery of earth and nature was radically transformed from organic to mechanistic.[52]

Feminist writers who are tracing the rise of historical and cultural *presumptions* about the links between women and nature argue that prior to the European scientific revolution, nature was conceptualized as a living, nurturing organism. Work, culture, nature, and daily life were interwoven in a seamless web, and a nurturing, female-identified earth was considered to be the root of all life. This imagery, some historians of science argue, served to constrain the abuse of nature: as long as the earth was considered to be alive and sentient, it could be considered a breach of human ethical behavior to carry out destructive acts against it. As Carolyn Merchant says, "one does not readily slay a mother, dig into her entrails for gold or mutilate her body."[53]

The metaphor of the earth as a nurturing organism was overwhelmed by the rise of modern Western science. Under the gaze of modern science, nature was reduced to a set of laws presumed to be knowable, and principles of physical properties, presumed to be universal and mono-lithic. Scientists set about to discover the "laws of nature," the mysteries of which could be penetrated by men of inquiring minds. In the new worldview, culture was sentient, nature was insentient; nature didn't act, it was a physical backdrop, to be acted upon. The identification of nature as female, though, was not lost as the scientific revolution transformed metaphors of nature. Rather, this identification was hardened by an emerging ideology that cast both nature and women as servants for the male scientific spirit. The subordination of (female-identified) nature was re-conceptualized as a struggle of male consciousness over female. Men of science struggled to subdue nature/woman, to know her secrets, to tame her wildness, and to put nature to work in the service of human enterprise. Modern scientists posited a hierarchy of life and consciousness, for which they found support in the Judeo-Christian ethic. On the pyramid of con-sciousness, God was considered to have dominion over all. Beneath "Him," but in His image, were men. Women and children followed next on the hierarchy, and below them, Nature. The "chain of being" was cast as a "chain of command," in which Nature and women were both subservient to men and God. This was reified as a "natural order."

The metaphor of "rational man" subduing an insentient female "Nature" has persisted—it is no mere historical quirk. The literature of (male) exploration, including contemporary accounts, is rife with metaphors of raping the wilderness, penetrating virgin lands, conquering a capricious Nature, mastering the wild, and subduing untamed lands. This is a view of nature that remains prominent in contemporary popular culture. (See Figure 5.2).

A number of feminists point out that, not unexpectedly, this too is the language and the controlling metaphor of modern scientific enterprise. As Merchant suggests, the "scientific method" is an interrogatory one, whose key feature "is the constraint of nature in the laboratory, dissection by hand and mind, and the penetration of hidden secrets—language still used today in praising a scientist's 'hard facts,' 'penetrating mind,' or the 'thrust of his argument."[54]

At the same time that nature was being reconceptualized as a feminized terrain for male exploitation and domination, justified by "rational" science in the service of cultural advance, the notion persisted of the earth as a nurturing mother. This odd dualism is paralleled by the "whore/Madonna" dichotomy into which women are cast by men who are nervous about women's power: Nature (and women) can be capricious, lustful, wild, and uncultured, needing to be tamed, subdued, and mastered; Nature (and women) can also be nurturing, life-giving, and supportive, an entity out of whom sustenance is to be coaxed, seduced, and wheedled.[55] The coexistence of these apparently contradictory paradigms is made possible by the existence of a prior metadichotomy: the nature/culture dualism.

The dichotomy between Nature and Culture was achieved by the reduction of nature to a feminized insentient *object*. As the unifying bonds of the older, organic, European cosmos were severed in the sixteenth century, European culture increasingly set itself above and apart from all that was symbolized by nature. The nature/culture dichotomy provided the rationale for the advance of culture at the expense of nature. It also provided the ideological underpinnings for the identification of women and animality with a lesser form of life.

Many feminist scholars agree that the subordination of nature to culture construed as a process of "natural law" and as a power relationship justified by science and religion is linked to the subordination of women. Feminists are *not* so clear about the implications of this for forging strategy to advance women's struggle for social equality; similarly, environmentalists are divided about the implications of this for forging strategy to protect environmental integrity. The assumption that there are particular connections between women and nature, and that women occupy a bridge-like position between nature and culture, poses a dilemma both for feminist and for environmental theory and practice. The "nature" issue is a thorny one for feminists.[56]

Feminist response to the presumption of a woman-nature bond has taken a number of tacks. In the first instance, and most noticeably in the

THE MOTOR-CYCLE THAT CONQUERED THE AFRICAN JUNGLE NOW COMES EQUIPPED FOR THE ASPHALT JUNGLE.

Introducing New West: for the man who conquers frontiers.

Figure 5.2. Real men conquer nature. Advertisments for a motorcycle (top—note also the Playboy sticker on the fuel tank) and for a men's cologne, 1989 and 1990.

1970s and early 1980s, many feminists seemed to agree that the way to advance the cause of feminism was to deny the potency of difference between women and men. This might be characterized as a "rationalist" or anthropological feminist position. Writers such as Simone de Beauvoir and Sherry Ortner argued that the woman-nature connection should be seen as a cultural artifact, the product of a particular historical period, with little contemporary relevance or value other than as a tool of the patriarchy in justifying the ongoing oppression of women (and of nature).[57] Ortner, for instance, argued that women should reject their presumed link with nature, and should seek to be integrated into the ("men's") world of culture, and that feminists should explore theoretical work that exposes the presumed woman-nature bond as a bankrupt male artifice.

In the late 1970s, other feminists, most notably Susan Griffin and Mary Daly, called for a feminist revaluation and reclamation of the woman-nature connection.[58] They argued that while the woman-nature bond had been defiled and denigrated by patriarchal culture, the bond of women with nature in fact represents a significant and empowering bridge for women—a bridge to their past, and a bridge to the natural cycles that can seem to have such significance in women's lives. For these feminists, a celebration and affirmation of women's distinctive culture offers an avenue out of and away from the dominant male culture. The essence of women's distinctive culture, they argue, is derived from women's close association with nature; in Griffin's poetical work, *Woman and Nature*, she writes, "We know ourselves to be made of this earth. We know this earth is made from our bodies. For we see ourselves. And we are nature."

The text of "ecofeminism," as it is defined in the writings of people who self-identify as ecofeminists, is located quite firmly in this second feminist camp.[59]

ECOFEMINISM

Ecofeminism, like feminism itself, is not a monolithic philosophy. As it is emerging in North America and Europe, it coheres from a combination of interests and ideas, most of which stem in some measure from an underlying assumption that women have a special relationship to the earth.[60] While ecofeminists often disagree among themselves on the specifics of ecofeminism, most women who identify themselves as "ecofeminists" would agree with Andree Collard's description of this relationship:

> The identity and destiny of woman and nature are merged. Accordingly, feminist values and principles directed towards ending the

oppression of women are inextricably linked to ecological values and principles directed towards ending the oppression of nature. It is ultimately the affirmation of our kinship with nature, of our common life with her, which will prove the source of our mutual well-being. . . . Good women have kept houses on the model of Mother Nature for as long as there have been mothers. . . . Women's experience with oppression and abuse, as well as their experience of mothering, can make them more sensitive to the oppression and abuse of nature, as well as better situated to remedy it.[61]

Most ecofeminists try to combine motherist politics with explicit feminist consciousness. Thus, on the one hand, ecofeminists, drawing on maternal affinities, revalue the identification of women with nature and affirm the continuity between the mothering behavior of women and the archetypal nurturer, Mother Earth, the celebration of which is rooted in an emergent women's spirituality movement. On the other hand, ecofeminists extend feminist analyses of the roots of women's oppression (and the role of male violence in such oppression) to a consideration of environmental destruction as another form of male violence, directed against the woman-identified earth. The allure of ecofeminism is in exactly this combination of feminist and ecological consciousness. Feminist analyses of social domination could be considerably sharpened and enriched by considering the environmental equivalent—the domination of nature. Conversely, a philosophy of ecology that does not take account of a feminist analysis of domination that reveals the interconnected roots of misogyny and nature domination, in the words of Ynestra King, remains an "incomplete abstraction."[62]

And yet, the combination of a feminist analysis of social domination with a motherist, spiritually based, Mother Earth-reverent philosophy is an uneasy ideological alliance. There are a number of unresolved contradictions in ecofeminism, as it is currently articulated by its Euro-American proponents, that limit its appeal and threaten its internal cohesion.

In the first instance, much ecofeminist writing comes perilously close to positing a universal "essentialism." In affirming the woman-nature association, many ecofeminists embrace the notion that women are *inherently* and *naturally* closer to the earth. Rather than construing this bond as a justification for the subordination of women though (as patriarchal cultures have done), ecofeminists argue that this is a position of virtue and strength for women—it locates women closer to the source of life and nurturance for all on the planet. The reification of the woman-nature bond, in fact, leaves men at a universalized disadvantage. Ecofeminism combines an ideology of the affinity of women to nature with a motherist ideology: that women carry within themselves the seeds of nurturing and earth-caring, the exemplar of which is the Mother Earth.

This essentialism—the argument that there are inherent and *universal* female values—ignores differences of class, culture, and race among women. While it is true that women alone mother (and men do not), not all women do it in the same way, with the same consciousness, under the same same conditions, or within the same kinds of social relations. A young, white academic mother in New York City and a Peruvian Indian mother might well have more in common with their *male* counterparts than with each other.[63] Not all women are naturally equity seeking and life enhancing, nor can it be assumed that they be would be so if removed from the constraints of patriarchy. Men may treat nature as a woman and a woman as nature, but this does not mean that all women actually experience a closeness to nature.

The essentialism of ecofeminism raises the same problems as does the maternalism of peace activism: the essentialist argument is indistinguishable from conservative arguments that women have an especially nurturing and maternal nature and that it is "natural" for men to dominate and rule. A feminist movement that aligns itself with centuries of rationalizations for male power and female subordination is skating on very thin ideological ice. It is dangerous to assume that we can embrace a patriarchal construct that men have used for centuries to oppress women, without embracing some of that oppression. The facility with which women's presumed "special relationship with nature" can be turned into a modern anti-feminist agenda is perhaps most recently illustrated by the startling meteoric rise of an antifeminist academic, Camille Paglia. Paglia, who has become the darling of the right wing, argues, among other things, that there is a biological basis for sex differences and that it is the inescapable force of a brutal pagan female nature that justifies male domination, violence, and superiority in Western culture.[64] Ecofeminism, while embraced by women seeking to free themselves from cultural definitions and restrictions, has a high potential for ideologically backfiring. It draws on the same source of evidence and rationale as patriarchal sex-role stereotyping that assumes that women have specifically womanly interests in preventing pollution, nurturing animals, or saving the planet.

Some critics find unwitting complicity with patriarchy even in the basic structure of the ecofeminist question. By implicitly framing the feminist debate over ecology in terms of the question, "Are women closer to nature than men?," ecofeminists are implicitly leaving unchallenged the legitimacy of the patriarchally constructed nature/culture dualism. That is, the idea that one group of people is closer to nature than others assumes the very nature-culture split that ecofeminism would deny. Thus, in the judgment of some critics, the root question of ecofeminism is itself flawed.[65]

The universalizing sweep of ecofeminism not only ignores social and cultural context, but by emphasizing the interrelatedness of women with nature as central to women's identity, it also runs the risk of leaving no room for considering people as actors outside their biological categories.[66] The universalism of ecofeminism echoes the patriarchal assumption that "biology is destiny"—a trap that robs women of the power of independent agency. It suggests that no matter who we are, or where we are, the essence of our being and the core of our identity is located in the fact that we are female—and, as a result, have an affinity to nature that men do not and cannot have. If this is true, if the essence of our identity is already fixed—fixed, indeed, at birth—then the structures of daily life that oppress women, such as patriarchy, classism, racism, poverty, and destructive bureaucracies, can be cast as derivative and superficial.

Indeed, one of the main critiques of ecofeminism is that it can lead to an apolitical ennui—it can be interpreted as undermining the rationale for women to take political action. If we humans are essentially or naturally dichotomized by sex-linked traits, then there is a certain futility in trying to change human cultural practices. Instead of arguing that gender and sex roles are socially constructed, and therefore could be socially reconstructed, ecofeminism gives us essential human natures that transcend political immediacies. Such a doctrine is at odds with the feminist conviction that men and women alike are (distorted) products of the psychological, social, and cultural practices of patriarchy.[67] Much ecofeminist philosophy ignores the extent to which women's oppression is grounded in concrete and diverse and culturally specific social structures.

The apolitical undercurrent of ecofeminism is reinforced by the New Age spirituality with which it is so closely associated, a spirituality that urges looking inward, not outward. Some proponents of ecofeminism, and of the women's spirituality movement, urge women to direct their energies inwards, to "find the goddess within," to nurture their inner strength. Critics of this stance argue that while it is, of course, important for women to nurture themselves and make inner peace, this cannot be at the expense of political action to improve their lot, or the lot of their less fortunate sisters. While many ecofeminists and women's spirituality advocates do not suggest a retreat from political action, others do.[68] Sonia Johnson, for example, a well-known American advocate for the women's spirituality movement, created a firestorm of controversy in the late 1980s by urging women to dissociate themselves from the patriarchy by withdrawing from all forms of engagement with structures of men—by which she meant, for example, that women working in battered women's shelters and rape crisis centers are tools of the patriarchy: "The system uses us to clean up after it, so it can go on battering and raping and maiming and wounding." She urged women to withdraw their support from those services, so that the patriarchy will die. She urged an "inner migration."[69]

To eschew political action directed at social change is a position that only privileged, Western, middle-class, white women can afford—if even they can. For the millions of women who feel their oppression directly and daily as survivors of imperialism, racism, and poverty, an argument for social transformation based on a "special relationship with nature" or on "Rapturing Out" is insulting.[70] For women who are struggling to change the social structures of domination, inward-turning personal transformation is an idyll. A number of women of color in the US have taken issue with Sonia Johnson's program for change by pointing out, for example, that if women stopped staffing rape crisis centers, the patriarchy may eventually die, but so will thousands of women:

> As activists and/or women of color, we have special concerns about participating in a feminist Rapture, a mass exodus of Johnson's followers away from such supposedly "easy" pursuits as legislative reform, advocacy, civil disobedience, work in battered women's shelters, and the like. . . . Can we really accept Johnson's highly individualistic, just-think-your-way-out-of-oppression program as the new true route to a liberated future? Or is this a devastating slip backwards into a cul-de-sac which can offer nothing—not consolation, nor subsistence, nor witnessing—to a significant part of the world's women?[71]

Johnson is not an ecofeminist, nor were her remarks directed at environmental concerns. However, there are many ecofeminists who adopt a similar "think-your-way-out-of-oppression" stance, a tendency that is reinforced in the works of those ecofeminists who embrace the contemporary women's spirituality movement. Irene Javors, for example, writes in *Reweaving the World* (one of the most comprehensive American ecofeminist anthologies), ". . . what we most fear in external reality—isolation, poverty, disease, loss of control, ugliness, death—are but the shadows and demons of those aspects of our inner worlds that are ruled by ego." Women who suffer from isolation, poverty, and disease might take issue with this analysis. In the same collection, Brian Swinne suggests that the "solution" to environmental catastrophe lies in "the perspectives, awareness, and consciousness found most clearly in primal peoples and women . . . for women are beings who know from the inside out what it is like to weave Earth into a new human being;"[72] a heightened primal awareness may be desirable in many ways, but it will not solve our environmental crises.

For many proponents of an ecofeminism rooted in an earth-based spirituality, political transformation appears to be primarily an internal process. A prominent ecofeminist, Starhawk, writes: "Feminist spirituality, earth-based spirituality is not just an intellectual exercise, it's a practice. For those of us called to this way, our rituals let us enact our visions. . . . Although the structures of war and domination are strongly entrenched, they must inevitably change, as all things change. We can become agents of that transformation, and bring a new world to birth." In another essay, she suggests: "We can begin our long-term commitment [to environmental change] by first getting together with people in a small way . . . Maybe you will form a circle where members take off their clothes and go to the beach and dance around and jump in the waves and energize yourselves that way. And then you'll all write letters to your congresspersons about the ozone condition."[73] Charlene Spretnak, a leading spokesperson for integrating spirituality and green politics, echoes a similar trust that political change will follow from personal transformation:

> How to save us all from annihilation, how to birth a just world of inner and outer peace and harmony? One essential element of our strategy should be achieving broad human awareness that the human race *does* have a heritage of long eras of peace among societies that lived by holistic values. . . . A second element must be our self-regeneration. To avoid being overwhelmed by the dimensions of our struggle, we should carefully maintain our most essential political resource: us. . . . Trust your body knowledge. Feed your natural tendencies toward multilayered perceptions. Feel your wholeness. Feel our oneness. . . . The authentic female mind is our salvation.[74]

Ecofeminism thus defined, (and while these writings may not be the rule, neither are they exceptions in the ecofeminist genre), is primarily an exercise in personal transformation,[75] one that does not necessarily suggest a political direction for a feminist re-thinking of the ecology question.

Even among ecofeminists who are not so closely rooted in an earth-based spirituality, the strategies for political transformation are not much clearer. A number of ecofeminists underscore the need for understanding the structures of domination—industrialism, militarism, male consciousness—that have created conditions of ecological destruction. But, perhaps because most ecofeminists share a belief that women are in a special position to "save the planet" (based on the presumption that women are more **of** nature than are men), political change is still typically cast as a change in consciousness:

> Replacing the unconscious and silent assumptions that have gov-
> erned our relationships for so long with new principles of respect,
> accountability, and harmony must become the core of our efforts to
> create change collectively. Evaluating whose interests are being
> served, and whether or not we are moving away from domination and
> control should be the basis of decisions. . . . By breaking the power
> taboo and naming the power dynamics that are operating . . . we can
> build new ties and ways of being together that heal us and reestablish
> communities that are once again sustained by the sanctity of life.[76]

The current wave of Western women's spirituality is rooted in the feminist rediscovery of forgotten women and of women's iconography that
male historians and record keepers have erased from the slate of history.
The reclamation of a distinctive women's spirituality from patriarchal
religion, one of the most recalcitrant bastions of male power and male
exclusivity, energized feminists throughout the 1960s and 1970s. Merlin
Stone's book, *When God Was a Woman*, was a watershed in this reclamation.[77] Startling discoveries followed, as women reconstructed a Western
history in which goddess-worshipping matriarchal cultures flourished,
living in assumed harmony with nature. Resuscitating goddess culture
from the obscurity to which it was consigned by male-defined religiohistor-
ical reconstructions has created a flourishing spiritual community within
the contemporary feminist movement.

In the 1980s and 1990s, the women's spirituality movement became
intertwined with "New Age" spirituality. Because both the women's and
New Age spirituality movements have a strong interest in other cultures
and other times in which the earth was seen as sacred, many ecofeminists
assert a strong affinity with Native American cultures. However, in practice,
this genuine interest has as often as not been contorted into a simplified
nostalgic romanticization. It is hopelessly idealistic to predicate an environmental philosophy on the assumption that Native Americans lived in
their environment without disturbing it in any lasting way. In fact, the
ecological record does not provide evidence of *any* culture living in
non-exploitative "harmony" with nature. Native cultures and goddess-
worshipping cultures alike have been active agents of ecological change
and ecological disruption throughout the millennia; all have left their
impress on the natural landscape.[78]

More distressingly, the genuine longing for historical exemplars of a
more harmonious, indigenous way of being has been cynically manipulated by some non-native New Age opportunists. The commercialization
of crystals, herbal medicine, and easy spirituality has reaped large fortunes
for a few women and men (and generated a large New Age publishing

empire), while reducing complex cultures to prayer-sticks hung with feathers and bags of incense worn around the neck. While many ecofeminists speak out against this opportunism, others remain sanguine about the complexities of predicating a Western and white movement on contemporary interpretations of non-Western, indigenous traditions.[79]

The divisions between feminists over ecofeminism are often bitter. One critic of ecofeminism writes:

> Especially in our present society—a society glutted with myths and tinsel images—we must seriously question the use of myth in alternative political movements. Myth cannot fight myth. Ruling classes have always encouraged confusions between illusion and reality in underclasses. The fact is that whether a goddess was worshipped in prehistory or not, she was an illusion then and she is an illusion now. In an age of manipulation and myth . . . we must restore to the ecology movement a realistic—not illusory—view of nature, and a political— not a religious—view of politics.[80]

A proponent of ecofeminism reflects on the relations between her movement and the larger feminist movement:

> Scorned, trivialized, or ignored by many in the academy, the strength of ecofeminism is in the streets. Here we see that the supposed antagonism between active political resistance and spiritual passivity does not seem to hold. Indeed, individuals who have been inspired and motivated by their beliefs in Earth-based spirituality have often been some of the most active defenders of the planet. Thus, it would seem that narratives that deny or ignore the spiritual threads of ecofeminism fail to acknowledge some of its most creative features . . . Many critics have claimed that ecofeminism is hopelessly mired in essentialism, reifying the female body and the essential femaleness of nature. While not denying the presence of essentialism within this complex constellation, we also believe that close attention to the practices of ecofeminism recovers what much academic discourse loses.[81]

These internal arguments among feminists are perhaps a tempest in a North American teapot—not particularly interesting or relevant to outsiders or non-Western feminists. However, the significance of ecofeminism is not to be judged primarily in terms of its relationship to feminism, but in its contribution to the environmental goals it professes. And it is here that ecofeminism fails most sharply. Despite the many ecofeminists who do not fall into this trap, there is a strong apolitical, acultural, and ahistorical undercurrent to ecofeminism that is especially limiting. Environmental destruction takes place in a political and politicized context. Environmentalism must remain a political movement. Such a movement, while it should certainly be concerned with the psychic well-being of its supporters, should not exist primarily to minister to their personal needs.[82] Conversely, ministrations of personal psychic well-being do not translate into platforms of political change.

The ecofeminist umbrella is a big one, and there are many important shades of difference among people who self-identify as ecofeminists. Not all ecofeminist writing is essentializing, nor do all proponents of ecofeminism urge an inward-turning self-nurturing. Many ecofeminists highlight the real-world overlapping structures of women's and nature's oppression, and cast these issues as integral to the struggle for social justice. However, the fact of identifying the problem doesn't in itself automatically suggest a solution, and a strategy for ecofeminist environmental action has *not* yet been clearly articulated.

Women *do* have specifically gendered experiences that can usefully contribute to understanding and reversing the assault on the environment. These experiences, though, are *culturally* conditioned and *historically* variable, as in Ynestra King's "embodied woman as social historical agent," or Vandana Shiva's female targets of maldevelopment. To the extent that ecofeminists rescue this gender-specific reality from enforced obscurity, they open up a new environmental dialogue.[83] Moreover, ecofeminism, inasmuch as it taps women's rage and despair at the destruction of our planet, may inspire women into environmental activism. But having catalyzed passion, it is not clear that ecofeminism provides a strategy for environmental change beyond one based on a personal response to the planetary crisis.

Chapter 6

Hysterical Housewives, Treehuggers, and Other Mad Women

THE GREEN CONSUMER

Personal responses to planetary crises seem to be in vogue in the 1990s. Many Europeans and Americans, especially those of us weaned on the social activism of the 1960s and 1970s, really do believe that the cumulative weight of small, individual efforts can turn the tide—whatever that tide may be. Entering the 1990s on the cusp of environmental crises that seem larger than all of us, facing what seem to be impossible environmental odds, we search for some way to shape, even in a small way, our environmental future; we strive to find ways to "make a difference." The "green consumer" movement, which started in Europe in the late 1980s, rolled across Canada, and is just beginning to show its colors in the US, gives direction and coherence to this inchoate individual impulse to reduce global environmental issues to a manageable scale. Many of us are comforted by the belief that we can change the world one kitchen at a time.

Green consumers buy "environmentally friendly" products—goods that have been produced without toxic chemical processes, usually without animal testing, goods that do not leave a toxic trail in either their use or disposal, and goods that are not produced at the expense of fragile and diminishing natural resources. Green consumers are people who read labels (on tins of food, on household cleaners, on wooden cutting boards . . .) as a conscious act of environmental awareness. Green consumers are trying to exert environmental influence through the power of the purse.

Consumer movements are not new. The 1980s–1990s green consumer movement can be seen as a continuation of a long and global tradition of women taking up activism for social change while remaining situated in their "proper" domestic sphere. In countries such as Japan (the Japan Housewives' Association), Malaysia (the Penang Consumer's Association) and in the Caribbean (the Housewives' Association of Trinidad and Tobago), consumer associations are—and have been for several decades— important political players.[1]

While the Eurocentric green consumer movement broadly resembles these "consumer-association" forerunners, it is much less politically potent. In the first place, the green consumer movement is not really a movement—it is the accumulation of individual trends that represent, in sum, shifts in collective consumption patterns. The Euro "green consumer" is typically white and middle-class; collectively organized consumer action groups are much more multiracial and mixed-class. Organized consumer groups rally collective support for specific health, safety, or consumer-rights campaigns; the momentum of the green consumer movement derives from an accumulation of unorganized individual acts.

In all cases, though, "consumers" are understood to be women. The green consumer movement is home-based and thus explicitly women-oriented. Advertisers and producers of green products are well aware that women remain the primary shoppers for home products, especially for mundane home products such as toilet-bowl cleansers and dish detergents. For the first time, it seems, the most trivialized and often scorned of women's traditional chores in the home—shopping for household products—is taking on the lustre of an "important" job. Such things as the environmental merits of one laundry detergent over another have become "serious" matters of public discussion, and, just as amazingly, it is *men,* men with political and career ambitions, who are trumpeting the significance of household-based consumer decisions. Women's household decisions are shown to have wider, even global, consequences, and thus women are suddenly seen to be in control of a much more serious economic-environmental unit (the home). Through the green consumer movement, women can put one foot into the political arena without leaving the kitchen—an alluring and empowering combination, though not without mixed political implications.

The green consumer movement is fueled not only by global environmental concerns but by pragmatic and immediate family-health issues. Just as feminists have revealed the extent to which violence against women occurs in the home and that the home is not the mythical refuge for women that it is supposed to be, so it is becoming evident that the home is environmentally hazardous as well. The conventional European and North American home is a highly toxic environment. A recent survey of 40 homes in the United States found from 20 to 150 hazardous chemicals in each home, with indoor concentrations of toxins at least 10 times higher than outdoor concentrations.[2] By asserting themselves as green consumers, women have the opportunity to safeguard the health of themselves and their families, while simultaneously contributing to planetary environmental health.

While there is good reason to be concerned about household toxins, and good reason to move quickly to clean up our domestic environmental act, the politics of environmental consumption are extremely complicated.[3] It *is* important to embrace the principles of environmentally responsible consumption—indeed, it might actually even be considered an environmental imperative that affluent consumers in the West do so. But we also need to be aware of and wary of the shadow politics behind the green consumer movement, women especially so. In other arenas of social activism and political change, women have found that when they don't control or shape the agenda of social movements that are made in their name, they end up with most of the blame for any failures, and little of the credit for any successes.

Healthy Homes, Then and Now

The current green consumer movement uncannily resembles an earlier consumer-based, home-based movement—the British and American "home healthy" drive of the 1910s and 1920s. Fueled by pressing concerns for reducing infant mortality and preventing the spread of disease in urban areas, reformers in the early twentieth century turned health concerns into a domestic orthodoxy that required a full-time housewife in every home to safeguard family health by meeting hysterically high standards of hygiene.[4] This was the era when the President of the United States wrote a monthly column in *Ladies Home Journal* warning women of the danger to national well-being if they did not take seriously their responsibilities to provide pure and ameliorative home environments for their families; this was the era that gave us the "Good Housekeeping Seal of Approval"; "correct" housekeeping was merged with "correct" consumerism in a bizarre mix of heavy blame and heady responsibility for housewives; advertisements warned women that the lives of their children and the health of their husbands depended on correct consumer choices, and on their constant vigilance against germs and unhealthful substances in the home—and any lapse invited ruination, ill health, possible death, and sure national catastrophe. Linking "correct consumerism" with broader social and political concerns was a lucrative twist for manufacturers of household goods and an ingenious marketing device; not surprisingly, manufacturers were among the most energetic in pushing the domestic reform agenda.

While many of the health issues raised by the early twentieth century reformers were legitimate, the "domestic reform" movement of the turn of the century had many hidden agendas, all of them conservative. In the United States, especially, the domestic reform movement was fueled by racism and anti-immigrant phobia—it presented a notion of what constituted a good "American" home, and tried to impose this standard across the board. Despite its many prominent women proponents, the movement also was used as a clever way to thwart early feminist activism by putting forward a compelling reason to privatize and "professionalize" women's work in the home, keeping women out of the public arena, out of waged workforce and in the home. (This turn-of-the-century ideal of women as non-wage earning, full-time housewives has stuck with us—it continues to set the tone for contemporary public policy debate on a range of issues from "family values," to equal wage issues, to child-care provision.) And the domestic-reform movement deflected attention away from real social inequities: social ills of all kind, from rising crime rates to homelessness, were blamed on poor standards of housekeeping. Social problems were presented as problems that could only be solved in the home and on a personal level. Since it was construed as women's responsibility to achieve the high standards set by the reformers, any failure was also thus women's. The propaganda from the 1910s and 1920s was blatant in laying blame on women for all kinds of social disorders.

GREEN CONSUMING: A
DOUBLE-EDGED SWORD

It is difficult to take the political pulse of the Eurocentered green consumer movement. There is considerable merit in the notion that individuals, by taking responsibility for their personal consumption patterns, can incrementally induce broader social change. There is also merit in empowering women and their domestic work (consuming) as agents of environmental change. But the contemporary green consumer movement is as complicated as the "good housekeeping" movement of the early twentieth century before it: woman-blaming is a recurrent undertone; it is a movement open to cynical manipulation by industrial interests; and it promotes individual solutions to what are in fact large structural environmental problems.

Woman-Blaming   Looking at the green consumer movement through feminist lenses affirms, in the first place, the crucial importance of "greening" the home environment. A consumer association in Southeast Asia points out that, globally, the home is a highly hazardous place, and especially so for women:

> Horror stories abound about hazardous products. . . . Gas cylinders may explode in the kitchen; your doctor may prescribe drugs which have been outlawed in another country because of their dangerous side-effects; chemicals sprayed in your garden can contain cancer-causing agents. In the US alone, about 20 million people are hurt in accidents in and around the home, 30,000 of whom will die. In many of these cases women—the main purchasers and users of home products—are the victims. Each year over one million of them seek hospital treatment for injuries arising from products like caustic cleaners, lawnmowers, sun lamps, and paint. It's the same elsewhere. Twenty percent of all the alerts sent out in the period 1982–1986 by the International Association of Consumer Unions' Consumer Interpol, a global programme for quick information exchange and action on dangerous products, were about items used exclusively or mostly by women. . . . Hazardous products are an area where women are especially vulnerable because they are more likely to be illiterate than men, have less access to information about product hazards, and be working in areas which expose them to danger.[5]

Women's role as primary caretakers of the home and family makes them vulnerable to consumer manipulation and subject to blatantly coercive advertising and sales techniques. In the 1910s, American manufacturers of products from refrigerators to Lysol "germ killing spray" all ran advertisements in women's magazines accusing mothers who did not purchase their products of neglect and virtual infanticide. In 1980s Asia, a tug at women's maternal responsibilities boosts the sale of insecticides: an advertisement in a Malaysian newspaper for insect spray shows a mother and her young daughter in loving closeness, with the words, "Love means never having to say you're sorry."[6] Many of the same manufacturers who in the 1960s and 1970s were persuading us that a "modern" and hygienic home required a different cleaning agent for each room, air freshening sprays, and scented toilet paper have now turned green and are pushing "environmentally friendly" products from laundry detergents to flea collars for the family dog.

While we like to think that we are not as susceptible today to the same kind of heavy-handed advertising and propaganda that women in the 1910s endured at the hands of the domestic reform campaigners, there are certain similarities in the way the green consumer movement is being framed. The chronology is much the same: important health and environment concerns raised for years by an activist minority slowly seep into mainstream consciousness, are legitimated by changing political tides, and then are suddenly seized upon as a lucrative marketing ploy by major industrial concerns. While the message of the green consumer movement is directed to women, as in the domestic reform movement before it, women's contributions to shaping the campaign and setting the terms of discussion are negligible—women are weeded out of the process very quickly when big business and big politics take over.

Blaming women for poor consumer choices is an undercurrent that runs throughout the green consumer movement. In defense of environmentally unsound production and packaging practices, industry typically turns the blame back on the consumer. Industry spokespeople often argue that they produce only what consumers demand—and that, therefore, if industries haven't yet done enough to detoxify their products and promote recycling, it's only because consumers haven't demanded it of them. But this is a specious argument, one that distorts the true relationship between consumers and producers. Consumer "demand" is more often than not a fictitious creation of industry and advertising executives,[7] and blaming "the consumer" (who, in abstract, is always a woman) for the environmentally unsound practices of industry does not ring true to the experience of most people in the aisles of the grocery store. One shopper in New York, for example, expressed her frustration with the trend to plastic packaging: "I have switched brand after brand of peanut butter to keep up with glass containers, but now it's at the point where everything is plastic."[8]

Consumer-based industries have set up a closed loop of women-blaming: in the first place, consumers (a.k.a. women) are blamed for not demanding more environmentally sound practices of industry. Then when putatively "green" products *are* made available, women who don't immediately jump on the green bandwagon are made to feel guilty for not caring enough for their families health and for the health of the planet—regardless of the fact that green products are typically priced beyond the reach of many family budgets. This push and pull of manipulation and blame is not only promulgated by industry; unfortunately, environmental groups themselves have created blame-filled campaigns directed at women, the most recent of which is the disposable-diaper campaign.

There is little question that disposable diapers are environmentally unsound: they are bleached, manufactured with a process that creates dioxins, produced with diminishing forest resources (it takes one tree to make about 500 disposable diapers), and the 16 billion disposable diapers that Americans throw out each year contribute to the garbage problem, a problem that is all the more urgent because throughout Europe and North America rubbish dumps are running out of space. In the early 1990s, several mainstream environmental groups waged a high-publicity campaign against the disposable diaper, highlighting it as a *major* contributor to the garbage problem, and turning disposable diapers into a symbol of all that is wrong with our throw-away society.

But taking aim at disposable diapers is a cheap shot. Diapers make up approximately 2 percent of America's garbage, a figure that is minor in comparison with the total composition of our garbage—which consists of, among other substances, 17 percent yard waste (grass clippings and the like), 30 percent plastics (by volume, 8 percent by weight), and 40 percent paper. When pressed, the environmental groups that are leading this campaign will reluctantly admit that diapers are a small part of the garbage problem, and may have been inappropriately inflated as a symbol of modern environmental crises.[9] Many garbage experts immediately dismissed the disposable diaper issue as a red herring, and suggested that environmentalists would be better off tackling the plastics and paper industries. But harried mothers who are trying to balance environmental concerns with parenting are easier targets. The public recriminations over disposable diapers aimed against parents (in abstract, always women) by environmental groups is not dissimilar from the women-blaming anti-fur campaigns of various animal rights groups. In both instances, while the issues are important for environmentalists to take on, the tactics adopted by environmental groups are anti-women and decidedly unfeminist.

The diaper example, in fact, provides good evidence that environmental campaigns can be handled differently when informed by feminist perspectives. While mainstream environmental groups focused on the "garbage crisis" created by mothers using disposable nappies, the Women's Environmental Network (London) produced a brochure on the problem that focused instead on the health threats to women and children posed by the use of bleaches and dioxin in the production of disposables. The brochure goes on to make the link between dioxin used domestically and the reproductive disorders caused by the chemical disaster in Seveso, Italy, and dioxin use in the Vietnam War.[10] Both the mainstream and the feminist disposable-diaper campaigns have the same end purpose—to stop production of environmentally unsound disposable products—but use very different means to that end.

Consumer Manipulation, Fraudulent Claims   The turning of industry's green leaf seems suspiciously sudden, and warrants close scrutiny. A number of feminist women's groups (while acknowledging the importance of changing consumer patterns) take a jaundiced view of the current marketing of green awareness. In fact, inasmuch as there has been a public critique of the green consumer movement, it has come primarily from women's organizations. In contrast, the mainstream environmental groups moved quickly to endorse the green wave in its early days—only to be subsequently burned by their eager embrace of industry.

In both Canada and the UK, major environmental groups have been subsequently embarrassed by companies they rushed to endorse. In early 1989, two Canadian environmental groups, Pollution Probe and Friends of the Earth Canada endorsed several of the new "green" products developed by Loblaws, a major grocery chain; in return, the environmental groups were to receive royalties from sales of those products. Within weeks of the endorsement deal, independent testing of the products cast doubt on the "green" claim, establishing, for example, that one of the products, a home garden fertilizer, contained toxic substances that were common in regular commercial fertilizer but that were supposed to have been eliminated from the "green" version.[11] In the UK, Friends of the Earth formed a promotional alliance with AEG, a major industrial group which was introducing water-efficient washing machines. It was subsequently revealed that the company is one of the largest UK electronic warfare systems producers. When asked about this apparent incompatibility in product philosophies, the managing director of AEG didn't see that there was any problem, saying, "In our view, supplying systems for advanced weaponry is of little relevance to us giving greater attention to developing household appliances with environmentally protective features."[12] Friends of the Earth withdrew from the promotional agreement.

This is not to suggest that environmental groups should *never* form alliances with industry; obviously there are mutual benefits to forging closer working relations between environmentalists and manufacturers. But environmentalists need to be cautious about cozying up to industries that have traditionally been their opponents, and they must be rigorous in their scrutiny of the principles of the *nouveau* green manufacturers. The green consumer movement has become industry-led; it is no longer the grassroots, "yogurt and nuts" movement that it once was. And under the thumb of big business, the green consumer movement is susceptible to all kinds of cynical manipulation.

Some green industries *are* genuinely motivated by environmental concerns. Most, though, are motivated by the same profit motive that drove them earlier to produce hazardous goods; "green" just happens to be the marketing wave of the moment, and industry is seizing the opportunity to turn a large profit by playing on environmental concerns. The manipulation of environmental concerns runs throughout industry, and there are endless examples of corporate "green duplicity." AEG promotes environmentally friendly products at the same time as they produce weapons of war. British Petroleum (BP), an oil company that was especially aggressive in marketing itself as newly-green, and which won an international award for conservation, is at the same time destroying hundreds of thousands of acres of Brazilian rainforest for open-cast mining.[13] Ironically, shortly after these environmental violations were made known, BP turned up eighth out of 20 on a list of companies most admired by British businessmen and financial analysts—proving the point, perhaps, that playing on both sides of the environmental fence is good for business.[14] Dozens of companies are now manufacturing "biodegradable" plastic bags, trying to cash in on our guilt about garbage; in truth, though, most "biodegradable" plastic bags don't really degrade, or they do so only under ideal conditions that are not possible to achieve in standard garbage dumps. And at the same time that industry is loudly turning over a green leaf, more and more consumer goods are being packaged in plastic containers—from jam jars to tampon applicators, containers that were recyclable or biodegradable just three or four years ago are now made of plastic.

Women's groups appear to have been more shrewd than mainstream environmental groups in their assessment of the green industrial wave— perhaps because they are more willing than the major environmental groups to make public judgments on the basis of values and moral reasoning, or because they feel less pressured to be "taken seriously" by corporate men. Bernadette Vallely of the London-based Women's Environmental Network explains that, "If a company is going to claim to be green, it has to have high morals—and that means not producing things that are worthless." Caroline Lucas of the British Green Party echoes this sentiment: "Companies have been selling goods for years which are changing our planet into a global dustbin, and using up scarce resources. To really go green, they need to change the products—and that means withdrawing the harmful ones which are not in the public eye, as well as those which are."[15] Lucas also put her finger on one of the great fallacies of the green consumer movement—that, after all, green or not, this *is* a consumer movement, and the purpose of green companies is to promote *more*, not less, overall consumption: "We believe the only way to protect our world is by consuming less, and companies which go green are hardly encouraging us to buy less of their goods. All their suggestions are aimed at increasing their share of the green market."

Women have learned a lot of hard lessons in recent years about the

manipulation and abuse of consumer products by multinational corporations that put profits ahead of principles. One of the more important things
we have learned is that it is crucial to inquire into the *gender politics* of
consumer abuse. Recent international consumer scandals—the Depo-
Provera birth-control controversy, the Dalkon Shield-IUD scam, the Nes-
tle's infant-formula abuses, the breast-implant scandal—have brought into
sharp relief the fact that there are serious consequences when multinationals that are run by men make research, design, and production decisions,
often literally life-and-death decisions, for a mostly women consumer
population. What we have learned, put simply, is that most men making
most decisions in most large corporations don't think about women seriously, don't take women's needs seriously, and inasmuch as they do think
about women, they perceive women as an easily manipulated consumer
pool with little political clout—and that this is true even when (or especially when), they are designing and marketing products designed mostly
or exclusively for women. Largely as a result of these international scandals, women around the world are comparing notes on consumer issues,
and the result is an emerging feminist analysis that is useful in assessing
the current European–North American green consumer movement.

The story of DepoProvera and of the Dalkon Shield, in which products
deemed unsafe for Western white women were dumped on Third World
women, underscores the necessity of looking at the global interconnected-
ness of industry practices. Speaking of Depo-Provera, Carol Downer of the
Los Angeles Feminist Women's Health Center said: "We can fight against
some problem here, only to see it exported to women overseas. But we're
not going to sit by while a victory at home turns into a tragedy abroad. We
have a responsibility to women all over the world."[16] The Depo-Provera and
Dalkon experiences suggest an important critique of the green consumer
movement: we must ask, "green for whom?" It is impossible not to notice
that industry has targeted its green goods campaign at white, middle-
class, and Western markets. Middle- and upper-class families are targeted
for green sales, and it is only they who can afford the expensive new green
products; their homes will be safe from toxics, but the poor will have to
struggle on as always with unsafe and unhealthy goods. The green consumer movement, as it has been framed in the UK and North America, is
thus hardening the lines between the "haves" and the "have nots," and
introduces a new *environmental* measure of privilege—the privilege of
being safe from household environmental hazards.[17]

The underlying philosophy of environmentalism—"that everything is
connected"—must be applied to an analysis of the green consumer movement itself. Environmental accountability must stretch across national
borders and *across class lines*. Feminists have learned to push cross-
cultural and cross-class accountability to the fore. It is not acceptable to

clean up one market and yet continue to produce hazardous products for sale to more vulnerable populations at home or abroad (among whom, both at home and abroad, women are the most vulnerable). Double-standard marketing is not only a morally untenable and usually racist practice, but it is environmentally unsound to produce hazardous products—full stop—no matter whose kitchen they'll end up in. (And ironically, of course, poisons exported "over there" always come home again; on a small planet there is no "away.")

More broadly speaking, feminists insist that the ethics of a company *as a whole* are important. A "green" company is not one that produces environmentally sound laundry detergents at the same time as it is clear-cutting tropical forests. A holistic critique of green marketing and consuming also needs to be informed by an explicit analysis of sexual politics. As Pat Hynes asks, "If a cosmetic is made with natural ingredients, but is advertised with pictures that objectify women and promote the message that aging in women in unattractive, is that product a 'green' product?"[18] The green movement should help *dismantle* the artificial compartmentalization that so many industries rely on to cloak unsound practices.

Personalizing the Environmental Crisis   In the late 1980s, public concern in North America about the thinning ozone layer reached a peak when it was announced that ozone loss was beginning to occur in the northern hemisphere, not just over Antarctica. The response to this crisis from the top levels of American government was a big shrug, and a suggestion from one of then-President Reagan's aides that the "solution" to this problem was for all Americans to wear hats and sunglasses!

The "sunglass solution" illustrates a fundamental fallacy that underlies much of the new green consumerism: green consumerism is often predicated on the assumption that individual actions can solve what are actually large structural problems. Green consumerism, if uninformed by a broader political analysis, leaves intact the institutional framework that is the cause of much environmental degradation. Real environmental solutions require a collective approach to establish appropriate parameters for technological choice and economic activity—not just a private solution based on individuals living "cleaner and greener."[19] As appealing as green consumerism is, collective action is the only effective mechanism for real environmental change.

OUT OF THE KITCHENS, INTO THE STREETS: GRASSROOTS ORGANIZING

Women are the backbone of virtually every environmental group around the world. With few exceptions, women constitute approximately 60 to

80 percent of the membership of most environmental organizations—averaging 60 percent of the membership of general- interest environmental groups, 80 percent or more of small grassroots groups and animal- rights groups. The "gender gap" in environmentalism turns up in other measures also. Around the world, women consistently express more progressive environmental values and attitudes than their male counterparts—international polls establish, for example, that women favor passing more stringent environmental protection laws, they favor spending more money on environmental protection, and they express more alarm over the state of the environment than their male equivalents.[20]

Clearly, many women see the environment as one of "our" issues. Despite the synergy between women's concerns and environmental issues, women continue to be excluded from the power tracks of ecomanagement (and remain alienated from the radical fringe of the movement). However, from this relegated position on the environmental sidelines and the social margins, women are creating a new culture of environmental activism. They are taking the initiative and assuming leadership roles in environmental alliances that lie outside the formal structure of green politics and environmental organizations—in grassroots, community-based organizing.

The prominence of women as catalysts and leaders of the grassroots environmental movement is a global phenomenon: in India and Kenya, it is women-led movements that are fighting to save forests. It was an ad-hoc group called the Mothers of the Aral Sea that helped to bring international attention to the catastrophe of the disappearing Aral in the former USSR. Women in the Ukraine and across Europe took the lead in forcing their governments to acknowledge the seriousness of the Chernobyl nuclear accident and its fallout. Women are in the forefront in virtually every community group in the US and Canada organizing around toxic wastes.

Grassroots Stories

Two grassroots initiatives stand out as leading examples of women's effective community-based environmental organizing: the Chipko forest-protection movement in India, and the toxics-waste movement in the United States.

USA   Lois Gibbs is often the first person who comes to mind when Americans think of grassroots environmental action. In 1978, Gibbs emerged from the obscurity of shy housewifery into the limelight as an outspoken environmental activist when she discovered that her bucolic suburban neighborhood in Niagara Falls, New York, was built on top of a huge chemical dump left behind by Hooker Chemical Company. "Love Canal," Gibbs's neighborhood, is now part of our collective environmental iconography in much the same way as are "Bhopal," "Chernobyl," and "Three Mile Island." But Bhopal, Chernobyl, and Three Mile Island were catastrophic and sudden disasters that literally exploded into public view. They were catapulted into our consciousness. Love Canal had to be dragged into prominence, against the mighty efforts of public officials and Hooker Chemical executives, who did everything in their power to minimize, trivialize, and bury the Love Canal story.[21]

Sexism was the first line of defense used by Hooker executives and recalcitrant public safety officials. Gibbs was constantly being dismissed as a "hysterical housewife." When Gibbs carefully collected health data from her neighbors, and mapped it—showing a pattern of deaths, birth defects, cancer, and serious illnesses that clearly overlapped with the boundaries of the chemical dump—her maps were discounted by health officials as "useless housewife data." It was only when she and some volunteer scientists put her "housewife data" into scientific form, "with all the pi-squareds and all that junk,"[22] that the officials acknowledged the accuracy of her findings. Eventually, a large part of the Love Canal neighborhood was designated an uninhabitable toxic zone, residents were evacuated, their houses were boarded up.

Without exaggeration, Gibbs has almost single-handedly shaped the nature of grassroots environmental politics in the US. Her unwavering stubbornness at Love Canal, which ultimately won grudging admiration even from her detractors, catalyzed local activists across the country, especially women. After evacuating Love Canal, Gibbs started the Citizen's Clearinghouse for Hazardous Waste (CCHW), an organization that serves as an umbrella resource center for community activists across the country, which now services almost 4,000 community-based groups. Gibbs herself has become something of a small "c" celebrity, and is a major figure in debates swirling around grassroots politics—such as whether grassroots groups should seek and accept government funding (CCHW does not), how to bring multiracial and multicultural coalitions together over environmental issues, and how to address the "jobs vs. the environment" argument that invariably arises when women take on environmental violations in their community.

The Love Canal battle ushered in a new era in citizen environmental activism in the US, dubbed the "NIMBY" (Not In My BackYard) movement. In unprecedented numbers across the US and Canada, NIMBY neighborhood coalitions are formed on short notice to battle environmentally hazardous plants and plans—whether the siting of a road or an incinerator, or expansion of zoned industrial districts. NIMBYs are small-scale, site-specific action groups, the largest proportion of which run on the volunteer energy of community women. The political and social impact of NIMBYs is considerable—across the country, NIMBY groups have been successful in stopping dozens of environmentally irresponsible projects, and NIMBYists have become a major irritant to corporate America.[23]

Hugging Trees in India   The Chipko movement is one of the best-known grassroots movements in the world, and is certainly what comes to mind when we think about grassroots environmentalism in India. The Chipko, or "tree hugging," movement was started by women in the Himalayan region of northern India in the early 1970s as a social protest movement to protect the fragile forests there from commercial exploitation by the timber industry.[24] To the women of this region, saving the trees is literally a matter of survival. The trees provide fuel, fodder, and fertilizer, and, just as important, they stabilize the hilly terrain, protecting villages from landslides and floods. Fed up with the commercial exploitation of their life-supporting forests, and drawing on a longstanding Indian tradition of nonviolent resistance, in March 1973, as a timber company headed for the woods above an impoverished village in northern India, women and children rushed ahead of them to *chipko* (or "hug") the trees. The Chipko forest movement inspired others, and throughout India women started to confront loggers in the forests by hugging the trees, linking arms, singing, and chanting. Their campaign has been successful in slowing the rate of deforestation, although the linked timber and mining industries that they face are powerful and not accustomed to brooking resistance.

The original Chipko movement has widened its influence, and Chipko-based resistance is now used throughout India to protest environmentally irresponsible road building, mining, and dam projects. In 1987, for example, Chipko activists formed a seven-month blockade at a limestone quarry that was destroying the ecosystem of an entire valley. Chipko has gone beyond resource protection to ecological management: the women who first guarded trees from loggers now plant trees, build soil-retention walls, and prepare village forestry plans.[25]

Further, the significance attributed to the Chipko movement has broadened. Chipko now symbolizes Third World resistance to misdirected "international development" aid in general, and resistance to nonlocal commercial exploitation of locally based resources in particular. Moreover, the movement has come to symbolize a struggle for autonomy from the stranglehold that Western reductionist science has come to have on resource management everywhere in the world. As Vandana Shiva, an advocate for Chipko, says,

> Forestry management practiced by forest dwellers such as tribals who see trees as living entities providing them with the conditions of life are declared unscientific even though tribals have conserved forests over centuries while foresters have destroyed them in decades through so called "scientific" management. . . . A non-reductionist ecological perception of nature such as that provided by Chipko leads to the awareness that reductionism is a particular way of looking at nature which picks certain facts of nature while denying the existence of others. Trees seen in terms of their dead product, wood, exclude the possibility of seeing them in their living function as providers of water, food, fertilizer, etc.[26]

The Indian Chipko and the American toxic-waste movement are particularly well-known examples of women's grassroots organizing, but they are literally just two of thousands of environmental initiatives taken up by women.[27] A quick survey of the environmental news of the past decade suggests the breadth of women's grassroots environmental activism around the world. Women everywhere seem to be on the march:

• A Boston newspaper started a story about protests over a Mexican nuclear power plant with the headline, "Mexicans Protest New Atomic Plant." The real story of this story, though, is clear in the first sentence: "About 60 *mothers and their children* marched to City Hall yesterday to protest the government's decision to open Laguna Verde, Mexico's first nuclear power plant [emphasis added]." Mexican women were instrumental in organizing resistance to construction of this nuclear facility—which is, by all accounts, poorly sited and shoddily built—and women continue to lead the fight against its operation.

• In upper New York State, women from the Mohawk Indian community are leading the battle against industrial pollution, including the dumping of PCBs (polychlorinated biphenyls—a highly toxic compound) by General Motors. GM has been charged with 19 violations of state environmental statutes, but it is the Mohawk women who have set up the only ongoing monitoring system.[28]

• Canadian activist Margherita Howe founded an organization in 1979 called Operation Clean to oppose the plans of an American chemical firm to dump millions of gallons of chemical waste into the Niagara River, a waterway shared by both Canada and the US. After years of hearings, a compromise plan was signed, one that represents the first case in which Canadian environmental groups became directly involved in regulating American pollution discharges.[29]

• In 1988, the Arab Women Committee, an organization of women representing Arab states, passed a special resolution calling on the Council of Arab Ministers to support women's environmental activities, especially in rural areas. The resolution was intended to draw attention to "the role played by women in the conservation of the environment and the biosphere."[30]

• Jessie Deer In Water, a Cherokee activist living in Oklahoma, co-founded a group called Native Americans for a Clean Environment (NACE) in 1985 in response to a plan by the nuclear energy conglomerate Kerr-McGee to dispose of nuclear waste in "injection wells" in the eastern part of the state. Kerr-McGee, a notorious nuclear polluter, has left 35 million tons of radioactive waste at various processing sites throughout the US, and has already contaminated sites in 11 states. Under Jessie's leadership, NACE successfully blocked the dumping plan.[31]

• In Kenya, a fuelwood crisis is looming—while almost 60 percent of Kenyan households rely on wood as their major energy source, deforestation is threatening the survival of many rural communities. The National Council of Women stepped in in 1977, and now sponsors tree-planting programs in rural areas. Without government funding or outside resources, the Council has already provided thousands of trees, and funding for tree caretakers, as part of its "Green Belt" scheme for the Kenyan Rift Valley.[32] Over a million trees in 1,000 greenbelts are now in place, 20,000 mini-greenbelts have taken root, and 670 community tree nurseries are in place. Meanwhile, Maendeolo Ya Wanawake, Kenya's largest women's development network, initiated a campaign in 1985 to construct improved, wood-saving cookstoves.[33]

• Women in Japan are organized: they have formed an estimated 13,000 consumer groups; the umbrella organization of housewives' associations claims a membership of almost 6 million.[34] In Japan over the past four decades the deceptively unassuming Japan Housewives Association has pressed causes like forcing manufacturers to reveal the true ingredients of fruit-juice drinks, warning the public about industrial pollution sites, and fighting new sales taxes.[35]

Some of the most contentious environmental issues in Japan have been battles over land use—the construction of the Narita airport, for example, or the appropriation of common land by the American military in Zushi and Shibogusa. Women have taken the lead in several of the most hotly contested of these land-use battles; they have formed Mothers' Associations that bring out hundreds of women to rally at a moment's notice. They have established protest camps, staffed by women 24 hours a day. They have sent representatives to Washington to argue over the US military presence.[36] One woman activist, commenting on the prominence of women in land struggles explained, "Women are more persevering than men. . . . The men quickly give in to the government. They are weak in the face of authority. But anonymous farm women like us can't be pushed down any further than we already are and we won't give up."[37]

Why Women? Why the Grassroots?

The prominence and distinctiveness of grassroots environmentalism as a women's activity, when combined with the evidence of differences between women's and men's attitudes to environmental issues, suggests the possibility that we are witnessing the emergence of a distinctive women's voice on the environment. This sense of carving out a niche of our own is a source of empowerment for many women. Without wanting to diminish that reality, though, it is important to address the complications raised by talking about a special "woman's voice."

Many feminists who are leery of ascribing "special" and inherent characteristics to women (or men) are uncomfortable with the idea that there may be a "women's perspective" or a "women's activism" on the environment. Further, given the obvious fact that there are class, race, income, and status differences among women within and between countries, it is at first glance difficult to imagine how we can talk about a distinctive "women's" contribution to environmentalism without falling into the simplistic essentializing and universalizing that the notion of a "women's perspective" often engenders.

But, in fact, the emergence of a women's voice on the environment does not imply the workings of biological destiny, nor does it suggest a universalized sex-specific behavior. Rather, a "women's voice" on the environment derives from materially-grounded facts about women's social location. Women's environmental activism occurs within the context of, and as a result of, their particular socially assigned roles—roles that in many key ways do transcend boundaries of race, ethnicity, and class. At the 1991 Global Assembly of Women for a Healthy Planet, Peggy Antrobus outlined the common consciousness that women bring to (and that brings women to) environmental activism:

> We are different women, but women nonetheless. The analysis and the perspectives that we get from women are certainly mediated by, influenced very profoundly, by differences of class, and race, and age, and culture, and physical endowment, and geographic location. But my hope and my optimism lies in the commonalities that we all share as women—a consciousness that many of us have, if we allow ourselves to have it, of the exploitation of our time and labor in

unremunerated housework, subsistence agriculture and voluntary work. Our commonality lies in the often conflicting demands of our multiple roles as caretakers, as workers, as community organizers. Our commonalities lie in our primary responsibility for taking care of others. Our commonality lies in our concern about relationships; the commonality that we share is the exploitation of our sexuality by men, by the media and by the economy. The commonality that we share is in our vulnerability to violence. Our commonality finally lies in our otherness, in our alienation and exclusion from decision-making at all levels.[38]

It is the linkage of "women's work" to the environments of lived ordinariness that explains why, globally, it is women who are usually the first to become environmental activists in their community. Everywhere in the world, women are responsible for making sure that their families are fed, housed, and kept healthy. They are, in many cases, responsible for creating the means of subsistence—globally, it is often women's work in the waged work force and in the family fields that provides the basics for family survival. And in all economies, women take raw materials and income, whether provided by themselves or by others, and fashion it into food, clothing, housing, health care, child care. In environmental management terms, we could say that women make the primary consumer and resource-use decisions for their families and their communities—that women in all cultures serve as managers of fixed resources.

Building a Politics from a Movement

Until recently, conventional wisdom in the environmental community cast grassroots activism as a sideshow. However, given the increasingly evident shortcomings of fast-track, establishment environmentalism, the center of gravity is now shifting, and all of a sudden everyone is becoming much more interested in grassroots activism. As Gerry Poje, an environmentalist with the National Wildlife Federation recently said,

A reordering of priorities, a rethinking of strategy and tactics is taking place throughout the entire environmental movement because of the increased activism by the very people who are most at risk. Here in Washington it is becoming increasingly obvious that true change will occur at the local level.[39]

Over the course of the past century, most radical movements for social change have been shaped by preexisting political agendas. In contrast, grassroots environmentalism is a movement of locally specific responses to particular circumstances, with no "core" doctrine; it is activism shaped more by internal realities than external politics. However, if grassroots environmentalism *is* to become the leading edge of eco/social transformation, we need to draw larger lessons from the experiences of thousands of locally-specific, often idiosyncratic grassroots actions.

Because of the gender skew in grassroots movements, we can only distill these larger lessons if we listen to women's voices and take seriously women's experiences. There is a certain repetitiveness to women activists' stories—particular patterns and themes turn up over and over again in the narratives of these women, almost no matter where in the world they live.

First and foremost, women who take the lead in community-based environmental causes do not generally have prior experience as political activists; many describe themselves in self-deprecating terms, as "mere housewives." In the great majority of cases, women become involved in environmental issues because of their social roles: as sustainers of families, it is often women who first notice environmental degradation. Many women grassroots activists report that their role as mothers and family caretakers is the key catalyst in their concern for protecting the environment. Because of those same social roles (and their related reproductive roles), women suffer the effects of environmental degradation first and longest—whether in Bhopal, the South African homelands, the Himalayas, or Vietnam, women are hardest hit by a diminished resource base, by exposure to toxins, and by localized pollution. Women, therefore, often have a vested interest in environmental protection.

Most women activists face tremendous resistance to their activism from the men they oppose (who enter the fray as representatives of industry, the military, the government), and also from men who are part of their daily lives, their husbands, sons, brothers, and friends. This resistance is patently based in sexist assessments of appropriate roles for women; for many men, the notion of a woman activist is an oxymoron. Women activists are stepping outside the bounds of sanctioned feminine behavior, and the techniques which men invoke to put women back in their place are often entirely based on sexist "policing"—there can hardly be a woman environmental activist in the world, for example, who has not been called a "hysterical housewife." It is clear that when women walk out of their homes to protest a planned clear-cutting scheme, toxic- waste dump, or highway through their community, their gender and sex identity goes with them—in a way that is not true for men. Many women take up grassroots environmentalism not just as activists, but self-consciously as *women* activists—often as mothers. But whether they encourage it or not, they are perceived by men not just as environmental activists, but as *women* activists.

These commonalities in women's grassroots experience underscore the extent to which priorities and practices in the environmental movement will have to shift if the grassroots model is going to become the new paradigm for environmentalism. And even though many grassroots women may not be individually feminist, their collective experience points the way to constructing a *feminist* set of principles for coming to terms with the environmental tasks ahead of us.

The experience of women on the environmental front-lines should help us change our notion of what environmental destruction looks like: It's usually not big, flashy, or of global proportions—or, if a global problem, it manifests itself locally. Environmental degradation is pretty mundane; it occurs in small measures, drop by drop, tree by tree. This fact is discomfiting to big scientific and environmental organizations whose prestige depends on solving "big" problems in heroic ways.

Because environmental destruction shows up in small ways in ordinary lives, we need to change our perception of who are reliable environmental narrators. Women, worldwide, are often the first to notice environmental degradation. Women are the first to notice when the water they cook with and bathe the children in smells peculiar; they are the first to know when the supply of water starts to dry up. Women are the first to know when the children come home with stories of mysterious barrels dumped in the local creek; they are the first to know when children develop mysterious ailments.

When the environment starts to suffer, signs of its degradation show up first in the water, food, and fuel women have to work with on a daily basis. Very early on, environmental decay starts to subtly impinge on the daily lives of women. In whatever context women are living, environmental degradation makes women's daily and ordinary lives more difficult. For Mrs. Woodman in Jacksonville Florida, the earliest sign of "something going wrong" was the fact that she couldn't get the clothes in her laundry to wash out white, no matter how hard she tried.[40] For Sithembiso Nyoni of Zimbabwe, environmental deterioration shows up first in her daily workload: "As a woman, it means that I have to walk long miles to fetch firewood and water. I have very little time, therefore, to grow vegetables and other food."[41]

Because women in all countries populate the most vulnerable segments of society, a disproportionate share of the impact of environmental problems falls on women. Women are confronted daily with having to care for children and go about their household chores without access to safe water, food, and sanitary facilities. It is often the case that community women are first spurred into environmental action when they perceive a threat to their children's health, and/or their own reproductive health. Reproductive disorders are an early-warning indicator of environmental deterioration. The presence of poisoned water, air and food often becomes evident first in unusually high numbers of miscarriages, still-births, low birth weights, birth defects, and occurrences of unusual diseases among infants and young children. Low birth rates, for example, are commonly found in the vicinity of hazardous waste sites; birth defects frequently occur when pregnant women are exposed to high levels of dioxins and other chemicals; childhood leukemia crops up in unusual clusters around nuclear power plants. These patterns of reproductive disorders and diseases are usually first detected not by a trained epidemiologist or local doctor, but by neighborhood women as they gather over communal coffee or child care and compare notes about sickness and health in their daily lives.

When women start to act on their environmental concerns, their actions often revolve around issues of the quality of daily life and the environmental safety of home and neighborhood. Knowledge of the wider links between "the local and the global" should not undercut a certainty, forged by women from their daily experience, that the appropriate forum for environmental action is often the local community itself.

But, the other side of the grassroots coin is that a home-based, local-defense environmentalism can also be mired in parochialism. The environmental report card on the American NIMBY movement, for example, gives a mixed review. While it is true that much environmental reform has come from people who face pollution in their own backyards and *then* go on to care about everyone else's, this ripple effect is not always present. Instead, the NIMBY movement often sets off an elaborate geographical shell game, wherein environmentally hazardous activities are shifted from one region to another (or one country to another). NIMBY movements do not necessarily present a radical challenge to the underlying structures of corporate and industrial enterprise—they challenge the right of businesses to locate dangerous activities in a *particular* neighborhood without necessarily challenging the presumed right of business to engage in environmentally unsound practices in the first place. Because of this, NIMBY movements thus often have the effect (perhaps unintended, but predictable) of shifting the burden of environmental hazards from the community that organizes the fastest—often predominantly white communities with an educated, middle-income population—to poorer, already-disadvantaged communities.[42] A number of recent studies of hazardous waste dumps in the US establish clearly that, among other things, the majority of the nation's largest hazardous waste dumps are located near African American or Hispanic communities; communities that host one toxic waste landfill or incinerator have, on average, a population that is 24 percent minority, while towns that have two or more dumps have an average minority population of 38 percent.[43]

In response to a NIMBY challenge, rather than change their environmentally unsound practices, corporations typically try to second-guess where hot spots of local activism are likely to flare up. In order to facilitate this sort of predictive planning, a large American waste disposal firm in the mid-1980s contracted with a research team to compile a profile of communities that were most and least likely to be resistant to the local siting of hazardous activities.[44] It is a tribute to the success of the NIMBY movement that "housewives" turned up in the "most resistant" occupational category! Not surprisingly, the communities described as "least resistant" to environmentally hazardous projects were small, rural communities in the South or Midwest with conservative politics and a middle-aged or elderly, low-income population—the sort of place like Emelle, Alabama.

The class and race subtext of the environmental agenda is written clearly on the landscape of Emelle, home of the largest hazardous dump in the US, probably in the world. Waste from 46 of the 50 states is transported (not all of it legally) to Emelle for "disposal." Emelle also receives wastes from other countries through shipments from military bases,[45] and current plans call for expanding the dump to include a hazardous waste incinerator. Sixty-nine percent of the families in the county live below the official poverty line; seven out of ten residents are African American; illiteracy and unemployment rates are high. And despite state assurances that no acutely toxic chemicals would be dumped at the site, residents have recently become aware of the dumping of benzene, PCBs, and possibly dioxins—substances with effects ranging from skin disorders to birth defects—and recent testing established that deadly chemicals have contaminated the groundwater around Emelle.[46]

The fact that the toxic trail often starts when relatively well-off communities prevent hazardous wastes and activities from locating in their "backyards" fuels accusations that the environmental community, as it is now constituted, is not taking seriously issues of racism in the environmental struggle. This tension then in turn undermines the effectiveness of environmental activists in making alliances with people of color in poorer communities—so their voices are missing from the environmental chorus, and the cycle of exclusion and privilege perpetuates itself.

Because of the nature of their daily lives, women often provide early warning on pollution problems and environmental degradation. They are also thus well placed to serve as agents for change—and because of their subordinate status they have the most to gain from movements for social, political, and environmental change. Successful environmental projects start with a recognition that women and men may have different priorities, agendas, and needs, and that men's needs must not be privileged over women's. One of the implications of this, which has yet to be fully acknowledged by the powers that be, is that women are crucial to environmental action plans—whether those plans involve changing household consumer patterns in London or developing water-supply projects in Zimbabwe. If women are not counted as serious environmental players, if their needs and their experiences are excluded from environmental decision-making processes, then plans for environmental action will fail—inevitably. The creation of acceptable living conditions for the world's people will depend on reversing environmental degradation on a large scale by introducing appropriate technology and appropriate use of resources on a small scale. Since the "appropriateness" of "appropriate" technology can only be measured in terms of quality-of-life impact, the inclusion of women's expertise and women's needs will be central to the success of changing technological paradigms.

Once roused into action, women report over and over again the extent to which they are ridiculed as know-nothing housewives, as hysterical and naive, as nosy busy-bodies.[47] In some cases, the fact that women activists are not taken seriously serves an initial advantage. Beth Gallegos, a Colorado woman who challenged military pollution in Denver, reported that she was given access to information that otherwise would have been restricted, "because they thought I was a nobody."[48] But the camouflage of being a "nobody" soon wears off, and then the challenge to women's authority starts in earnest.[49]

Women environmental activists often face the most resistance to their new role from their families, and especially from the men in their families. In Lois Gibbs's words:

> Women are still seen as child-bearers, caretakers, homemakers, while the men are seen as the protectors, providers and heads of the household, as well as anything else they decide they, as men, want to be. In many families, the woman who becomes active is seen as a threat to the "strong" male. He feels that he is losing control over "his woman" and might feel that he is being out-done or "out-shined" by his mate, a problem that is exacerbated if she is successful. . . . Pressure begins to build on her as she tries to balance her commitment to the cause with the conflicting demands that come out of her [male] mate's emotional needs.[50]

In many cases, Gibbs continues, the male spouses of women leaders become the target of ridicule from their male co-workers. This triangle of tension between the woman activist, her male partner, and *his* male buddies is heightened if environmental issues are perceived as a threat to jobs—i.e., to men's jobs. Women activists are often accused of undermining the community's economic base, of destroying men's livelihoods.

Because women get caught in the crossfire of the "jobs vs. environment" debate, feminists need to develop more clear-headed analyses of the presumed "trade-off" between jobs and a clean environment. It is *not* true that doing business in an environmentally sound manner costs jobs. What *is* true is that the pursuit of environmentally unsafe practices endangers the health of a community, and often costs lives. If the economy and the environment are on a collision course, it is not because of environmental "extremism"; it is more the result of what might be called economic extremism—the pursuit of short-term gains at any environmental cost.[51] Worldwide, it is clear that the primary environmental issue is not "jobs vs. the environment," but "poverty vs. the environment." Environmental issues need to be framed in terms of the everyday, local, health and safety trade-offs that they really represent, and cannot be allowed to be misrepresented as far-off, far away, maybe-someone-somewhere-else-in-the-world-will-suffer issues. People are dying and people are being disabled every day in communities around the world because of environmentally inappropriate, unsound, and dangerous practices. It is precisely this sense of immediacy that fires up women community activists, and that men shy away from.

Many men, when talking about the environment, adopt a tone of measured reasonableness; they don't like to admit panic, urgency, despair, or emotionalism about the state of the earth. In the eyes of many men, women who are anxious about the health of their families, women who are emotional about environmental destruction, or women who raise questions with a sense of urgency are inappropriately intruding in an environmental discourse that should, properly, be the province of men of reason and men of science—the experts. Contemporary environmental issues are often scripted as the drama of "the hysterical housewife meets the man of reason."

The use of science and scientific expertise in environmental impact assessment is becoming particularly contentious, and it may be the single most important factor in widening the schism between women grassroots activists and men in the environmental mainstream. As grassroots environmental watchdogs, women see it as an advantage that they have been socialized to listen to "their gut feelings"; men are socialized to veer away from intuition, and to formulate judgements only on the basis of irrefutable "facts." On an academic level, there is a growing gap between feminist and masculinist presumptions about the appropriate applications of scientific inquiry to environmental issues. Feminists have formulated sophisticated critiques of science, specifically focusing on the extent to which science *is* a values-based activity, arguing that there is no "neutral," objective science. Many scientists, especially male scientists, resist this premise; many male environmentalists are reluctant to cast aside the prop of presumed scientific neutrality and expertise.

It is very difficult to pinpoint "scientifically" or "absolutely" the particular cause of an illness or death. To "neutral" observers, just because a family's water supply is inadvertently poisoned by leachate from a chemical dump doesn't mean that the child they lost to cancer was poisoned by that water; even if it is proven that toxins did enter the water supply, it can seldom be proven that *those* substances *directly* caused the cancer or illness of *that* child. Corporations rely on this scientific uncertainty to save them from culpability. Many families and many community groups have fought local environmental battles through the media and the courts for years, only to be denied justice in the end because "the experts" can't "prove" causality.

Not surprisingly, many women activists have come to mistrust the role of science—and of scientific experts—in arbitrating environmental disputes. Margherita Howe, the Canadian activist, speaks for many women activists when she says "I didn't know anything about chemical waste then, but I sensed that it wasn't right. It was a gut feeling." As Howe goes on to point out, in settling environmental disputes there are seldom irrefutable facts:

> I gradually became aware that you don't have to be an expert. There are very few experts for one thing. If you have 12 scientists looking at a report, for instance, you'll probably have 12 different interpretations; and depending on who the scientists or experts worked for, that determines the policy.[52]

Howe actually *is* expressing an expertise—but it is an expertise grounded in daily experience, not externally ratified knowledge, a "women's" expertise usually scorned and ignored.

Which brings us back to thoughts about "women's voice." Representing the environment as somehow especially a "woman's issue" is troubling for feminists who fear that in a world where women are the second sex, this association will only be used to trivialize them both. To some extent, the self-proclamations of mothers and mothering in women's grassroots organizing engenders this conundrum. It is clear that maternalism brings

SYLVIA   **by Nicole Hollander**

Figure 6.1.   Men and women often express different views on the appropriate role of science in setting the environmental agenda. (© Nicole Hollander, by permission.)

many women into environmental politics (although almost no one assumes that mothers are intrinsically more environmentally responsible than other people).[53] For reasons both banal and deep, it "matters" what mothers say and do, and women can often bring attention to their cause if they speak as mothers. But a maternalism-based activism that is not informed with a broader feminist analysis can paint women into a corner—or, rather, keep women in the corner that society has cordoned off for them. It allows women to sneak onto the wings of the political stage without broadening the role for women in the script of the political play as a whole. It reinforces the notion that women's most useful and natural role is "bearing and caring," and that women's public activities are primarily appropriate only insofar as they remain rooted in this maternalism. A movement based on a sentimentalized motherhood is obviously a potent political force, but it is a tenuous basis on which to construct a new environmental politics. The "moral mother" argument is a poor organizing tool: it does not challenge us to think in complex ways about the nature of environmental threat, nor about women's—or men's—consciousness and social activity.[54]

The sexual division of emotional and intellectual labor on the environment—women care about it, men think about it—may be somewhat descriptive of present realities; without guarding against it, this could also become *pre*scriptive. If women are consigned to "care" about the environment, this lets men off the hook (why do men have to care about the environment if women are doing so?). If we are looking to build a new environmental agenda based on the lessons from grassroots activism, we must be willing to critically examine the assumptions about gender roles *before* they get inscribed in social movements. If a grassroots-informed environmentalism tells us anything, it is that both men and women must *care*—and think—about the environment.

One absolutely crucial lesson to take from grassroots organizing is that environmentalists need to be aware of gender inequalities in all guises in which they appear. For example, environmental organizing in India, and especially the Chipko movement, provides a number of lessons that are more widely applicable.[55] Independent groups in India report that the most reliable grassroots organizers are middle-aged mothers: they have a good rapport with villagers, especially other women, and they are likely to stay put, while younger people often migrate to cities. There cannot be social or gender discordance between the community and the organizer for the community. Another lesson on gender relations from Chipko organizing is that social stratification can make all-inclusive working groups counterproductive: lumping male farmers with landless women, for example, virtually guarantees that the men will reap the bulk of the rewards. Many government-led community-development programs, in developing and industrial countries alike, ignore the realities of social inequities, and treat all residents as essentially equal in interests and status, thus allowing the more powerful (usually men) to co-opt projects for their own benefit.

The participation of women in social-justice movements (of which the environmental movement is one) must be allowed to develop free from male attitudes about women's appropriate place. Specifically, male violence against women is often used to "keep women in their place." It is often the *form* of women's activism—women acting as independent agents outside the home—not necessarily the *content* of women's activism which sparks male resentment and violence in the first place. If there is a single lesson to take from the world of grassroots organizing, it is this: women are absolutely key players in the environmental arena, but for women to act as agents of environmental change, they must be freed from narrow male assumptions about appropriate gender behavior, and they must be free to act without the threat of male violence. Men need to change what is considered to be the "normal" pattern of male exercise of power. The struggle to forge an environmentally just and sound future is inextricable from the struggle for gender justice and equality. Environmental relations are embedded in the larger gender relations that shape modern life.

Conclusion

The grassroots environmental movement expands our sense not only of what is possible, but of what is necessary. It is a movement that is fueled by persistence, resistance, stubbornness, passion, and outrage. Around the world, it is the story of "hysterical housewives" taking on "men of reason"—in the multitude of guises in which they each appear.

In preparation for the Global Assembly of Women for a Healthy Planet held in Miami in 1991, the organizers compiled a dossier of more than 200 "success stories" of women's community-based environmental projects, ranging from forest protection schemes in Tanzania to waste education programs in California. The published report, which provides a one-page summary of each project, is inspirational.[1] It maps out the realities, in rich countries and poor, of women struggling to stop the pollution of their water supplies, to stop the sale of foods contaminated by the nuclear accident at Chernobyl, to grow food on depleted soils, to clean up garbage dumps, to recycle wastes, to harness solar energy for their cooking, to plant trees against the encroaching desert—and in doing so, to challenge the governments, corporations, and militaries which act, so often and so clearly, against the interests of ordinary people.

But, inspirational as they are, these projects alone will not solve the problems of massive environmental degradation. Peggy Antrobus, at that same Miami meeting, laid out the broader agenda that the environmental crisis demands:

> The primary task for us as women is to formulate analyses which will help us to identify the root causes of our environmental problems.

> We must clarify the links between environmental degradation and
> the structures of social, economic and political power . . . the links
> between decisions made in boardrooms, parliaments and military
> command centers and the conditions under which we live . . . the
> links between the structure of our own subordination as women and
> the processes by which this subordination serves to perpetuate all
> other systems of oppression.

Antrobus exhorts us to ask what I call "questions of agency": questions about who is doing what to the environment and why.

In the 1990s, we are bombarded with environmental information. But mostly it is information that masks agency—and thus, it is hollow information, information without knowledge. Environmentalists are caught in the bind that many problem-solving groups and individuals find themselves in—faced with urgent, often life-threatening, assaults on human health and natural habitats, there is often not the time or energy to ask big questions about root causes. And yet, if we don't ask those questions, we commit ourselves to a treadmill of one crisis intervention after another, with fewer and fewer respites in between. Crisis response and community resistance are essential; they must be guided, though, by knowledge about the links between environmental degradation and the structures of social, economic, and political power. Those structures are not gender-neutral.

In mixed-sex, action-oriented environmental groups, it is hard to ask questions about the relationships between men and women and our environmental crisis. It is not easy to ask about militarism and masculinity, sovereignty and sex, and it's not considered polite to point out that men have been far more implicated in the history of environmental destruction than women. But there's too much at stake to stick to the easy questions and polite conversation.

◑

For most of this century, problems in the environment have been assumed to be problems for the scientists—we turn to scientists to tell us when things have really gone wrong, and we are made complacent by the unrealistic expectation (one which many scientists themselves *dis*courage) that scientists will find the magic bullet that will Make Things Better. We are lulled by the hope that there is a technological fix that will dissipate greenhouse gases, close the ozone hole, purify our water and air, perhaps even regenerate species driven into extinction, or create space colonies to which we can escape when this planet is no longer habitable.

But the causes and solutions to our environmental crisis are more complex than "techno-boomerism" would suggest.[2] In the early 1990s, a small sea-change started to ripple through American environmental and academic circles: a number of powerful funding agencies (which have a de facto influence in setting the direction of environmental research) and an increasingly vocal group of academics started to pay more attention to the proposition that non-scientists—especially humanists, people who usually think about literature, history, art, and culture—might have something to contribute to environmental understanding. Leo Marx, a well-known American cultural historian, signaled the turn: "Although the work of scientists and engineers is indispensable for coping with the most urgent environmental problems, the problems themselves are invariably the result of social practices. They are quintessentially *social* problems whose roots lie deep in a long-standing matrix of cultural proclivities. Hence, they are bound to remain intractable until we find ways to integrate 1) scientific analyses of their nature; 2) an adequate understanding of their social, cultural, or behavioral genesis; 3) a plan to change the behavior, or institutional structures necessary to resolve them."[3]

In point of fact, this is precisely the sort of analytical work that feminists have been doing for years, although not in the environmental arena, and the new interest in humanities-informed environmentalism holds great promise. Now that it has the stamp of approval from think tanks, funders, and universities, its influence in environmental circles will expand during the 1990s. Unfortunately, it is not at all clear that the "culture clubbers" are any more open to thinking about gender relations than their "techno-boomer" brothers. And if we go down this road without a feminist road map, we will once again end up at a dead end. It *is* possible to pursue an analysis of the cultural basis of our ecological crisis *without* a feminist sensibility—but it will not be insightful, truthful, useful, or, even, very interesting.

◑

Feminist analysis starts with the insight that "the environment" is not somehow a separate realm that exists beyond the ordinariness of everyday life; to start to figure out what's wrong with the environment, one doesn't have to be, as they say, a rocket scientist (indeed, it especially helps if one is *not* a rocket scientist). Rather, the relationships of power and institutional control that shape our environmental affairs are extensions of the ordinary and everyday relationships between men and women, and between them and institutions—and this is something about which feminists have a lot to say, and about which most women have a certain "lived expertise."

Feminist analysis is not the *only* tool necessary for making sense of our environmental crisis. Indeed, it might be thought of as one among many spotlights of concern and inquiry illuminating the environmental arena.[4] But by focusing on questions about gender and the environment, feminist analysis highlights aspects of the problem that other analyses do not even begin to shed light on. Given the severity and complexity of our environmental problems, we can't afford to sit in the dark.

Notes

Introduction

1. Pam Simmons, "The challenge of feminism," editorial introduction to the special issue of *The Ecologist*, "Feminism, Nature, Development," January/February 1992.

2. Quotes taken from taped transcripts of the Global Assembly of Women for a Healthy Planet conference, Miami, 1991; see also the report on the conference in Elayne Clift, "Reclaiming the sanity agenda," *On the Issues*, Spring 1992.

3. This sentence paraphrases a comment by Lorene Cary in "Why it's not just paranoia," *Newsweek*, April 6, 1992.

4. J. Ann Tickner, *Gender in International Relations: A Feminist Perspective on Achieving Global Security*. NY: Columbia University Press, 1992, p. xi.

5. R. W. Connell. *Gender & Power*. Stanford: Stanford University Press, 1987; this point is also made especially well in Ann Tickner's book.

6. A point made by Ann Tickner. *Gender in International Relations*, p. xii.

7. "Patriarchy," here means the systematic and systemic dominance of men in society, and the structures that support and further this dominance.

8. Simone de Beauvoir. *The Second Sex*. NY: Random House, 1968.; Sherry Ortner, "Is Female to Male as Nature is to Culture?" in Michelle Zimblast Rosaldo & Louise Lamphere eds. *Woman, Culture & Society*. Stanford: Stanford University Press, 1974.

9. See, for example, Mary Daly. *Gyn/Ecology: The Metaethics of Radical Feminism*. New York: Harper & Row, 1980; Susan Griffin. *Woman and Nature: The Roaring Inside Her*. New York: Harper & Row, 1980.

10. The growing literature by women who identify themselves as "ecofeminists," and who self-consciously identify themselves as shaping the course of ecofeminism, includes

284

the following: Judith Plant, ed. *Healing the Wounds: The Promise of Ecofeminism*. Philadelphia: New Society Publishers, 1989; Andree Collard with Joyce Contrucci. *Rape of the Wild*. London: The Women's Press, 1988; Bloomington: Indiana University Press, 1988; Irene Diamond & Gloria Orenstein. *Reweaving the World: The Emergence of Ecofeminism*. San Francisco: Sierra Club Books, 1990. The ecofeminist articles published in journals and newsletters over the past three or four years are too numerous to be cited here, but references to many of them are scattered throughout this book. The term "ecofeminism" is not, of course, the exclusive property of the women mentioned above, and there are no doubt women who call themselves "ecofeminists" who would take issue with some of the above-mentioned authors. Nonetheless, it is important to note that the term increasingly *is* identified with a spiritually-based and goddess-centered philosophy, and its meaning is increasingly specific to that philosophy.

11. This emergent strand of feminism relies on the academic work of historians such as Joan Kelly and Carol Smith Rosenberg who identify the separate culture of women in other historical periods, and the groundbreaking work by Carol Gilligan on the construction of women's and men's moral and behavioral development. Ynestra King is a key figure in building bridges between a "transformative feminism" and environmental issues. See her recent collection of articles in *What Is Ecofeminism?* New York: Ecofeminist Resources, 1990.

12. Ynestra King in Judith Plant, ed., *Healing the Wounds: The Promise of Ecofeminism*, p. 23. King suggests that this is the direction promised by "ecofeminism," but semantics are confusing here. The direction of what is now widely called "ecofeminism" is clearly away from this middle ground, and more emphatically towards "essentialist" feminist analysis and a philosophy of goddess-worshipping self-realization.

13. Audre Lorde, "The Master's Tools Will Never Dismantle the Master's House," in Cherrie Moraga and Gloria Anzaldua, eds., *This Bridge Called My Back: Writings By Radical Women of Color*. NY: Kitchen Table Press, 1981.

1 Up in Arms against the Environment

1. Donald Snow, ed. *Inside the Environmental Movement*. Washington, DC: Island Press, 1992, p. 53.

2. Excellent overviews of the environmental costs of militarism include: Arthur Westing, ed. *Herbicides in War: The Long-term Ecological and Human Consequences*. London: Taylor & Francis, 1984; Arthur Westing, ed. *Cultural Norms, War and the Environment*. NY: Oxford University Press, 1988; Stockholm International Peace Research Institute (SIPRI).*Warfare in a Fragile World: Military Impact on the Human Environment*. London: Taylor & Francis, 1980; Anne Ehrlich and John Birks. *Hidden Dangers: Environmental Consequences of Preparing for War*. San Francisco, CA: Sierra Club Books, 1992.

3. The ecological disaster in Vietnam has been extensively studied. For readers interested in pursuing this further, I suggest: Patricia Norland, "Vietnam's Ecology: Averting Disaster," *IndoChina Issues 66*, June 1986; Phan Nguyen Hong, "Vietnam's Mangroves—Going, Going. . ." *Asian-Pacific Environment*, Vol. 5, 3, 1988. Stockholm International Peace Research Institute. *Ecological Consequences of the Second IndoChina War*. Stockholm: Almquist & Wiksell, 1976; Arthur Westing, ed. *Herbicides in War*; Environmental Project on Central America. *Militarization: The Environmental Impact*. Green Paper 3, 1988.

4. The classic study of the gendered politics of food and famine is Lisa Leghorn & Mary Roodkowsky. *Who Really Starves? Women and World Hunger*. NY: Friendship Press, 1977.

5. This story comes from personal discussions with Peggy Perri, an American nurse stationed in Vietnam during the war, who is now involved in setting up exchanges with and American assistance to Vietnamese women. See also, "The Effects of Dioxin Poisoning on Maternal and Infant Health in Vietnam," paper presentation by Peggy Perri, "Women, War and the Environment Conference," University of Massachusetts, Boston, March 1990, and, Nick Malloni, "Agent of destruction," *Far Eastern Economic Review*, December 7, 1989.

6. Quoted in Nick Malloni, "Agent of Destruction."

7. For an extended analysis of the militarized construction of masculinity, see: Cynthia Enloe. *Does Khaki Become You?* London and New York: Pandora Press/Harper Collins, 1988 and "Bananas, Bases & Patriarchy," *Radical America*, Vol. 19 4, 1985; and Eva Isaksson,ed.*Women and the Military System*. NY: Harvester, 1988.

8. I am aware of only one study that attempts to trace direct links between the feminization of nature and the assault on nature by militaries: Mehmed Ali, Kathy Ferguson and Phyllis Turnbull, "Gender, Land and Power: Reading the military in Hawaii," paper presented at the XVth World Congress of the International Political Science Association, Buenos Aires, 1991.

9. Carol Cohn, "A Feminist Spy in the House of Death," in Eva Isaksson, ed., *Women and the Military System*.

10. Helen Caldicott. *Towards a Compassionate Society*. Westfield, NJ: Open Media, 1991.

11. Ibid, p. 4.

12. For fuller discussions of the environmental costs of the Gulf War, see: Joni Seager, "Operation Desert Disaster: Environmental Costs of the War," in Cynthia Peters, ed. *Collateral Damage: The 'New World Order' At Home and Abroad*. Boston: South End Press, 1992 and in Haim Bresheeth & Nira Yuval-Davis, *The Gulf War and the New World Order*, London: Zed Books, 1991; Ramsey Clark, ed. *War Crimes: A Report on US War Crimes Against Iraq*. Washington DC: Maisonneuve Press, 1992; "The Impact of War on Iraq," A Report to the Secretary General of the United Nations on Humanitarian Needs in Iraq in the Immediate Post-Crisis Environment, March 20, 1991 (published as a pamphlet by Open Magazine Pamphlet Series, Westfield, NJ, 1991).

13. See, for example, several of the essays in Arthur Westing, ed. *Environmental Hazards of War: Releasing Dangerous Forces in an Industrialized World*. London: Sage Publications, 1990.

14. John Horgan, "Why are data from Kuwait being withheld?" *Scientific American*, July 1991, p. 20.

15. For further information on the Central American military/environment connection, I recommend the publications of the California-based Environmental Project on Central America (EPOCA), especially "Nicaragua: An Environmental Perspective" (Green Paper 1, 1987), "Central America: Roots of Environmental Destruction" (Green Paper 2, 1988), and "Militarization: The Environmental Impact" (Green Paper 3, 1988). Also: Mike Friedman, "Nicaragua: War, Injustice fuel nation's environmental woes," *The Guardian (US*, June 15, 1988; Joshua Karliner & Daniel Faber, "The other revolution: Nicaragua's environmental crisis," *Utne Reader*, Jan/Feb. 1988; and William Weinberg. *War on the Land: Ecology and Politics in Central America*. London: Zed Books, 1991.

16. Alan Riding, "The Struggle for Land in Latin America," *NY Times*, March 24/89; Environmental Project on Central America, *Update*, Spring 1990; Daniel Faber, "The intimate

link: environment and justice in Central America," *Central America Reporter*, May/June 1992.

17. Environmental Project on Central America, *Update*, Spring 1990.

18. Bianca Jagger, "Save the rain forest in Nicaragua," *New York Times*, November 12, 1991.

19. Sonia DeMarta, "War in El Salvador took its toll," *Central America Reporter*, May/June 1992.

20. "Panama: Ecological Consequences of the Invasion," Environmental Project on Central America, *Update*, Spring 1990.

21. Todd Steiner, "Southern African counterrevolutions awash in elephant blood," *In These Times*, November 23, 1988; Craig Van Note, "Reagan's Freedom Fighters in South Africa are Killing Elephants to Buy Arms," *Earth Island Journal*, Fall 1988; Patrice Greanville, "Ivory for Arms," *The Animals' Agenda*, January 1989.

22. Quoted in Patrice Greanville, "Ivory for Arms."

23. Quoted in Alan Durning, "Apartheid's Environmental Toll," *Worldwatch Paper 95*, May 1990, p. 32.

24. Todd Steiner, "Southern African counterrevolutions."

25. Alan Durning, "Apartheid's Environmental Toll."

26. Quoted in "The Pentagon's War on Drugs: The ultimate bad trip," *The Defense Monitor*, Vol. XXI, Number 1, 1992, p. 3.

27. Tracy Cohen,"US Defoliates Guatemalan Rain Forest," *Science for the People*, November/December, 1987; Daniele Rossdeutscher, "US sprays defoliants on Guatemalan forests," *EPOCA Update*, Summer 1986, p. 6.

28. "US puts Guatemala on its drug blacklist," *San Diego Union*, April 1, 1990.

29. Ibid.

30. Quoted in "Guatemala: A political Ecology," Environmental Project on Central America (EPOCA), Green Paper 5, 1990, p. 13.

31. "US, 3 nations agree to use military in drug war," *Boston Herald*, January 15, 1990.

32. Susanna Rance, "Growing the stuff," *New Internationalist*, October 1991, p. 11; Alfonso Serrano. "Fighting cocaine in Peru," *Nuclear Times*, Winter 1991–92, p. 45.

33. Ruth Conniff, "Colombia's dirty war, Washington's dirty hands," *The Progressive*, May 1992; "An appeal from the Guatemalan human rights commission," *Earth Island Journal*, Summer 1988, p. 32.

34. Saul Landau, "The enemy is us," *The Progressive*, May 1992.

35. Quoted in James Brooke, "Fighting the drug war in the skies over Peru," *New York Times*, March 28, 1992.

36. "Chemical Warfare Plans Off-Target," *Greenpeace Magazine*, October 1988; Pamela Constable, "In Peru's coca valley, herbicide is reviled," *Boston Globe*, June 26, 1988.

37. "SPIKE: Uncle Sam's deadly remedy," *New Internationalist*, July 1989, p. 15.

38. "Eli Lilly just says no," *In These Times*, June 8–21, 1988, p. 4; "Chemical warfare plans off target," *Greenpeace Magazine*, September/October 1988, p. 4.

39. Bertil Lintner, "The deadly deluge: Opium eradication program linked to deaths," *Far Eastern Economic Review*, November 12, 1987.

40. An excellent summary of domestic military environmental problems is Seth Shulman's

book, *The Threat at Home: Confronting the Toxic Legacy of the US Military.* Boston: Beacon Press, 1992.

41. Transcript of "Poison and the Pentagon: A Frontline Report." Boston: WGBH Educational Foundation, 1988, p.2.

42. Lois Gibbs, "Women and Burnout," Citizen's Clearinghouse for Hazardous Wastes, 1988.

43. Transcript of "Poison and the Pentagon: A Frontline Report." Boston: WGBH Educational Foundation, 1988, p.4.

44. "Report criticizes cleanup at atomic arms plants," *New York Times*, February 11, 1991, p. A20.

45. *The Defense Monitor,* Vol. XVIII, No. 6, 1989.

46. Ted Olson, "Appetite for land growing," *Army Times*, April 9, 1990.

47. Michael Renner, "Assessing the military's war on the environment," in Worldwatch Institute, *The State of the World 1991*. NY: Norton, 1991.

48. Kathy Ferguson, Phyllis Turnbull & Mehmed Ali. *Rethinking the Military in Hawaii.* Honolulu: University of Hawaii, 1991, p. 8.

49. Michael Renner, "Assessing the military's war on the environment," in Worldwatch Institute, *The State of the World 1991*. NY: Norton, 1991.

50. "Military animal abuse," *The Animals' Agenda*, November 1990, p. 38.

51. Cited in Stephanie Pollack and Seth Shulman, "Pollution and the Pentagon," *Science for the People*, May/June 1987.

52. William Vaughan, Assistant Secretary of Energy during President Reagan's first term in office, quoted in *The Defense Monitor,* Vol. XVIII, No. 4, 1989.

53. Quoted in Will Collette, "Dealing with Military Toxics," Citizen's Clearinghouse for Hazardous Waste, 1987.

54. Barbara Ann Scott, "Help wanted: Women defense experts and decision-makers," *Minerva: Quarterly Report on Women and the Military*, Vol. VI, 4, Winter 1988.

55. For example, see Hedrick Smith's description of the workings of US government policy, Chapter 6 in *The Power Game: How Washington Works*. NY: Random House, 1988.

56. Worldwide, women represent on average about 2 to 4 percent of government officials. In Scandinavian countries, where women represent approximately 25 percent of elected officials, there does seem to be a different concensus emerging about the nature of "national security," and one must wonder whether there is a relationship between these two facts.

57. Hedrick Smith. *The Power Game,* pp. 71–72, 78.

58. Ibid., p. 185.

59. Cynthia Enloe. *Bananas, Beaches & Bases: Making Feminist Sense of International Politics.* London: Pandora Press, 1989 and Berkeley: University of California Press, 1989, pp. 11–12.

60. For a fuller discussion, see Hedrick Smith. *The Power Game,* esp. chapter 8; Gordon Adams. *The Iron Triangle.* NY: Council on Economic Priorities, 1981; Helen Caldicott. *Missile Envy.* NY: Bantam, 1985, esp. the chapter on the "Iron Triangle."

61. Cited in Rosalie Bertell, *No Immediate Danger,* London: The Women's Press, 1985, p. 201.

62. The US Navy has since volunteered to try to meet most of the provisions of this plastics pollution agreement, but they are not bound by compliance.

63. National Toxics Campaign Fund, *The US Military's Toxic Legacy* Boston: National Toxic Campaign Fund, 1991.

64. There are several good exposes of deliberate military and governmental malfeasance in obscuring and altering the environmental records of militaries. See, for example, Rosalie Bertell. *No Immediate Danger*, London: Women's Press, 1985; Nigel Hawkes, et. al *The Worst Accident in the World*. London: Pan Books, 1986; Daniel Ford. *MeltDown: The Secret Papers of the Atomic Energy Commission*. NY: Touchstone, 1982; and, Paul Loeb. *Nuclear Culture: Living and Working in the World's Largest Atomic Complex* Philadelphia: New Society Publishers, 1986.

65. Ruth Sivard. *World Military and Social Expenditures, 1987- 88*. Washington DC: World Priorities Institute, 1987.

66. Comparisons from: Ruth Sivard, annual updates; Larry Tye, "Pulling resources from the military," *Boston Globe*, April 9/89, p. A1; World Commission on Environment and Development. *Our Common Future*. New York: Oxford University Press, 1987; Michael Renner, "Who Are the Enemies?" *Peace Review*, Spring 1989; *The New Economy*, (National Commission for Economic Conversion & Disarmament, Washington DC) Vol. 1, No. 1, August/September 1989; and, *Plowshare*, The Center for Economic Conversion, Vol. 14, 2, Spring 1989. An excellent compilation of information on the effect of military spending on women and women's programs in the US is put out by the Women's International League for Peace and Freedom: *The Women's Budget*, Third Edition, 1988.

67. Figures from Ruth Leger Sivard. *World Military and Social Expenditures, 1991*. Washington DC: World Priorities Institute, 1991; K. S. Jayaraman, "Poor and buying weapons," *Panoscope*, No. 14, September 1989; Jim Hollingworth, "Global militarism and the environment," *Probe Post (Toronto)*, Fall 1990, p. 42; "The cost of the Persian Gulf war," Massachusetts SANE/FREEZE fact sheet, 1991.

68. On masculinity and war, see: Diana Russell, ed. *Exposing Nuclear Phallacies*. NY: Pergamon Press, 1989; Leonie Caldecott & Stephanie Leland, eds. *Reclaim the Earth: Women Speak Out for Life on Earth*. London: The Women's Press, 1983; Helen Caldicott. *Missile Envy*; Cynthia Enloe. *Does Khaki Become You?* ; Birgit Brock-Utne. *Educating for Peace: A Feminist Perspective*. Oxford: Pergamon Press, 1985; Judith Stiehm, ed. *Women and Men's Wars*. Oxford: Pergamon Press, 1983. (Originally published as a special issue of *Women's Studies International Forum*, Volume 5, 3/4, 1982); Virginia Held, "Gender as an influence on cultural norms relating to war and the environment," in Arthur Westing, ed. *Cultural Norms, War and the Environment*. Caldicott, in her 1985 book, puts forward a biologically-based argument about men as aggressors and women as nurturers; some of the most persuasive 'social structure' arguments can be found in the Diana Russell anthology.

69. Penny Strange, "It'll make a man out of you: A feminist view of the arms race," in Diana Russell, ed. *Exposing Nuclear Phallacies*.

70. "More Pollution from the Pentagon," *Science for the People*, September/October, 1988.

71. As described in David Beach, "Ohioans Are Learning to Hate the Bomb," *In These Times*, Feb. 1–7, 1989, p. 8.

72. Much of this information is from a personal conversation with Lisa Crawford, March 31, 1989.

73. David Beach, "Ohioans Are Learning to Hate the Bomb."

74. Quoted in Stephen Hedges, " Bomb makers' secrets," *US News & World Report*, October 23, 1989.

75. Information from Stephen Hedges, " Bomb makers' secrets," *US News & World Report*, October 23, 1989; personal communications with Lisa Crawford, FRESH; and, presentation by Lisa Crawford at the "Women, War & the Environment" conference, University of Massachusetts, Boston, March 1990.

76. This information comes from: Press reports; Radioactive Waste Campaign. *Deadly Defense: Military Radioactive Landfills*, NY: Radioactive Waste Campaign, 1988; Anne Ehrlich & John Birks. *Hidden Dangers*.

77. Campbell Reid, "13000 nuclear guinea pigs," *The Sunday Sun (Australia)*, July 15, 1990; "GE's Radioactive & Toxic contamination at Hanford," INFACT fact sheet, Boston, 1991.

78. Quoted in Karen Dorn Steele, "Hanford's Bitter Legacy," *Bulletin of the Atomic Scientists*, Jan/Feb 1988.

79. Karen Steele, "Tracking down Hanford's victims," *Bulletin of the Atomic Scientists*, 47:7–8, May 1991.

80. The following descriptions of early life in Hanford are from Paul Loeb, *Nuclear Culture*.

81. "Hanford Blues" National Public Radio, April 11 1988.

82. Keith Schneider, "Scientist who managed to shock the world on atomic workers' health," *New York Times*, May 3, 1990.

83. An excellent overview of the "downwinders" plight is: Howard Ball. *Justice Downwind: America's Atomic Testing Program in the 1950s*. NY: Oxford, 1986.

84. Table is from Keith Schneider, "Opening the record on nuclear risks," *New York Times*, December 3, 1989.

85. Bill Keller, "Soviet city, home of the A-Bomb, is haunted by its past and future," *New York Times*, July 10, 1989; Mark Hertsgaard, "From here to Chelyabinsk," *Mother Jones*, January/February 1991.

86. Mark Hertsgaard, "From here to Chelyabinsk," p. 55.

87. Quoted in Mark Hertsgaard, "From here to Chelyabinsk."

88. Gail Bradshaw is the spokeswoman, quoted in Matthew Wald, "US Waste Dumping Blamed in Wide Pollution at A-Plants," *New York Times*, December 8, 1988.

89. Quoted in Stephen Kurkjian, "Dilemma of US Nuclear weapons program," *Boston Globe*, Nov. 13,1988.

90. Quoted in Dan Reicher & S. Jacob Scherr, "The bomb factories: Out of compliance and out of control," in Anne Ehrlich & John Birks. *Hidden Dangers: Environmental Consequences of Preparing for War*. San Francisco, CA: Sierra Club Books, 1992.

91. Matthew Wald, "Whistleblower at nuclear laboratory was disciplined, Labor Department rules," *New York Times*, February 5, 1992; Keith Schneider, "Inquiry finds illegal surveillance of workers in nuclear plants," *New York Times*, August 1, 1991.

92. Keith Schneider, "Inquiry finds illegal surveillance of workers in nuclear plants," *New York Times*, August 1, 1991.

93. Ibid., p. A18.

94. In the film "The Life and Times of Rosie the Riveter," which chronicles the entry of American women into "men's professions" during the labor shortages caused by World War II, one of the women workers tells of her surprise when she found out that welding

was not a mystical and unattainable skill. "All these years," she exclaims, "men have been telling us how difficult welding is, how you have to study for years and years to get the skills. . . .It's just not true."

95. Brian Easlea. *Fathering the Unthinkable: Masculinity, Scientists and the Nuclear Arms Race*. London: Pluto Press, 1983.

96. For excellent commentaries on the masculinist underpinnings of science, see: Evelyn Fox Keller. *Reflections on Gender & Science*. New Haven, Conn: Yale University Press, 1984; Sandra Harding. *The Science Question in Feminism*. Ithaca: Cornell University Press, 1986.

97. A point suggested by Lisa Greber, "The Unholy Trinity: Physics, Gender & the Military," unpublished BA Honors Thesis, Massachusetts Institute of Technology, 1987.

98. Probably the best study of the culture of the nuclear weapons project (though written with no particular gender awareness) is: Richard Rhodes. *The Making of the Atomic Bomb*. NY: Simon & Schuster, 1986.

99. Ibid., p. 429.

100. Ibid., p. 571.

101. Ibid., p. 564.

102. Readers familiar with the Chipko movement in India will immediately recognize the similarities.

103. Richard Rhodes, *The Making of the Atomic Bomb*, pp. 564–565.

104. For contemporary accounts of the world of weaponeers, see: Robert Del Tredici. *At Work in the Fields of the Bomb*. NY: Harper & Row, 1987 and Rosenthal, Debra. *At The Heart of the Bomb: The Dangerous Allure of Weapons Work*. Reading, MA: Addison-Wesley, 1990.

105. This point is also made in Paul Loeb's book, *Nuclear Culture*. Further readings on the world of weapons designers and makers include: Helen Caldicott, *Missile Envy*; Debra Rosenthal, *At the Heart of the Bomb*; Robert Del Tredici, *At Work in the Fields of the Bomb*; and the classic, Richard Rhodes, *The Making of the Atomic Bomb*.

106. Carol Cohn, "A Feminist Spy in the House of Death," in Eva Isaksson, ed., *Women and the Military System*. See Cohn's Bibliography in this article for a good list of feminist critiques of Western male rationality in the military system.

107. For more information on this mothering/fathering debate, see Brian Easlea, *Fathering the Unthinkable*, Chapter 4.

108. W. H. Auden, "Moon Landing," in Charles Muscatine et al., eds. *The Borzoi College Reader*. NY: Knopf, 1971.

109. From Cynthia Enloe. *Bananas, Beaches and Bases: Making Feminist Sense of International Politics*. London: Pandora Press, 1989 and Berkeley, CA: University of California Press, 1989.

110. Winona LaDuke, "Nitassinan: Innu women say no to NATO," *Listen Real Loud* (AFSC), Vol. 10, 1–2, 1990.

111. National Defence Department. *Goose Bay EIS*. Ottawa, July 1989.

112. Peter Armitage, ed., *Compendium of Critiques of the Goose Bay EIS*, Sheshatshit, Labrador, February 1990.

113. Frank Carroll and Stephen Mather, "Turkey favored for NATO centre," *The Labradorian,* March 20, 1990.

114. "Moruroa Mon Amour," *Greenpeace Magazine,* January/February 1989.

115. In April of 1992, the French government announced a temporary suspension of nuclear testing in the Pacific; most observers believe that the moratorium will not be long-lived.

116. Michael Hamel-Green, "A future for the South Pacific," *Peace Dossier 8,* Victorian Association for Peace Studies, Australia, December 1983, p. 2.

117. An excellent report on the British nuclear testing program in Australia is Robert Milliken's. *No Conceivable Injury.* Victoria (Australia): Penguin Books, 1986.

118. Michael Hamel-Green, "A future for the South Pacific," p. 1.

119. "Maralinga clean-up imperative," *The Canberra Times,* July 4, 1990, p. 14.

120. Cited in Peter Hayes, Lyuba Zarsky and Walden Bello. *American Lake: Nuclear Peril in the Pacific.* London: Penguin Books, 1986.

121. Quoted in Helen Caldicott, *Missile Envy,* pp. 160–61.

122. Rosalie Bertell, *No Immediate Danger* p. 74; excellent overviews of the environmental and social havoc wreaked in the South Pacific include: Jane Dibblin. *Day of Two Suns: US Nuclear Testing and the Pacific Islanders.* London: Virago, 1988; NY: New Amsterdam Books, 1990; Eric Weingartner. *The Pacific: Nuclear Testing and Minorities.* London: Minority Rights Group, 1991.

123. Quoted in Dianne Dumanoski, "The South Pacific: decades after testing, the legal fallout," *Boston Globe,* November 10, 1985, p. A25.

124. Michael Hamel-Green, "A future for the South Pacific."

125. Reprinted with permission from Women Working For a Nuclear Free Pacific. *Pacific Women Speak: Why Haven't You Known?* London: Green Line, 1987, pp. 6–9. In England, the contact address for Women Working for a Nuclear Free Pacific is: c/o 82 Colston St., Bristol BS1 5BB, Avon, UK.

126. Michael Hamel-Green, "A future for the South Pacific," p. 3.

127. James Lerager, "Soviet victims of nuclear testing," *Earth Island Journal,* Fall 1990.

128. Michael Bedford, "Collision Course in Fiji," *Cultural Survival Quarterly,* 11(3), 1987, p. 55.

129. Quoted in Jennifer Scarlott, "US offers Palau dollars or democracy," *Bulletin of the Atomic Scientists,* November 1988, p.33; see also, Glenn Alcalay, "Murder, arson force women to drop antinuke suit," *The Guardian (US),* September 23, 1987; "Belauan women stand firm," *Spare Rib* (London), July 1988, p. 20; Roger Clark & Sue Rabbitt Roff. *Micronesia: The Problem of Palau* (London: Minority Rights Group, Report 63, 1987); Bernadette Keldermans and Lynn Wilson. *Belau and the Compact: Storms of Distrust.* Novato, CA: Times Change Press, 1990.

130. Quoted from Women Working for a Nuclear Free Pacific, *Pacific Women Speak,* p.24.

131. Information on the household waste plan was provided by Jane Dibblin, and is mentioned in the update to her book, *Day of Two Suns;* also, "US waste dumping firm," *Greenpeace Waste Trade Update,* July 15, 1989.

132. Jennifer Tichenor, "Japan's Proposed High-Level Rad Dump," *Waste Paper* (newsletter of the Radioactive Waste Campaign, New York) Spring 1989.

133. The Marshall Islands' chief of mission in Washington, quoted in Jennifer Tichenor, "Japan's Proposed High-level Rad Dumps," p. 10.

2 Business as Usual

1. Walter Robinson, "Worries on hazardous gases waft over W. Va.'s 'Chemical Valley'," *The Boston Globe,* December 7, 1984.

2. Ibid.

3. Alexandra Allen, "Poisoned Air," *Environmental Action,* January/Feb. 1988.

4. Walter Robinson, "Worries on hazardous gases waft over W. Va.'s 'Chemical Valley'."

5. "The chemical corridor," *Newsweek,* April 16, 1990, pp. 77 –80; Pat Costner & Joe Thornton. *We All Live Downstream—The Mississippi River and the National Toxic Crisis.* Greenpeace, December 1989.

6. An excellent survey of the literature on environment/cancer links is: Rita Arditti & Tatiana Schreiber, "Breast Cancer: The environmental connection," *Resist Newsletter,* (Boston) 246, May/June 1992; see also Ralph Moss. *The Cancer Industry—Unraveling the Politics.* NY: Paragon House, 1989.

7. Alexandra Allen, "Poisoned Air," *Environmental Action,* January/Feb. 1988.

8. National Wildlife Federation. *The Toxic 500.* 1987.

9. James Bellini. *High-Tech Holocaust.* San Francisco: Sierra Club, 1986.

10. See, for example, the now-classic study, David Weir & Mark Schapiro. *Circle of Poison.* San Francisco: Food First, 1981.

11. John Elkington. *The Poisoned Womb.* London: Penguin, 1985.

12. Excerpts from "Form 10-K" (annual report), American Vanguard Corp., December 31,1977, cited in David Weir & Mark Schapiro, *Circle of Poison.*

13. Russell Mokhiber, "Crime in the suites," *Greenpeace,* V. 14, 5, Sept/Oct, 1989.

14. For a compelling description of the overseas asbestos market, see: Ruth Norris, et al. *Pills, Pesticides and Profits: The International Trade in Toxic Substances.* NY: North River Press, 1982, and Barry Castleman & Manuel Vera Vera, "The Selling of Asbestos," *The Ecologist,* Vol. 11, 3, May/June 1981.

15. Mokhiber, "Crime in the Suites."

16. Minard Hamilton, "Workplace, environment both need cleaning up," *In These Times,* January 16–22, 1991, p. 16.

17. David Johnston, "Weapon plant dumped chemicals into drinking water," *New York Times,* June 10, 1989.

18. Matthew Wald, "Rockwell threatens to close nuclear weapons plant," *New York Times,* September 16, 1989, p. 9. In early 1992, Rockwell agreed to plead guilty to criminal violations of hazardous waste laws arising from its operation of Rocky Flats, which will end the investigation in exchange for a fine.

19. Quoted in Jack Todd, "Reliving the terror," *Montreal Gazette,* December 9, 1989.

20. In state socialist countries, state-set production quotas operate as the same trigger mechanism as does profit in capitalist economies. As Eastern European state economies switch, over the next decade, to a profit economy, Eastern Europeans will experience major social and political changes—but a healthier environment won't be one of them.

21. Nicholas Freudenberg. *Not in Our Backyards: Community Action for Health and the Environment.* NY: Monthly Review Press, 1984, p. 37.

22. Ibid., p. 39.

23. Andree Collard. *The Rape of the Wild.* London: Women's Press, 1988, pp. 81–82.

24. Karen Wright, "Going by the numbers," *New York Times Magazine*, December 15, 1991.

25. Sandra Postel, "Toward a new 'eco-nomics'" *Worldwatch Magazine*, September/October 1990, pp. 20–28.

26. Sandra Postel, "Toward a new 'eco-nomics'"; for a further analysis of the environmental costs of contemporary economics, see: David Pearce, ed. *Blueprint 2: Greening the World Economy.* London: Earthscan, 1991.

27. Cynthia Enloe. *The Morning After: Sexual Politics At the End of the Cold War.* Berkeley: University of California Press, forthcoming, 1993.

28. The analyses of the relationships between capitalism and patriarchy are too complex to explore here in any depth. For readers interested in this literature, I recommend: Anne Witz. *Professions and Patriarchy.* New York: Routledge, 1992; Z. R. Eisenstein, ed. *Capitalist Patriarchy and the Case for Socialist Feminism.* NY: Monthly Review Press, 1979, especially the articles in this collection by Heidi Hartmann and by Eisenstein; Heidi Hartmann, "The Unhappy Marriage of Marxism and Feminism," in Lydia Sargeant, ed., *Women and Revolution.* NY: Monthly Review Press, 1981; Cynthia Cockburn. *Brothers: Male Dominance and Technological Change.* London: Pluto Press, 1983; Sophie Walby. *Patriarchy at Work.* Cambridge: Polity Press, 1986; and, Sophie Walby. *Theorizing Patriarchy.* Oxford: Basil Blackwell, 1990.

29. "Women make slow progress up the corporate ladder," *The Economist*, March 14, 1987; "Few Women at the Top," *Executive Female*, Sept/Oct. 1988; Benjamin Forbes, et al. "Women Executives: Breaking Down Barriers?" *Business Horizons*, Nov/Dec 1988; "White males still hold most executive posts," *Los Angeles Times*, August 14, 1990. The use of male pronouns throughout this chapter to refer to industry executives and corporate management is, thus, intentional.

30. Wess Roberts. *The Leadership Secrets of Attila The Hun.* Buffalo, NY: Peregrine Pubs., 1986.

31. Quoted in Robert Jackall. *Moral Mazes: The World of Corporate Managers.* NY: Oxford University Press, 1988.

32. Terrence Deal and Allan Kennedy. *Corporate Cultures: The Rites & Rituals of Corporate Life.* Reading, MA: Addison-Wesley, 1982.

33. Kathleen Hirsch, "A new vision of corporate America," *The Boston Globe Magazine*, April 21, 1991.

34. Quoted in Jackall, *Moral Mazes*, p. 127.

35. Ibid., p. 136–37.

36. Ibid., p. 35.

37. Suzanne Berger, et. al. "Toward a new industrial America," *Scientific American*, June 1989, Vol. 260, 6, pp. 39–47.

38. Robert Jackall, *Moral Mazes*, p. 149.

39. Ibid., p. 156–57.

40. Quoted in *Newsweek*, April 23, 1990, p. 17.

41. Quoted in Bill Walker, "Green like me," *Greenpeace*, May/June 1991, p. 10.

42. Lynette Lamb, "Deceptive associations," *Utne Reader*, January/February 1992, p. 18.

43. Given the corporate record of casting environmental problems as "failures of management," it is particularly worrisome that environmentalists seem to be increasingly adopting management metaphors such as "managing planet earth," "harvesting resources," etc.

44. This is particularly true of European and North American managers. By contrast, Japanese executives are not rewarded on short-term profits, but they are even more strongly company-identified, and under pressure from a government economic policy that puts priority on continual growth.

45. A term described by prominent business theorists, Irving Janus & Leon Mann in *Decision Making: A Psychological Analysis of Conflict, Choice, and Commitment*. NY: Free Press, 1979.

46. Thomas Horton. *What Works for Me: 16 CEOs Talk About Their Careers and Commitments*. NY: Randon House, 1986.

47. Quotes from CEOs, cited in Thomas Horton *What Works for Me.*

48. Harro Von Senger. *The Book of Stratagems*. NY: Viking, 1991.

49. "The toughest companies in America," *US News & World Report*, October 28, 1991.

50. Information on the ozone case-study comes from: Philip Shabecoff, "Consensus on the threat to the ozone," *New York Times*, December 7, 1986; Dick Russell, "Politics of ozone: delay in the face of disaster," *In These Times*, August 17–30, 1988; Dick Russell, "'Beyond Sunglasses', but still too little, too late," *The Guardian* (US), October 7, 1987; Transcript of "The Hole in the Sky," *Public Broadcast Service*, 1987; Transcript of "The Ozone War," *CBS Television News*, 1987; Sharon Roan. *The Ozone Crisis*. New York: John Wiley, 1990.

51. Dick Russell, "Politics of ozone."

52. Art Kleiner, "Brundtland's legacy: can corporations really practice environmentalism while fattening their profit margins?" *Garbage Magazine*, September/October 1990, pp. 58–62.

53. See, for example, the essay by Michael Meacher, a British Member of Parliament, in *New Statesman & Society*, May 1, 1992.

54. David Moberg, "Oversight of workplace hazards," *In These Times*, April 25–May 1, 1990, p. 3.

55. Jolie Solomon, "Taking a stand," *Boston Globe*, July 21, 1991, pp. 45–47.

56. Charles Perrow. *Normal Accidents: Living with High Risk Technologies*. NY: Basic Books, 1984.

57. Hazelwood's alcohol blood level was above the Coast Guard legal limit even nine hours after the accident. Timothy Egan, "Elements of tanker disaster," *New York Times*, May 22, 1989, and other press reports.

58. Sarah Chasis & Lisa Speer, "How to avoid another Valdez," *New York Times*, May 20, 1989.

59. David Nyhan, "Exxon, not Hazelwood, is the culprit," *Boston Globe*, March 25, 1990.

60. William Coughlin, "Exxon curbing overtime on ships," *Boston Globe*, April 30, 1989; William Coughlin, "Exxon cutting size of crews," *Boston Globe*, April 23, 1989.

61. Quoted in *Newsweek*, April 24, 1989.

62. William Coughlin, "Three managers fault Exxon for tight-fisted ship policies," *Boston Globe*, May 13, 1989.

63. William Coughlin, "House panel sifts charge of illegal surveillance in Exxon case," *Boston Globe*, August 17, 1991, p. 4.

64. Of the extensive literature on Bhopal, these studies are among the most comprehensive: Anil Agarwal, Juliet Merrifield, and Rajesh Tanden. *No Place To Run: Local Realities and Global Issues of the Bhopal Disaster.* New Market, Tennessee. The Highlander Center, 1985; Padma Prakash, "Neglect of Women's Health Issues," *Economic and Political Weekly* (Delhi), December 14, 1985; Sanjoy Hazarika. *Bhopal: The Lessons of a Tragedy.* New Delhi: Penguin Books, 1986. A more journalistic account is Dan Kurzman, *A Killing Wind: Inside Union Carbide and the Bhopal Catastrophe,* (NY: McGraw Hill, 1987).

65. Quoted in Anna DeCormis, "Union Carbide ducks paying Bhopal victims," *Guardian (US),* August 3, 1988.

66. Quoted in Anna DeCormis, "Carbide still evading Bhopal compensation," *Guardian* (US), December 31, 1986.

67. David Bergman, "Tragedy without end," *New Statesman,* May 20, 1988.

68. Cited in Stuart Diamond, "Doing Business in the Third World," *New York Times,* December 16, 1984.

69. Cited in Sehdev Kumar, "The three legacies of Bhopal," *Alternatives,* 1989.

70. Cited in David Bergman, "Surviving Bhopal," *New Statesman and Society,* November 16, 1990, p. 5.

71. Ibid.

72. Anil Agarwal et al., *No Place To Run.*

73. Anna DeCormis, "Union Carbide Ducks."

74. "Union Carbide and the devastation of Bhopal," *Multinational Monitor,* April 6, 1987.

75. David Bergman, "Tragedy Without End."

76. Quoted in Timothy Egan, "Oil cleanup brings sad prosperity to village," *New York Times,* September 18, 1989, p 16.

77. Padma Prakash, "Neglect of Women's Health Issues," *Economic and Political Weekly* (Delhi), December 14, 1985.

78. Padma Prakash, "Neglect of Women's Health Issues," *Economic and Political Weekly* (Delhi), December 14, 1985. See also, Dr. Sathyamala, "Lest we forget the women of Bhopal," *Isis/Women's World,* June 1987.

79. Steven R. Weisman, "Bhopal suit marches on, in circles," *New York Times,* August 5, 1988.

80. "Bhopal women fight on," *Spare Rib,* January 1988.

81. Padma Prakash, "Neglect of Women's Health Issues."

82. Sanjoy Hazarika, "Bhopal victims still wait for Carbide money," *New York Times,* January 30, 1989.

83. "Bhopal women fight on."

84. See, for example, Kathleen Hirsch, "A new vision of corporate America," *The Boston Globe Magazine*, April 21, 1991.

85. Harris Sussman, "Women and the workplace," *Boston Business Journal*, November 11, 1991.

86. See, for example, Rosabeth Moss Kanter, *Men and Women of the Corporation*, NY: Basic Books, 1977 and, by the same author more graphically, *The Tale of O, On Being Different in an Organization*. NY: Harper Collins, 1986. See also Sue Freeman. *Managing Lives: Corporate Women and Social Change*. Amherst: University of Massachusetts Press, 1990.

87. Kathy Ferguson. *The Feminist Case Against Bureaucracy*. Philadelphia: Temple University Press, 1984, p. 186.

88. Ann Hornaday, "The most powerful women in corporate America," *Savvy Woman*, May 1989.

89. Ibid.

90. Jeremy Laurance, "Lost for words," *The European*, May 17–19, 1991, p. 9.

91. Cynthia Cockburn. *In the Way of Women: Men's Resistance to Sex Equality in Organizations*. Ithaca, NY: ILR Press, 1991.

92. Cited in Cynthia Cockburn, *In the Way of Women*, p. 71.

93. There is a growing literature that explores these differences. Probably the best known research is that of Carol Gilligan (*In A Different Voice*, Cambridge, MA: Harvard University Press, 1982), who details the differences in women's and men's moral judgments. Women's moral judgements, she finds, are informed by an ethic of nurturance, responsibility, and care, and women interpret complexity as a problem of inclusion rather than one of balancing competing claims. Men express greater need for distinctive and individual achievement, and separate identity.

94. Gilligan, *In A Different Voice*, pp. 164–65

95. Linda Weltner, "Even Mother Earth has limits," *Boston Globe*, April 28/89.

96. Jean Bethke Elshtain. *Private Man, Public Woman*. Princeton, NJ: Princeton University Press, 1981; and Sara Ruddick. *Maternal Thinking: Toward a Politics of Peace*. Boston: Beacon Press, 1989.

   See also Kathy Ferguson, *The Feminist Case Against Bureaucracy*.

97. An excellent analysis of the secretarial role is Rosemary Pringle. *Secretaries Talk: Sexuality, Power and Work*. London: Verso, 1988.

98. Jamaica Kincaid, "The Tongue" *The New Yorker*, October 9/89. My thanks to EJ Graff for bringing this essay to my attention.

99. Quoted in Robert Jackall, *Moral Mazes*, p.51.

100. Nicholas Freudenberg. *Not in Our Backyards!*. New York: Monthly Review Press, p. 26; a compelling expose of hazards in one particular workplace is Steve Fox. *Toxic Work: Women Workers at GTE Lenkurt*. Philadelphia: Temple University Press, 1991.

101. Keith Schneider, "Study finds link between chemical plant accidents and contract workers," *New York Times*, July 30, 1991.

102. Cynthia Daniels, ed., *Women in the Workplace Conference Proceedings*, Commonwealth of Massachusetts, May 1986; see also Maureen Hatch, "Mother, Father, Worker: Men and women and the reproduction risks of work," in Wendy Chavkin, ed. *Double*

Exposure: Women's Health Hazards on the Job and At Home. NY: Monthly Review Press, 1984.

103. Darren John Lewis, "The workplace isn't safe for baby," *Alternatives,* November 1986.

104. An excellent study of workplace exclusionary policies is: Judith A. Scott, "Keeping women in their place: Exclusionary policies and reproduction," in Wendy Chavkin, ed. *Double Exposure.*

105. Eileen McNamara, "Factory and fertility," *Boston Globe,* October 11/89.

106. Cynthia Daniels, ed., *Women in the Workplace* , p. 8.

107. Cynthia Daniels et al., "Health, Equity and Reproductive Risks in the Workplace," forthcoming, *Journal of Public Health Policy*

108. Although the Cyanamid policy was overturned in subsequent lawsuits in the early 1980s, a ruling on a similar case in 1989 reinforced the rights of companies to implement "fetal protection" employment policies. Citation from Cynthia Daniels et al., "Health, Equity and Reproductive Risks in the Workplace."

3 Governments

1. Alejandro Bendana, "Latin America after the Gulf war,"*Alternatives: Social Transformation & Humane Governance,"* Volume 16, 3, Summer 1991, p. 345.

2. Ruth Sivard. *World Military and Social Expenditures 1991.* Washington DC: World Priorities, 1991.

3. Generalizations about military influence in government admit the possible exception of a few countries with no standing military, such as Iceland and many of the Caribbean island nations, or the Scandinavian countries in which military influence appears to be somewhat contained. (The absence of a military sector can be documented for only a total of 19 sovereign nations; another 9 have very limited military commitment, as indicated by reported military expenditures. As reported in Arthur Westing, ed. *Environmental Hazards of War.* London: Sage Publications, 1990, p. 70). It may not be coincidental that these "under" or "un-militarized" countries are also prominent on the short list of those nations where women have gained the most access to formal political power.

4. The term "Third World" is extremely problematic, and is not the first choice of this author. Nonetheless, it is a designation widely used by others, including international governance bodies, and where it is used as a category of analysis by others, it is repeated here. Quote from Ruth Sivard, *World Military and Social Expenditures,* 1991, p. 19.

5. Ibid.

6. Ibid.

7. "Chilean road opening up rugged frontier," *Christian Science Monitor,* March 9, 1988.

8. Nicholas Kristof, "In North Korea, dam reflects 'Great Leader's' state of mind," *New York Times,* July 5, 1989.

9. Reports on the state of the environment in the former USSR and Eastern Europe are still sketchy, and press reports are still the most reliable sources for up-to-date information. The Worldwatch Institute and Greenpeace International are among the best sources for continuing environmental coverage on the Eastern bloc. In addition to the several books on the Chernobyl accident, a good general source is: Murray Feshbach and Alfred Friendly, Jr. *Ecocide in the USSR: Health and Nature Under Seige.* NY: Basic Books, 1991.

10. "Breast milk risk," *The Independent* (London), May 23, 1991; "Firms go east to clean up

their act," *The European*, May 24–26, 1991; Paul Hockenos, "East Germany opens its nuclear Pandora's box," *In These Times*, April 11–17, 1990; Marlise Simmons, "In Bulgarian town, the killers are lead and arsenic pollution," *New York Times*, March 28, 1990; Ian Traynor, "Welcome to the valley of death," *The Guardian*, May 17, 1991; Shane Cave, "Wrestling with the Soviet environment," *Our Planet: The Magazine of the United Nations Environment Programme*, Vol. 3, 5, 1991; "A nice red afterglow," *The Economist*, March 14, 1992.

11. Marlise Simmons, "Rising Iron Curtain exposes haunting veil of polluted air," *New York Times*, April 8, 1990; Hilary French, "Industrial wasteland," *WorldWatch*, November/December 1988; Don Hinrichsen, "Gloom: pollution is poisoning the future of Eastern Europe," *New Internationalist*, July 1990.

12. Quoted in Murray Feshbach & Alfred Friendly Jr. *Ecocide in the USSR: Health and Nature Under Seige*.

13. Quoted in Shane Cave, "Wrestling with the Soviet environment," p. 5.

14. Quoted in Marlise Simmons, "Rising Iron Curtain," p. 14.

15. Henry Kamm, "Prague chore: Cleaning up after Soviets," *International Herald Tribune*, July 25, 1990; Renate Walter, "Military pollution in East Germany," *Peace Magazine*, Vol. VIII, Issue II, March 1992; John Budris, "Evict Russia's squatter army from the Baltics,"*Wall Street Journal Europe*, June 30, 1992.

16. Keith Schneider, "The Soviets show scars from nuclear arms production," *New York Times*, July 16, 1989.

17. "Kazakhstan chief orders the closing of a nuclear test range," *New York Times*, August 30, 1991; James Lerager, "Soviet victims of nuclear testing," *Earth Island Journal*, Fall 1990; Nick Thorpe, "Cry for the poisoned, hairless children," *The Observer* (London), September 17, 1989; "A nice red afterglow," *The Economist*, March 14, 1992; Patrick Tyler, "Soviets' secret nuclear dumping raises fears for Arctic waters," *The New York Times*, May 4, 1992.

18. Paul Hockenos, "Ecology sinks in swamp of bureaucracy," *In These Times*, July 18–31, 1990.

19. I am using "Burma" rather than "Myanmar" throughout the text because it is more familiar to readers, and because many of the events described took place before the government changed the country's name.

20. Lindsay Murdoch, "A forest dies as teak loggers put guerillas to flight," *The Age* (Sydney, Australia), July 10, 1990; John C. Ryan, "War and teaks in Burma," *Worldwatch*, September/October, 1990, pp. 8–9; Steven Erlanger, "Burmese teak forest falls to finance a war," *New York Times*, December 9, 1990, p. 6..

21. Deforestation figures from Joni Seager,ed. *The State of the Earth Atlas*. New York: Simon & Schuster, 1990.

22. In Cambodia (Kampuchea) in the late 1980s, similarly, the Khmer Rouge government opened up its rainforest to Thai loggers to get money to finance its war effort. The Thai Prime Minister justified the logging by saying that "if Thailand does not engage in logging in Myanmar, other countries will move in anyway." (Quoted in Kevin Meade, "Six vie for Khmer rainforest deal," *The Sydney Age*, July 1, 1990.)

23. All quotes and figures from Lindsay Murdoch, "A forest dies as teak loggers put guerillas to flight," *The Age* (Sydney, Australia), July 10, 1990; John C. Ryan, "War and teaks in Burma," *Worldwatch*, September/October, 1990, pp. 8–9 .

24. Cynthia Enloe, "Patronage," unpublished manuscript, 1991.

25. United Nations Environment Programme, *Our Planet*, Vol. 3, 4, 1991, p. 11.

26. "The dwindling forest beyond Long San," *The Economist*, August 18, 1990, pp. 23–24.

27. Cited in *Greenpeace Magazine*, Jan/Feb. 1991, p. 5.

28. Quoted in *New York Times*, March 23, 1991.

29. "Philippine rainforest activists arrested," *Rainforest Action Network*, Action Alert 59, April 1991.

30. Judi Bari, "For FBI, back to political sabotage?' *New York Times*, August 23, 1990; Dennis Bernstein, "Earth First bombing: blaming the victim," *In These Times*, June 6–19, 1990.

31. Chip Berlet, "Activists face increased harassment," reprinted in *Utne Reader*, January/February 1991, pp. 85–88.

32. "Going to jail for the earth," *Environmental Action*, January/February 1988; Dianne Dumanoski, "Groups to campaign for release of Malaysian activists," *Boston Globe*, November 18, 1987.

33. Cited in Dianne Dumanoski, "Groups to campaign for release of Malaysian activists," *Boston Globe*, November 18, 1987.

34. Even restrained official assessments of the environmental crisis now point to poverty as a major environmental factor. See, for example, the World Commission on Environment and Development report, published as *Our Common Future*. Oxford: Oxford Univ Press, 1987, commonly known as the "Brundtland Report," named after Gro Brundtland, prime minister of Norway and Chairwoman of the Commission.

35. Bill Hall, "Natural catastrophes: Central America's ecological crisis is rooted in politics," *The Guardian (USA)*, July 15, 1987.

36. Bill Hall, "Natural catastrophes: Central America's ecological crisis is rooted in politics."

37. Personal correspondence, 1/2/87.

38. Paul Handley "The land wars," *Far Eastern Economic Review*, October 31, 1991.

39. "Repression and eco-devastation in Indonesia," *Utne Reader*, January/February 1987, p. 9.

40. Cited in Marcus Colchester, "Paradise promised," *New Internationalist*, November 1987, p. 25.

41. Marcus Colchester, "Paradise promised," p. 19.

42. Marcus Colchester, "Paradise promised," p. 25.

43. Marcus Colchester, "Paradise promised," p. 25; see also Thomas Leinbach, John Watkins & John Bowen, "Employment behavior and the family in Indonesian transmigration," *Annals of the Association of American Geographers*, 82 (1), 1992.

44. Marcus Colchester, "Paradise promised," p. 25.

45. Lynn White's article, "The historical roots of our ecologic crisis," (*Science*, Vol. 155, March 10, 1967) is among the more highly respected research in this area. Quote is from p. 1205.

46. Lynn White, p. 1204; Social ecologist William Leiss was another writer who, in the 1970s pointed out that modern science was centrally concerned with mastery over nature, but, like White, Leiss overlooks the gender implications of such dominance: William Leiss. *The Domination of Nature*. NY: Brazilier, 1972.

47. Carolyn Merchant. *The Death of Nature*. New York: Harper & Row, 1980.

48. Quotes from Carolyn Merchant, *The Death of Nature*, Chapter 7, "Dominion Over Nature."

49. See, for example: Sandra Harding. *The Science Question in Feminism*. Ithaca, NY: Cornell Univ Press, 1986; Evelyn Fox Keller. *Reflections on Gender & Science*. New Haven, CT: Yale University Press, 1984.

50. The report on Mandela's free-market speech is carried in Will Hutton, "Free market call by Mandela," *Manchester Guardian Weekly*, February 9, 1992; advertisement text from the *Boston Globe*, February 19, 1992.

51. Bill Keller, "In polluted Russian city, a new fear," *The New York Times*, November 26, 1990;

52. Bill Keller, "In polluted Russian city, a new fear"; Susan Greenberg, "Morale victories," *The Guardian*, May 17, 1991; Ian Traynor, "Welcome to the valley of death," *The Guardian*, May 17, 1991; Alex Duval Smith, "Triumphs and Trabants," *The Guardian*, May 17, 1991.

53. As described in Dick Russell, "We Are All Losing the War," *The Nation*, March 27, 1989, and Bob Gottlieb & Margaret Fitzsimmons, "New Approach to Green Issues," *In These Times*, April 16, 1989.

54. Dick Russell, "We Are All Losing the War," *The Nation*, March 27, 1989.

55. For more information the Council on Competitiveness, see Joni Seager, "The Environment: It's time to trash the environmental president," *The Village Voice*, April 28, 1992.

56. Geoffrey Baker, "The water privateers," *New Internationalist*, May 1990.

57. Geoffrey Baker, "The water privateers," p. 8.

58. Geoffrey Baker, "The water privateers," p. 9.

59. Quoted from a speech given by Peggy Antrobus to the Global Assembly of Women for a Healthy Planet, Miami, November 1991.

60. Alejandro Bendana, "Latin America after the Gulf war," p. 348.

61. This point is especially clear in the essays in Edward Goldsmith & Nicholas Hildyard, eds. *The Earth Report: The Essential Guide to Global Ecological Issues*. London: Price Stern Sloan, 1988.

62. James Sterngold, "7 rich nations to help Soviets plan economy," *New York Times*, October 14, 1991, p. 1.

63. For a fuller analysis of international development policies, and of their ecological consequences, see: Edward Goldsmith & Nicholas Hildyard, eds. *The Earth Report*; David Pearce, ed. *Blueprint 2: Greening the World Economy*. London: Earthscan Publications, 1991; Irene Dankelman and Joan Davison. *Women and Environment in the Third World*; The World Commission on Environment & Development. *Our Common Future*; The Bank Information Center. *Funding Ecological and Social Destruction: The World Bank and the IMF*. Washington DC: Bank Information Center, 1991.

64. The Bank Information Center, *Funding Ecological and Social Destruction*, p. 1.

65. Dianne Dumanoski, "World Bank to be more ecology minded," *Boston Globe*, May 10, 1987; Philip Shabecoff, "World lenders facing pressure from ecologists," *New York Times*, October 20, 1986; Paul Lewis, "Aid agencies defend policy for Third World," *New York Times*, July 10, 1988; The Bank Information Center. *Funding Ecological and Social Destruction: The World Bank and the IMF*. Washington DC: Bank Information Center, 1991.

66. Fred Pearce, "The giants that are stalking the earth," *Guardian Weekly*, July 8, 1990, p. 11.

67. "Greener faces for its greenbacks," *The Economist*, September 2, 1989.

68. Bruce Marshall, ed. *The Real World*. Boston: Houghton Mifflin Co., 1991, p.111.

69. Vandana Shiva, "Development as a new project of Western patriarchy," in Irene Diamond and Gloria Orenstein, eds. *Reweaving the World*. San Francisco: Sierra Club Books, 1990.

70. Barbara Rogers. *The Domestication of Women: Discrimination in Developing Societies*. New York: St. Martin's, 1980, p. 10.

71. The gender breakdown given for US AID directors and deputy directors only counts the total personnel identifiable by name as either male or female. Listed in *Front Lines*, USAID, February 1992.

72. The literature on women and development is too extensive to adequately cite here. For readers interested in this topic, I suggest starting with the following: Women's International Information and Communication Service (ISIS) ed. *Women and Development: A Resource Guide for Organization and Action*. Philadelphia: New Society Publishers, 1984; Vandana Shiva. *Staying Alive:Women, Ecology & Development*. London: Zed Books, 1989; Susan Bourque and Kay Warren, "Technology, Gender & Development," *Daedalus*, Fall, 1987; Eleanor Leacock & Helen Safa. *Women's Work: Development & the Division of Labor by Gender*. South Hadley, MA: Bergin & Garvey, 1982; Barbara Rogers. *The Domestication of Women: Discrimination in Developing Societies*. New York: St. Martin's, 1980; Gita Sen & Caren Grown. *Development, Crisis, and Alternative Visions: Third World Women's Perspectives*. Development Alternatives for Women in a New Era (DAWN), 1985; Janet Momsen, ed. *Women and Development in the Third World*. London: Routledge, 1990; Jeanne Vickers, ed. *Women and the World Economic Crisis*. London: Zed Books, 1991; Jill Gay and Tamara Underwood, "Women in danger: A Call for Action," *Background Paper*, NY: National Council for International Health, 1991.

73. For an extended discussion of the class/gender implications of privatization, see Bina Agarwal, "The gender and environment debate: lessons from India," *Feminist Studies*, Vol. 18, 1, Spring 1992.

74. For further information, see Rachel Kamel. *The Global Factory*. Philadelphia: American Friends Service Committee, 1990; Cynthia Enloe. *Bananas, Beaches & Bases: Making Feminist Sense of International Politics*. London: Pandora Press, 1989, especially Chapters 6 & 7.

75. Cynthia Enloe, *Bananas, Beaches & Bases*, p. 160.

76. Michael Goldman, "Green revolution has sown class conflict," *In These Times*, May 18–24, 1988; compelling analyses of the political, social , and environmental impacts of the green revolution can be found in: Francine Frankel. *The Political Challenge of the Green Revolution*. Princeton, NJ: Princeton Univ. Press, 1972; Vandana Shiva. *The Violence of the Green Revolution*. London: Zed Books, 1991.

77. Cynthia Enloe, *Bananas, Beaches & Bases*, p. 128.

78. Edward Goldsmith & Nicholas Hildyard, eds. *The Earth Report: The Essential Guide to Global Ecological Issues*. London: Price Stern Sloan, 1988, p. 116.

79. Jan Rocha, "Slave virgins up for auction at Amazon mining camps," *Manchester Guradian*

Weekly, February 23, 1992. See also Ximena Bunster & Elsa Chaney. *Sellers and Servants*. NY: Praeger, 1985.

80. The classic study of sex tourism is Kathleen Barry, Charlotte Bunch & Shirley Castley, eds. *International Feminism: Networking Against Female Sexual Slavery*. NY: International Women's Tribune Center, 1982; see also Kathleen Barry. *Female Sexual Slavery*. NY: Prentice-Hall, 1979 and Thanh-Dam Troung. *Sex Money and Morality: Prostitutionn and Tourism in South-east Asia*. London: Zed Books, 1990.

81. William Burke, "The toxic price of fre trade in Mexico," *In These Times*, May 22, 1991, p. 2.

82. Working Group on Canada-Mexico Free Trade, "Que Pasa? A Canada-Mexico 'Free' Trade Deal" *New Solutions*, Summer 1991, pp. 10–24.

83. Dianne Dumanoski, "Free-trade laws could undo pacts on environment," *Boston Globe*, October 7, 1991, p. 25; "International Trade Agreement threatens rainforests," *Rainforest Action Network Action Alert 66*, November 1991; David Morris, "Trading our way to devastation," *Utne Reader*, March/April 1991.

84. Quotes from: Stephen Powell, "Brazil issues plan to shield rain forests," *Boston Globe*, April 7, 1989; Marlise Simons, "Brazilians tell of the forest and the fears," *New York Times*, January 22, 1989; Marlise Simons, "Brazil, smarting from the outcry over the Amazon, charges foreign plot," *New York Times*, March 23, 1989, p. A14.

85. Marlise Simons, "Brazil, smarting from the outcry over the Amazon, charges foreign plot."

86. "Eight Amazon nations denounce pressure to save rain forest," *New York Times*, March 9, 1989.

87. For an overview of environmental issues in China, see: Baruch Boxer, "China's environmental prospects," *Asian Survey*, Volume 29, 7, July 1989; Katharine Forestier, "The degreening of China," *New Scientist*, July 1989.

88. Colin Nickerson, "China copies worst polluters," *Boston Globe*, December 20, 1989, p. 16.

89. "China's environment pays for growth, paper says,"*Boston Globe*, December 4, 1988; Kazuhiro Ueta, "Dilemmas of pollution control policy in contemporary China," *Capitalism, Nature, Socialism (CNS)*, Number 3, November 1989, pp. 109–127.

90. Colin Nickerson, "China copies worst polluters," *Boston Globe*, December 20, 1989, p. 16.

91. Colin Nickerson, "China copies worst polluters."

92. Colin Nickerson, "China copies worst polluters."

93. Quoted in Colin Nickerson, "China copies worst polluters," p. 16.

94. Marlise Simons, "Brazil, smarting from the outcry over the Amazon, charges foreign plot," *New York Times*, March 23, 1989, p. A14.

95. Michael Renner, quoted in the *New York Times*, May 29, 1989.

96. Mark Valentine of the US Citizens' Network, cited in Julia Preston, Eugene Robinson & Joel Achenbach, "Old Fashioned Nationalism in Control," *The Guardian Weekly*, June 14, 1992.

97. Jessica Tuchman Mathews, "Redefining security," *Foreign Affairs*, Vol. 68, 2, Spring 1989, p. 176.

98. Patricia Mische, "Ecological security and the need to reconceptualize sovereignty," *Alternatives: Social Transformations & Humane Governance*, Vol. XIV, 4, October 1989, p. 415.

99. Among the most eloquent pleas for reconsidering sovereignty/environment relations is a book now largely forgotten: Harold Sprout & Margaret Sprout. *Toward a Politics of the Planet Earth*. NY: Van Norstrand Reinhold, 1971; recent discussions of the need for reconsidering sovereignty include Jessica Tuchman Mathews, "Redefining security," and Patricia Mische, "Ecological security."

100. Some of the best writing on masculinity/sovereignty comes from feminist political scientists. I recommend especially: Rebecca Grant & Kathleen Newland, eds. *Gender and International Relations*. Bloomington: Indiana University Press, 1991; Wendy Brown. *Manhood and Politics*. NY: Rowman & Littlefield, 1988; V. Spike Peterson,ed. *Gendered States: Feminist (Re)Visions of International Relations Theory*. Boulder, Colorado: Lynne Rienner Pub., 1992; J. Ann Tickner. *Gender in International Relations*. NY: Columbia University Press, 1992. On nationalism, see: Kumari Jayawardena. *Feminism and Nationalism in the Third World*. London: Zed, 1986; Cynthia Enloe. *Bananas, Beaches & Bases* ; Andrew Parker, Mary Russo, Doris Summer, Patricia Yeager, eds. *Nationalisms and Sexualities*. NY: Routledge, 1992.

101. Christine Sylvester, "The emperor's theories and transformations: looking at the field theough feminist lenses," in Dennis Pirages and Christine Sylvester, eds. *Transformations in the Global Political Economy*. Basingstoke: MacMillan, 1989.

102. Robert O. Keohane, "International relations theory: contributions of a feminist standpoint," in Rebecca Grant and Kathleen Newland, eds. *Gender and International Relations*, p. 43.

103. J. Ann Tickner, *Gender in International Relations*, especially Chapter 4, "Man Over Nature."

104. Fred Halliday, "Hidden from international relations: women and the international arena," in Rebecca Grant & Kathleen Newland, eds. *Gender and International Relations*, p. 164.

105. Cynthia Enloe, *Bananas, Beaches and Bases*, p. 64.

106. Alejandro Bendana, "Latin America after the Gulf war," p. 346.

107. Marlise Simons, "Brazil, smarting from the outcry," p. A14.

108. Anil Agarwal, "A case of environmental colonialism," *Earth Island Journal*, Spring 1991, pp. 39–40.

109. Steven Greenhouse, "Europe's failing effort to exile toxic trash," *New York Times*, October 16, 1988; "The toxic waste trade," *Environmental Action*, November/December 1988.

110. A comprehensive report on the international traffic in hazardous waste is, Bill Moyers & Center for Investigative Reporting. *Global Dumping Ground*. Washington DC: Seven Locks Press, 1990.

111. "Combating Toxic terrorism," *Worldwatch*, Sept./Oct. 1988, p. 6.

112. "Toxic waste as recycled products," *Environmental News Digest*, Vol. 9, 1, 1991, p. 18

113. Ron Chepesiuk, "From ash to cash," *E, The Environmental Magazine*, July/August 1991.

114. "Combating Toxic terrorism."

115. James Brooke, "African nations barring toxic waste," *New York Times*, September 25, 1988.

116. Ibid.

117. "Dumping on the developing world," *Rachel's Hazardous Waste News*, 126, April 25, 1989.

118. Wana Leba, quoted in "Africa's shining example," *Greenpeace*, May/June 1991, p. 5.

119. "Toxic waste as recycled products," *Environmental News Digest*, Vol. 9, 1, 1991, p. 18.

120. "Combating Toxic terrorism," p. 7; Ron Chepesiuk, "From ash to cash."

121. *Rachel's Hazardous Waste News*, 57, December 28, 1987.

122. Quoted in *Environmental Action*, November/December 1988, p. 28.

123. Quoted in "Dumping toxic wastes on the Third World," *Utne Reader*, Sept./Oct. 1988.

124. Quoted in, "EPA respects Mexican sovereignty," *In These Times*, May 6–12, 1987, p. 5.

125. "Toxic Waste for PNG," *Environmental News Digest*, Vol. 9, 1, 1991, p. 21.

126. "Dumping toxic wastes on the Third World,"p. 89.

127. Pesticide Action Network, June 1991.

128. Ibid.

129. "Toxic trade," *New Internationalist*, February 1988, p. 27.

130. One of the early studies of shifts in the industrial economy, motivated by the search for unregulated sites, is: Barry Bluestone & Bennett Harrison. *The Deindustrialization of America: Plant Closings, Community Abandonment and the Dismantling of Basic Industry*. NY: Basic Books, 1982; more recent, activist-oriented, studies are: Annette Fuentes and Barbara Ehrenreich. *Women in the Global Factory*. Boston: South End Press, 1983 and Rachel Kamel. *The Global Factory*. Philadelphia: American Friends Service Committee, 1990.

131. "Let them eat pollution," *The Economist*, February 8, 1992, p. 66; Doug Henwood, "Toxic banking," *The Nation*, March 2, 1992, p. 257.

132. Jane Juffer, "A Mexican Wasteland," *Environmental Action*, November/December 1988, p. 25; "Maquiladoras: A different brand of toxic export," *Greenpeace*, November/December 1988, p. 10.

133. "Maquiladoras,"p. 10.

134. "EPA respects Mexican sovereignty," *In These Times*, May 6–12, 1987, p. 5.

135. Ibid.

136. "Maquiladoras", p. 10.

137. EPA respects Mexican sovereignty," p. 5

138. For more information on the environment in South Africa, see: Alan Durning. *Apartheid's Environmental Toll*. Worldwatch Paper 95, May 1990, Worldwatch Institute, Washington DC; Judy Christrup, "Of apartheid and pollution," *Greenpeace*, May/June 1990; Margaret Knox, "South Africa polluted by more than just apartheid," *In These Times*, February 7–13, 1990.

139. Alan Durning. *Apartheid's Environmental Toll*. Worldwatch Paper 95, May 1990, p. 8

140. Ibid.

141. Ibid, p. 10.

142. Judy Christrup, "Of apartheid and pollution," *Greenpeace*, May/June 1990, p. 18.

143. Ibid.

144. Alan Durning, *Apartheid's Environmental Toll*, p. 14.

145. Bernard Nietschmann & William Le Bon, "Nuclear weapons states and Fourth World nations," *Cultural Survival Quarterly*, 11 (4), 1987, pp. 5–7.

146. Larry Tye, "Indian reservations are targeted for nation's waste," *Boston Globe*, June 23, 1991, p. 12.

147. Michael Williams, "Serious reservations: Native Americans fight dumpers," *Everyone's Backyard* , (Citizen's Clearinghouse for Hazardous Wastes), June 1991, p. 3.

148. Larry Tye, "Indian reservations are targeted for nation's waste," p. 12; Chuck McCutcheon, "Enterprising tribe weighs risks of going to waste," *Albuquerque Journal* (New Mexico), November 3, 1991.

149. Larry Tye, "Indian reservations are targeted for nation's waste"; Mary Hager, "Dances with Garbage: Reservations as Toxic Dumping Grounds," *Newsweek*, April 29, 1991.

150. This heading is taken from the book by Robert Bullard, *Dumping in Dixie: Race, Class and Environmental Quality*, Boulder, Co: Westview Press, 1990.

151. Cited in Barbara Day, "Third World communities serve as toxic dumps," *The Guardian (US)*, June 22, 1988, p. 17; see also, Dana Alston, ed., *We Speak for Ourselves: Social Justice, Race and the Environment*, Panos Institute, 1990 and the CRJ study, Charles Lee, ed.*Toxic Wastes and Race in the United States: A National Report on the Racial and Socio-Economic Characteristics of Communities Surrounding Hazardous Waste Sites* United Church of Christ, April 1987.

152. Robert Bullard, *Dumping in Dixie.*

153. Charles Lee, "The integrity of justice," *Sojourners*, February/March 1990, p. 24.

154. John McCormick. *British Politics and the Environment*. London: Earthscan Publications, 1991.

155. Walter Rosenbaum, "The Bureaucracy and environmental policy," in James Lester, ed., *Environmental Politics and Policy: Theories and Evidence*. Durham NC: Duke University Press, 1989.

156. Ibid., p. 232.

157. "Banning books at EPA?" *Environmental Action*, May/June 1991, pp. 6–7.

158. "Free the book!" *Environmental Action*, May/June 1991, p. 33.

159. Poll cited in *Newsweek*, December 16, 1991, p. 30.

160. See, for example, Barbara Rogers. *Fifty-Two Percent: Getting Women's Power into Politics*. London: Women's Press, 1983.

161. Center for the American Woman and Politics, Rutgers University, New Brunswick, NJ. *The Impact of Women in Public Office*. 1991.

4 The Ecology Establishment

1. This chapter was informed by discussions with Francesca Lyman (former editor at *Environmental Action*); Sylvia Tesh, (Political Science, Yale University); Janet Marinelli (formerly with *Environmental Action)*; Ida Koppen, (European University, Florence); Priscilla Feral, (Friends of Animals, USA); Marina Alberti (Cooperativa Ecologia, Milan);

Mary Fillmore (formerly with the US EPA); and a number of other women currently working within environmental organizations who wish to remain anonymous.

2. Ann Holt, "French Women Gain," *Everywoman*, May 1989, p. 10.

3. *CREW Reports* (Brussels), July/August 1989.

4. Ynestra King, "Coming of Age With the Greens," *Zeta Magazine*, February 1988, p. 16.

5. Ibid, p. 19.

6. Charlene Spretnak and Fritjof Capra. *Green Politics: The Global Promise*. Santa Fe, NM: Bear & Co., 1986, pp. 50–51.

7. Ibid.

8. Cynthia Enloe. *Bananas Beaches & Bases: Making Feminist Sense of International Politics*. London: Pandora Press and Berkeley: Univ of California Press, 1989. See esp. the chapter on "Nationalism" for a discussion of the ways in which men trade off issues of nationalism with issues of gender equality.

9. Quoted in Flora Lewis, "Red-Green Tide in Germany," *New York Times*, June 18, 1989, p. E27.

10. "An Interview With Jutta Ditfurth," *Green Perspectives: A Left Green Publication*, Number 9, August 1988.

11. Cited in Charlene Spretnak and Fritjof Capra, *Green Politics*, p. 109.

12. Merle Hoffman, "The Greening of the World: An Exclusive Interview with Petra Kelly," *On The Issues*, Vol. IX, 1988, pp. 20–23.

13. Diana Johnstone, "Spring brings a touch of Green to France," *In These Times*, April 5–11, 1989.

14. "The Women of Green," *The Guardian* (UK) January 27, 1987, p. 26.

15. Michael Tronnes, "The Greening of Mexico," *Utne Reader*, March/April 1988.

16. Diana Johnstone, "Spring brings a touch of Green to France," p. 11.

17. Reported in *The New Catalyst*, Fall 1988.

18. Ruth Wallsgrove, "Greens, Browns, Purples: Motherhood and Party Politics in West Germany," *Off Our Backs*, Vol. xvii, 9, October 1987, p. 1.

19. Ibid.

20. Merle Hoffman, *op. cit.*, p.21.

21. "Kathryn S. Fuller Named President of World Wildlife Fund and the Conservation Foundation," *WorldWIDE News: World Women in Environment*, April 1989, p. 1.

22. In personal conversation, May 5, 1989.

23. An independent survey I conducted in 1989.

24. *The Economist*, April 21, 1990, p.31; and *The Economist*, October 20, 1990, p. 93.

25. *Maclean's*, April 3, 1989.

26. The groups constituting the Group of Ten are: National Wildlife Federation; Natural Resources Defense Council; Wilderness Society; Izak Walton League; National Audubon Society; Sierra Club; Defenders of Wildlife; National Parks & Conservation Association; Environmental Defense Fund; Environmental Policy Institute. As listed in Dick Russell, "We are all losing the war," *The Nation*, March 27, 1989, pp. 403–408.

27. All figures as of 1990.

28. Quoted in Beth Burrell, "Men Ride the Recycling Wave," *New Haven [Connecticut] Register*, May 21, 1989.

29. Ibid

30. Karen Zagor, "Rubbish industry smells new profit," *Financial Times*, May 29, 1991.

31. Donald Snow, ed. *Inside the Environmental Movement*, Washington DC: Island Press, 1992, p.55.

32. Stephen Kellert & Joyce Berry, "Attitudes, knowledge and behaviors toward wildlife as affected by gender," *Wildlife Society Bulletin*, Vol. 15, 3, Fall 1987.

33. Ibid., p. 365.

34. Figures from the Boston Women's Community Cancer Project.

35. For more information on domestic pollution, see: B.J. Roche, "Monitoring household pollution," *Boston Globe*, June 8, 1984; Lisa Belkin, "Experts say air pollution in the home is a growing risk," *New York Times*, March 7, 1985; "Stepping lightly on the earth: Everyone's guide to toxics in the home," *Greenpeace Action Bulletin*.

36. Philip Shabecoff, "Environmental Groups Told they Are Racists in Hiring," *New York Times*, February 1, 1990, p. A20; and Clay Carter, "Pulling Down the Barriers," *Everyone's Backyard*, (newsletter of the Citizen's Clearinghouse for Hazardous Waste) Vol. 8, 2, March/April 1990, pp 6–8; Holly Brough, "Minorities redefine environmentalism," *Worldwatch*, Sept/Oct 1990, pp. 5–7; Friends of the Earth USA, "Eco-groups asked to hire more minorities," *Not Man Apart*, April/May 1990, p.14; John Lancaster, "Role of minorities in environmental movement remains limited," *The Washington Post*, Nov. 23, 1990.

37. Cited in Dorceta E. Taylor, "Blacks and the environment," *Environment & Behavior*, Vol. 21, 2, March 1989, p. 193.

38. Julian Agyeman, "A snail's pace," *New Statesman & Society*, February 3, 1990, pp. 30–31.

39. Bob Ostertag, "Greenpeace takes over the world," *Mother Jones*, March/April, 1991, p. 32.

40. Salim Muwakkil, "US Green Movement needs to be Colorized," *In These Times*, May 2–8, 1990, p 8; Julian Agyeman, "A snail's pace."

41. Cited in Dorceta E. Taylor, "Blacks and the environment," p. 176.

42. Anne Witz. *Professions & Patriarchy*. NY: Routledge, 1992, p. 62.

43. See the discussion of professionalization in Anne Witz. *Professions & Patriarchy*. NY: Routledge, 1992, pp. 60–63, and also the work of J. Hearn ("Notes on patriarchy, professionalization, and the semi-professions," Vol. 16, 1982), whose work was instrumental in theorizing professionalization as a patriarchal process.

44. Sarah Benton, "The new party politics: Is there room for us?", *Everywoman*, October 1990, pp. 12–13.

45. The academic history of women in the physical sciences offers a particularly good case in point. For a description of the marginalizatioin of women with professionalization of the sciences, see Margaret W. Rossiter. *Women Scientists in America*. Baltimore: Johns Hopkins University Press, 1982.

46. "Porritt Power," *The Guardian* (UK), January 7, 1986.

47. Mark Muro, "Quiet: Environmental Politics at Work," *Boston Globe*, September 18, 1988.

48. Bob Gottlieb & Margaret Fitzsimmons, "New Approach to Green Issues," *In These Times*, April 16, 1989.

49. Quoted in Helen Ingram and Dean E. Mann, "Interest Groups and Environmental policy," in James P. Lester, ed. *Environmental Politics & Policy*. Durham: Duke University Press, 1989, p. 153.

50. Ibid.

51. Sally Ranney, "Heroines and Hierarchy: Female leadership in the conservation movement," in Donald Snow, ed *Voices from the Environmental Movement*, Washington DC: Island Press, 1992.

52. Michael Tobias. *Deep Ecology*. San Marcos, CA: Avant Books, 1988; John Davis & Dave Foreman, eds. *The Earth First! Reader*. Salt Lake City: Peregrine Smith Books, 1991.

53. Andrew Dobson, ed. *The Green Reader*. San Francisco: Mercury House, 1991.

54. Sandy Irvine. *A Green Manifesto*. London: Optima, 1988. Reviewed by Sara Parkin in *The Ecologist*, Vol. 19, 1, 1989.

55. Pam Simmons, "The challenge of feminism," Editorial introduction to *The Ecologist*, January/February 1992.

56. *Greenpeace Magazine*, Vol. 15, 1, Jan/Feb. 1990, pp. 4–8.

57. Quoted in Scott Sonner, "Dissident members lament Sierra Club's growing pains," *Middletown Press (Connecticut)*, April 1, 1991, p. 7.

58. Quoted in Helen Ingram & Dean E. Mann, "Interest Groups and Environmental policy," p. 145.

59. Dick Russell, "We Are All Losing the War," *The Nation*, March 27, 1989, p.404; Eve Pell, "Oiling the works," *Mother Jones*, March/April, 1991, p. 40.

60. Matthew Wald, "Exxon head seeks environmentalist to serve on board," *New York Times*, May 12, 1989.

61. Quoted in Kirkpatrick Sale, "The forest for the trees: Can today's environmentalists tell the difference?" *Mother Jones*, November 1986, p. 25.

62. John McCormick. *British Politics & the Environment*. London: Earthscan Publications, 1991.

63. Dick Russell, "We Are All Losing the War," p.405.

64. David Patrice Greanville, "Whither the World Wildlife Fund?" *The Animals' Agenda*, May 1989, pp. 32–33, and "Dateline: International," *The Animals' Agenda*, November 1990, pp. 42–43.

65. John Green, "Reilly given environmental portfolio," *In These Times*, January 18, 1989, pp. 4–5.

66. Eve Pell, "Oiling the works," p. 40.

67. David Beers & Catherine Capellaro, "Greenwash!" *Mother Jones*, March/April 1991, p. 88.

68. Quoted in Dick Russell, "We Are All Losing the War," p.404.

69. Ibid.

70. As described in Dick Russell, "We Are All Losing the War," and Bob Gottlieb and Margaret Fitzsimmons, "New Approach to Green Issues."

71. Dick Russell, "We Are All Losing the War."

72. An interesting discussion of the polemical pull in environmentalism is in: Mark Sagoff. *The Economy of the Earth*. Cambridge: Cambridge Univ. Press, 1988.

73. Rob Edwards, "Spirit of Outrage," *New Statesman & Society*, July 29, 1988.

74. The fact that women are a small minority among the ranks of scientists barely needs documenting, but for readers interested in this I recommend Stephen G. Brush, "Women in science and engineering," *American Scientist*, Volume 79, Sept/Oct. 1991.

75. Ruth Bleier, ed. *Feminist Approaches to Science*. NY: Pergamon Press, 1986.

76. Two excellent books on this topic by non-Western writers are: Ziauddin Sardar, ed. *The Revenge of Athena: Science, Exploitation and the Third World*. London: Mansell Publishing Co., 1988 and Susantha Goonatilake. *Aborted Discovery: Science & Creativity in the Third World*. London: Zed Books, 1984. My thanks to Sandra Harding for bringing these to my attention.

77. Readers interested in this literature should start with Carolyn Merchant. *The Death of Nature*. New York: Harper & Row, 1980; Vandana Shiva. *Staying Alive*.; and Susan Bordo, "The Cartesian Masculinization of Thought" *Signs: Journal of Women in Culture & Society*, Vol. 11, 3, Spring 1986.

78. Vandana Shiva, *Staying Alive*.; see also Michael Curry, "Beyond nuclear winter: On the limitations of science in political debate," *Antipode: Journal of Radical Geography*, Vol. 18, 3, 1986.

79. For fuller discussions of this, see especially J. Bandyopadhyay and V. Shiva, "Science and Control: Natural Resources and their Exploitation," in Ziauddin Sardar, ed.. *The Revenge of Athena;* and also,Vandana Shiva. *Staying Alive;* and Vandana Shiva. *The Violence of the Green Revolution*. London: Zed Books, 1991.

80. Khor Kok Peng, "Science and Development: Underdeveloping the Third World," in Ziauddin Sardar, *The Revenge of Athena*.

81. See, for example: Sandra Harding, *The Science Question in Feminism*. Ithaca, NY: Cornell Univ Press, 1986; Evelyn Fox Keller, *Reflections on Gender & Science*. New Haven, CT: Yale University Press, 1984; Ruth Bleier, ed. *Feminist Approaches to Science*; *Women's Studies International Forum*, Special Issue, "Feminism and Science," Vol. 12, 3, 1989.

82. Wendy Mishkin, a Canadian activist in the struggle against the expansion of a NATO base in Labrador, is one of the few people actually developing a feminist critique of EIA: Wendy Mishkin, "Towards a Feminist Perspective on Environmental Impact Assessment in Canada," unpublished, March 1990.

83. Margrit Eichler. *Nonsexist Research Methods*. Boston: Allen & Unwin, 1988.

84. This example was provided by Marina Alberti, a founding member of the Italian Green movement, and an environmental planner in Italy, in personal conversations, March 1989.

85. Pam Sommons, Editorial introduction to *The Ecologist*, January/February 1992.

86. Quoted in *Lynx Magazine*, Spring 1989.

87. I compiled this information from a number of sources, including: Personal communications from the American Fur Industry; Tom Davis, "The Killing Season: Pelting Tima at a Mink Ranch," *Animal's Agenda*, December 1987; Lynx (London) "Fur Trade Fact Sheet;" Trans-Species Unlimited (USA) "Fur Fact Sheets."

88. Personal communication from the American Fur Industry offices, New York, January 1990.

89. There is a large, and growing, literature on women and animal rights, including: the "Feminist for Animal Rights Newsletter" (bi-annual, available from FAR, Box 10017, North Berkeley Stn., Berkeley, CA 94709); Josephine Donovan, "Animal Rights and Feminist Theory," *Signs: Journal of Women in Culture & Society*, Winter 1990; Aviva Cantor, "The Club, the Yoke, and the Leash," *Ms. Magazine*, August, 1983; Carol Lansbury. *The Old Brown Dog: Women, Workers, and Vivisection in Edwardian England*. Madison: University of Wisconsin Press, 1985; Marti Kheel, "Meat and Misogyny: Why Animal Rights is a Feminist Issue," *ISIS: Women's World*, March 1987; Marjorie Spiegel. *The Dreaded Comparison: Human and Animal Slavery*. Philadelphia: New Society Publishers, 1988; Roberta Kalechofsky, "Metaphors of Nature: Vivisection and Pornography," *On the Issues*, Vol. 9, 1988; Carol Adams. *The Sexual Politics of Meat*. NY: Continuum, 1990.

90. The best study of women in the early anti-vivisection movement is Carol Lansbury, *The Old Brown Dog*.

91. Ibid., p. 111.

92. *Feminists for Animal Rights Newsletter*, Volume IV, Summer/Spring 1988.

93. Aviva Cantor, "The Club, the Yoke, and the Leash," *Ms. Magazine*, August, 1983

94. The peak times for calls to battered women's shelters occurs during the hunting season and, secondly, during the football Superbowl: Tim Peek, "Unnecessary roughness," *Vermont Times*, January 23, 1992.

95. See especially Peggy Sanday. *Female Power and Male Dominance*. New York: Cambridge University Press, 1981.

96. Russell Baker quoted in Carol Adams. *The Sexual Politics of Meat*, p. 32.

97. Quoted in Carol Adams *The Sexual Politics of Meat* , p. 34.

98. Sue Kirchoff, "Study urges cut in meat output to save environment," *Boston Globe*, July 14, 1991, p. 15.

99. Carol Adams, "Ecofeminism and the eating of animals," *Hypatia*, Vol. 6, 1, Spring 1991; for a broader discussion of the costs of meat production, see Frances Moore Lappe. *Diet for a small planet*. New York: Ballantine Books, 1982 (Tenth anniversary edition). Additional figures from the Worldwatch Institute *1992 Information Please Environmental Almanac*.

100. For more information on the toll of recreational hunting, see: Joni Seager, ed., *The State of the Earth Atlas*, and Lee Durrell. *The State of the Ark*. London & New York: Doubleday, 1986.

101. For more information on the wild trade, see Joni Seager, ed., *The State of the Earth Atlas*, London: Unwin Hyman, 1990; and the annual reports of the World Resources Insitute (Washington DC).

102. An international monitoring body, the Convention on International Trade in Endangered Species (CITES) collects statistics on the wildlife trade. Figures are available from CITES, and from secondary sources, such as the World Resources Institute annual reports (*World Resources*, Washington DC).

103. Kathleen Barry. *Female Sexual Slavery*. NY: Prentice-Hall, 1979.

104. Figures from *The New Internationalist*, Population Issue, October 1987; and, Joni Seager, ed., *The State of the Earth Atlas*, London: Unwin Hyman, 1990.

105. This becomes quite a complicated argument of environmental economics. For readers

interested, I recommend Barry Commoner, "Rapid population growth and environmental stress," *International Journal of Health Services*, Vol. 21, 2, 1991.

106. Christa Wichterich, "From the Struggle against 'Overpopulation' to the Industrialization of Human Production," *Reproductive and Genetic Engineering*, Vol. 1, 1, 1988.

107. A particularly compelling analysis of the abusive exercise of reproductive control is Betsy Hartmann. *Reproductive Rights & Wrongs*. NY: Harper, 1987.

108. Christa Wichterich, "From the Struggle."

109. Angelina Makwavarara, quoted in Irene Dankelman & Joan Davidson. *Women & Environment in the Third World*. London: Earthscan Publications, 1988, p. 132.

110. See, for example, the description of the feminine life-force in Indian cosmology in Vandana Shiva, *Staying Alive.*.

111. Patrick Murphy, "Sex-Typing the Planet," *Environmental Ethics*, Vol. 10, 1988.

112. W.J. Lines, "Is 'Deep Ecology' Deep Enough?" *Earth First*, May 1987.

113. An interview on National Public Radio, "All Things Considered," May 19, 1989.

114. Linda Weltner, "Even Mother Earth has Limits," *Boston Globe*, April 28, 1989.

5 The EcoFringe

1. Some of the debate between ecofeminists and deep ecologists can be followed through a series of "deeper than thou" articles that appeared in the late 1980s. The trail starts with Ariel Kay Salleh, "Deeper than Deep Ecology: The Eco-Feminist Connection," *Environmental Ethics*, Vol. 6, Winter 1984; Bill Devall, "Deep ecology and its critics," *Earth First!*, December 22, 1987; Jim Cheney, "Eco-Feminism & Deep Ecology," *Environmental Ethics*, Vol. 9, Summer 1987; Sharon Doubiago, "Deeper than Deep Ecology," *The New Catalyst*, No. 10, Winter 1987/88; Janet Biehl, "Ecofeminism and Deep Ecology: Unresolvable Conflict?" *Green Perspectives* 3 ("Green Perspectives" is a newsletter of the 'social ecologists', and is available from P.O. Box 111, Burlington, Vt. 05402, USA); Karen Warren, "Feminism and ecology: making connections," *Environmental Ethics*, Vol. 9, Spring 1987; Warwick Fox, "The deep ecology/Ecofeminism debate and its parallels," *Environmental Ethics*, Vol. 11, Spring 1989; Karen Warren, "The power and the promise of ecological feminism," *Environmental Ethics*, Vol. 12, Summer 1990.

2. Arne Naess, "The shallow and the deep, long-range ecology movement: a summary," *Inquiry*, Vol. 16, 1973.

3. Bill Devall & George Sessions. *Deep Ecology: Living as if Nature Mattered*. Salt Lake City, UT: Peregrine Smith Books, 1985. Readers interested in exploring the literature of deep ecology might also read: Rik Scarce. *Eco-Warriors: Understanding the Radical Environmental Movement*. Chicago: Noble Press, 1990; George Bradford. *How Deep is Deep Ecology?* Ojai, CA: Times Change Press, 1989; Michael Tobias, ed. *Deep Ecology*. San Marcos, CA: Avant Books, 1988 (2nd. revised printing); John Davis & Dave Foreman eds. *The Earth First Reader*. Salt Lake City, Utah: Peregrine Smith Books, 1991.

4. This notion is loosely based on James Lovelock's "Gaia hypothesis."

5. John Davis & Dave Foreman eds. *The Earth First Reader*. Salt Lake City, Utah: Peregrine Smith Books, 1991.

6. Brian Tokar, "Exploring the New Ecologies: Social Ecology, Deep Ecology and the future of green political thought," *Alternatives*, Vol. 15, No. 4, 1988. ("Alternatives" is an

environmental magazine published from the University of Waterloo, Ontario). This article also appeared as "Ecological Radicalism," *Zeta Magazine* (Boston), 1988.

7. Quoted in Janet Biehl, "Ecofeminsm and Deep Ecology: Unresolvable Conflict?" *Green Perspectives*, 3.

8. Brian Tokar, "Exploring the new ecologies."

9. Correspondence from Edward Abbey in *Utne Reader*, March/April 1988, p.7.

10. Judi Bari, "The Feminization of Earth First," *MS.* , May/June 1992.

11. Quoted in Jim Robbins, "The environmental guerillas," *Boston Globe Magazine*, March 27, 1988.

12. Dolores LaChapelle, "No I'm Not an Ecofeminist: A Few Words in Defense of Men," *Earth First!* March 21, 1989. LaChapelle frequently is cast in the role of 'token woman' for the deep ecology movement. She is the only woman, for example, included in a 22-author anthology, edited by Michael Tobias, *Deep Ecology*.

13. Sharon Doubiago, "Mama Coyote Talks to the Boys," in Judith Plant, ed. *Healing the Wounds: The Promise of Ecofeminism* (Philadelphia PA: New Society Publishers, 1989)

14. Annette Kolodny. *The Lay of the Land* (Chapel Hill: University of North Carolina Press, 1975) and *The Land Before Her* (Chapel Hill: University of North Carolina Press, 1984).

15. Vera Norwood & Janice Monk, eds. *The Desert Is No Lady: Southwestern Landscapes in Women's Writing and Art* . New Haven: Yale University Press, 1987.

16. A quote from noted American historian, Roderick Nash, *Wilderness and the American Mind* (New Haven: Yale University Press, 1967). Nash's book continues to be used as the standard source for understanding the American encounter with wilderness. Nash himself now appears to acknowledge the andro-centric bias in his earlier work (see, for example, his article, "Rounding out the American revolution," in the Michael Tobias anthology cited above), but he does not revise his basic presumptions about the nature of the American wilderness experience.

17. See, for example, Janet Biehl, "Ecofeminism and Deep Ecology"; Murray Bookchin, "Social Ecology versus Deep Ecology," *Socialist Review*, 88/3, 1988; and, Ynestra King, "Coming of Age With the Greens," *Zeta Magazine*, February 1988.

18. A point made by Ynestra King, quoted in "Social ecology vs. deep ecology," *Utne Reader*, Nov/Dec. 1988, p. 135.

19. Dave Foreman and Nancy Morton, "Good luck, darlin', it's been great," in John Davis and Dave Foreman eds. *The Earth First Reader*.

20. Paul Watson in *Earth First!*, December 22, 1987, p. 20.

21. Ramachandra Guha, "Radical American environmentalism and wilderness preservation: a Third World critique," *Environmental Ethics*, Volume 11, Spring 1989, p. 72.

22. Ibid., p. 75.

23. David Johns, "The relevance of deep ecology to the Third World: Some preliminary comments," *Environmental Ethics*, Vol. 12, Fall 1990, p. 238.

24. This call for severe population reductions has, surprisingly, recently been taken up by some representatives of the UK Green Party. A conference paper at the 1989 annual conference proposed a reduction of the population of the UK to 'between 30 and 40 million.' These population reductions, it was proposed, could be achieved through public education (to reverse attitudes that portray childless people as unfulfilled, for

example), and through family planning. Victor Smart, "Greens aim to halve population," *The Observer* (London) Sept. 17, 1989.

25. Quoted in Bill Devall and George Sessions, *Deep Ecology*, p. 76.

26. Miss Ann Thropy, "Population and AIDS," *Earth First!* May 1, 1987.

27. Ramachandra Guha, "Radical American environmentalism," p. 72.

28. See the population discussion in the preceding chapter

29. Quoted in Chris Reed, "Wild men of the woods," *The Independent* (UK), July 14, 1988. Similar sentiments were expressed in an article by Tom Stoddart, "On Death," and an accompanying editorial in *Earth First!* February 6, 1986.

30. For example, the Brundtland Commission on the Environment estimated that the Ethiopian government could have reversed the advance of desertification threatening its food supplies in the mid-1970s by spending no more than about $50 million a year to plant trees and fight soil erosion. Instead, the government in Addis Ababa pumped $275 million a year into its military machine between 1975 and 1985 to fight secessionist movements in Eritrea and Tigre. Figures cited in Michael Renner, "What's Sacrificed when we arm," *World Watch Magazine*, Vol. 25, September/October 1989.

31. Quoted in George Bradford, *How Deep Is Deep Ecology?* Ojai, CA: Times Change Press, 1989, p. 32.

32. See again, Tom Stoddard, *Earth First!*, February 1986; see also Miss Ann Thropy, "Overpopulation and industrialism," originally published in *Earth First!* 1987, reprinted in John Davis and Dave Foreman, eds. *The Earth First Reader*.

33. The original essay putting forward this idea is Miss Ann Thropy, "Population and AIDS," *Earth First!* May 1, 1987, p. 32. This was followed by a deluge of letters to the journal, and a very long article expanding the original argument by Daniel Conner, "Is AIDS the answer to an environmentalists prayer?" *Earth First!* December 22, 1987. The question was picked up in the progressive press, where it has been circulating since. See, for example, Brad Edmondson, "Is AIDS good for the earth?" *Utne Reader*, Nov./Dec. 1987, and letters to Utne Reader on this topic for the following three years.

34. Miss Ann Thropy, "Population and AIDS," p. 32

35. A theory developed over the past 20 years by a British atmospheric scientist, James Lovelock, and an American microbiologist, Lynn Margulis, that the earth itself is akin to a single, living organism in that it has self-regulation capabilities, and is self-monitoring. The Gaia hypothesis, initially discounted, is now enjoying a serious resurgence. Neither Lovelock nor Margulis have publically commented on the AIDS/Gaia link proposed by deep ecologists, but it is a proposal that distorts the original Gaia hypothesis.

36. A term coined in the 1940s by an American naturalist, Aldo Leopold, and adopted by deep ecologists. A current short-hand for a sense of 'oneness with nature'.

37. Nancy Hartsock was one of the earliest commentators to make this point in "Rethinking modernism," *Cultural Critique*, Vol 7, Fall 1987; and Janet Biehl, "Ecofeminism and Deep Ecology."

38. On feminist critiques of postmodernism and of the relativism argument, see: Frances Mascia-Lees, Patricia Sharpe and Colleen Ballerino Cohen, "The postmodernist turn in Anthropology: Cautions from a feminist perspective," *Signs: Journal of Women in Culture & Society*, Vol. 15, 1, 1989; and Sandra Harding, "Introduction: Is there a feminist method?" in Sandra Harding ed., *Feminism and Methodology*, Bloomington: Indiana University Press, 1987.

39. Janet Biehl, "Ecofeminism and Deep Ecology."

40. With regards to Sandra Harding, I have borrowed the style of this chapter title from her book, *The Science Question in Feminism*.

41. For more information on the Women's Strike for Peace in particular, and women's peace camps, readers should turn to essays in *Rocking the Ship of State: Toward a Feminist Peace Politics* (ed. Adrienne Harris & Ynestra King, Boulder, CO: Westview Press, 1989).

42. Sara Ruddick. *Maternal Thinking*. NY: Ballantine Books, 1989, p. 222.

43. Adrienne Harris & Ynestra King, "Introduction," in *Rocking the Ship of State*.

44. There are strong elements, though, of a non-maternal women's politics of peace. In the Harris and King collection, see especially articles by Carol Cohn and Nancy Hartsock. In Eva Isakkson, ed. *Women & the Military System*. NY: Simon & Schuster, 1988, see articles especially by Wendy Chapkis, Cynthia Enloe, and the essays in the section on the military and economy.

45. Sara Ruddick. *Maternal Thinking*, p. 137.

46. Amy Swerdlow, "Pure Milk, Not Poison," in Adrienne Harris & Ynestra King, eds. *Rocking the Ship of State*, see also a critique of maternalist politics in Wendy Chapkis, "Sexuality and militarism," in Eva Isaksson, ed. *Women and the Military System*.

47. Quoted in Alice Schwartzer, "Simone de Beauvoir Talks About Sartre," *MS.*, 1983.

48. Indeed, the articulation of a maternal-based politics raises a challenge to the contemporary feminist movement by claiming a territory for feminism that many feminists would consider to lie beyond its purview. The feminist movement needs to take seriously this challenge.

49. The literature on this debate is rich. I suggest only a few articles where readers can follow it: Janet Radcliffe Richards, "Why the Pursuit of Peace is No Part of Feminism," in Jean Bethke Elshtain and Sheila Tobias, *Women, Militarism & War*, Totowa, NJ: Rowman & Littlefield, 1990; Birgit Brock-Utne. *Educating for Peace: A Feminist Perspective*. New York: Pergamon Press, 1985; and a number of the selections in Eva Isakkson, ed. *Women & the Military System*. NY: Harvester, 1988.

50. The controversy over the Greenham Common peace camp, in particular, is not only over issues of "motherist" peace politics. The issue of maternal politics organizing has been overshadowed on many occasions by debate over the appropriateness of women's separatism, and especially of lesbian separatism. The "lesbian issue" often lies just beneath the surface of both feminist and non-feminist arguments about women's separate peace organizing.

51. Ynestra King, "The Eco-Feminist Imperative," in Leonie Caldecott and Stephanie Leland, eds. *Reclaim the Earth: Women Speak Out For Life on Earth*. London: The Women's Press, 1983.

52. Feminist historians of science have contributed the most to the exploration of the history of the linked subordination of women and nature. See, in particular, Carolyn Merchant. *The Death of Nature*. San Francisco: Harper & Row, 1980, and Sandra Harding. *The Science Question in Feminism*. Ithaca: Cornell University Press, 1986. Other important works are: Mary Daly. *Gyn/Ecology: The Metaethics of Radical Feminism*. New York: Harper & Row, 1980; Susan Griffin. *Woman and Nature: The Roaring Inside Her*. New York: Harper & Row, 1980.

53. Carolyn Merchant, *The Death of Nature*, p.3.

54. Carolyn Merchant, *The Death of Nature*, p. 171. The use of masculinist metaphors in the pursuit of science is also explored in Chapter 3. Articles by Carol Cohn, and the book by Brian Eslea are especially helpful.

55. Feminist historians and theologians reconstrue the European and American witch-hunts of the sixteenth and seventeenth centuries as the persecution of nature-identified women and as an expression of the attempt by men to control the perceived chaos of woman-identified nature.

56. There are a number of discussions about the treatment of the nature issue in different feminist ideologies. A very good overview is: Ynestra King, "Healing the Wounds: Feminism, Ecology, and Nature/Culture Dualism," in Alison Jaggar and Susan Bordo, eds, *Gender/Body/Knowledge*. Rutgers University Press, 1989.

57. Simone de Beauvoir. *The Second Sex*. NY: Random House, 1968.; Sherry Ortner, "Is Female to Male as Nature is to Culture?" in Michelle Zimblast Rosaldo & Louise Lamphere eds., *Woman, Culture & Society*. Stanford: Stanford University Press, 1974.

58. Mary Daly. *Gyn/Ecology: The Metaethics of Radical Feminism*. New York: Harper & Row, 1980; Susan Griffin. *Woman and Nature: The Roaring Inside Her*. New York: Harper & Row, 1980.

59. The term ecofeminism was first coined, by all accounts, by a French writer, Francoise d'Eaubonne in 1974. The growing literature by women who identify themselves as "ecofeminists," and who self-consciously identify themselves as shaping the course of ecofeminism, includes the following: Judith Plant, ed. *Healing the Wounds: The Promise of Ecofeminism*. Philadelphia: New Society Publishers, 1989; Andree Collard with Joyce Contrucci. *Rape of the Wild*. London: The Women's Press, 1988; Bloomington: Indiana Univ Press, 1988 ; Irene Diamond & Gloria Orenstein, *Reweaving the World: The Emergence of Ecofeminism*. San Francisco: Sierra Club Books, 1990. The large number of ecofeminist articles published in journals and newsletters over the past three or four years cannot be cited here, but references to many of them are scattered throughout this chapter. The term "ecofeminism" is not, of course, the exclusive property of the women mentioned above, and there are no doubt women who call themselves "ecofeminists" who would take issue with some of the above-mentioned authors. Nonetheless, it is important to note that the term increasingly IS identified with the philosophies identified by these women, and its meaning is increasingly specific.

60. I am deliberately leaving out of this discussion the ecological challenge raised by Third World women, such as the Chipko activists of India or the Greenbelt workers of Kenya, and emergent postcolonial analyses of ecological destruction, such as that raised by Vandana Shiva. "Ecofeminism," while very dependent on cross-cultural and panhistorical influences, is a particular First World philosophy, and it is not necessarily recognized by women working in other cultures. For discussion of related, but distinctive, women's environmental movements outside the Euro-American sphere, see Chapter 6.

61. Andree Collard with Joyce Contrucci. *Rape of the Wild,* pp. 137–138.

62. Ynestra King, "The Ecology of Feminism and the Feminism of Ecology," in *Healing the Wounds*.

63. Wendy Brown. *Manhood and Politics*. NY: Rowman & Littlefield, 1988, p. 190–191.

64. Camille Paglia, as reviewed in Teresa Ebert, "The politics of the outrageous," *Women's Review of Books*, Vol. IX, 1, October 1991.

65. Karen Warren, "Feminism and Ecology: Making Connections," *Environmental Ethics*, Vol. 9, Spring 1987.

66. Michael Zimmerman, "Feminism, Deep Ecology, & Environmental Ethics," *Environmental Ethics*, Volume 9, Spring 1987.

67. Ibid.

68. In *Healing the Wounds* (Judith Plant, ed.), Anne Cameron, for example, appeals to women to work for political change by drawing on a renewed inner strength.

69. Johnson's program for social change based on personal transformation is outlined in her two most recent books: *Going Out of Our Minds*, Freedom, CA: Crossing Press, 1987 and *Wildfire: Igniting the She/volution*, Albuquerque: Wildfire Books, 1989.

70. Susan Prentice, "Taking Sides: What's Wrong with Eco-Feminism?," *Women & Environments*, Spring 1988.

71. Angela Bowen et al., "Taking Issue With Sonia," *Sojourner*, February 1988.

72. Irene Javors "Goddess in the Metropolis," and Brian Swinne, "How to heal a lobotomy," in Irene Diamond and Gloria Orenstein, *Reweaving the World*.

73. The first quote is from Starhawk, "Feminist, earth-based spirituality and ecofeminism," in Judith Plant, ed. *Healing the Wounds*, the second from "Power, authority, and mystery: Ecofeminism and earth-based spirituality," in Diamond & Orenstein, *Reweaving the World*..

74. Charlene Spretnak, "Toward an Ecofeminist Spirituality," in Judith Plant, ed. *Healing the Wounds*. pp. 131–132. Charlene Spretnak also authored a book on spirituality and environmentalism, one of the few to make the connection explicit: *The Spiritual Dimension of Green Politics*. Santa Fe: Bear & Co., 1986.

75. Janet Biehl, "What is Social EcoFeminism?" *Green Perspectives*, 11, October 1988.

76. Margo Adair & Sharon Howell, "The subjective side of power," in Judith Plant, ed. *Healing the Wounds*, p. 226.

77. Merlin Stone. *When God Was a Woman*. NY: Harcourt, Brace, Jovanovich, 1976.

78. Hunting wild animals to the point of extinction, for example, is a very old story. Aleut hunters wiped out populations of sea otters around Amchitka Island over 2500 years ago; the New Zealand Maoris killed off the last of the maos hundreds of years before the Europeans arrived. The early North Americans profoundly changed the ecosystems of the new continent they colonized, killing off dozens of animal species, reducing forest land to grassland, and grassland to the deserts of the Plains.

79. An excellent feminist analysis of the exploitation of Native American culture by (some) New Agers is: Andy Smith, "For all those who were Indian in a former life," *Sojourner* (Boston), November 1990. Judith Plant, in "The Circle is gathering. . ." warns against the cynical exploitation of indigenous traditions, in Judith Plant, ed. *Healing the Wounds*, as do Irene Diamond and Lisa Kuppler, "Frontiers of the imagination: women, history, and nature," in *Journal of Women's History*, Vol. 1, 3, Winter 1990.

80. Janet Biehl, "The Politics of Myth," *Green Perspectives*, 7, June 1988.

81. Irene Diamond and Lisa Kuppler, "Frontiers of the Imagination."

82. Janet Biehl, "The politics of myth."

83. Jan Clausen, "Rethinking the world," *The Nation*, September 23, 1991.

6 Hysterical Housewives, Tree Huggers, and Other Mad Women

1. A good introduction to and summary of women's consumer movements is: Troth Wells and Foo Gaik Sim. *Till They Have Faces: Women as Consumers*. Rome: ISIS International, 1987.

2. Lisa Belkin, "Experts say air pollution in the home is a growing risk," *New York Times,* March 7, 1985; B.J. Roche, "Monitoring household pollution," *Boston Globe,* June 8, 1984; "Stepping lightly on the earth: Everyone's guide to toxics in the home," *Greenpeace Action,* 1988;

3. For further discussions and critiques of green consumerism, I recommend: H. Patricia Hynes, "The Pocketbook and the Pill: Reflections on Green Consumerism and Population Control," *Issues in Reproductive and Genetic Engineering,* Vol. 4, 1, 1991; Sandy Irvine, "Consuming Fashions? The Limits of Green Consumerism," *The Ecologist,* Vol. 19, No. 3, 1989; "Going for the Green," *Environmental Action,* Vol. 22, 3, Nov/Dec 1990.

4. The domestic reform movement is a fascinating chapter in the social history of the UK and North America. Several good histories are now available: Ruth Schwartz Cowan. *More Work for Mother.* New York: Basic Books, 1983; Gwendolyn Wright. *Moralism and the Model Home.* Chicago: Univ. of Chicago Press, 1980 and *Building the Dream* (New York: Pantheon Books, 1981); Dolores Hayden. *The Grand Domestic Revolution,* Cambridge MA: MIT Press, 1981; Joni Seager. *Father's Chair: Domestic Reform and Housing Change.* Clark University unpublished dissertation, 1987.

5. Troth Wells and Foo Gaik Sim, *Till They Have Faces,* pp. 90–91.

6. Ibid., p. 93.

7. I recommend Stuart Ewen's book, *Captains of Consciousness: Advertising and the Social Roots of the Consumer Culture* (New York: McGraw Hill, 1976) for an insightful discussion of the relationships between industry, advertising, and consumers, though Ewen only touches briefly on an analysis of gender issues in consumer culture.

8. Michael deCourcy Hinds, "In sorting trash, householders get little help from industry," *New York Times,* July 29/89.

9. Michael deCourcy Hinds, "Do disposable diapers ever go away?" *New York Times,* December 10, 1988.

10. "Clean and soft and white and dangerous," Campaign brochure, Women's Environmental Network (London), 1989.

11. "A divisive alliance: the controversy over product endorsements," *Maclean's,* (Canada) July 17/89; Michael Freitag, "Luring green consumers," *New York Times,* August 6/89.

12. "How green is my wallet?" *City Limits* (London), March 30–April 6/89.

13. "'Green' BP fells rainforest," *The Sunday Times,* June 18, 1989; "The greening of industry," *The Guardian,* June 24, 1989.

14. "Britain's most admired companies," *The Economist* (London), September 9, 1989.

15. "How green is my wallet?"

16. Quoted in Wells and Sim, *Till They Have Faces,* p. 5.

17. The international marketing of tobacco is another good example of this double-edged marketing strategy. Cigarette companies are marketing "low nicotine" tobacco to their newly health-conscious Western consumers while methodically marketing *high* nicotine cigarettes to Third World consumers. Women, and especially Third World women, are specifically targeted as the next big consumer group for cigarette marketers to reach.

18. H. Patricia Hynes, "The Pocketbook and the Pill," p. 48.

19. Sandy Irvine, "Consuming Fashions?" p. 92.

20. For example the Lou Harris/United Nations Environmental Program International survey of environmental opinion, the first coordinated effort of its kind to survey public opinion

on environmental issues internationally, established that across the board women's views on the environment differed from men's. The countries polled are a diverse group, including Argentina, Hungary, Japan, Jamaica, Saudi Arabia, and Zimbabwe, among others. Lou Harris Associates/UNEP, "Public and leadership attitudes to the environment in 14 countries" 1989.

21. The best description of the Love Canal struggle is in Lois Gibbs' autobiography, *Love Canal: My Story*, Albany, NY: SUNY, 1982. See also, H. Patricia Hynes, "Ellen Swallow, Lois Gibbs and Rachel Carson: Catalysts of the American Environmental Movement," *Women's Studies International Forum*, Vol. 8 4, 1985.

22. Quoted in Lois Gibbs, *Love Canal: My Story*.

23. For a description of the corporate response to the NIMBY movement, see William Glaberson, "Coping in the Age of 'Nimby'," *New York Times*, June 19, 1988.

24. Vandana Shiva, an Indian scientist, has emerged as the unofficial spokesperson of the Chipko movement, and it is her writing that initially sparked international interest in Chipko. I recommend Vandana Shiva's book, *Staying Alive*, and: Vandana Shiva, "Ecology movements in India," *Development: Seeds of Change*, 3, 1985; Ann Spanel, "Indian women and the Chipko movement," *Woman of Power Magazine*, 9; Jaya Jaitley, "If nobody fights, we will," *Manushi* 41, 1987; Bina Agarwal, "The gender and environment debate: lessons from India," *Feminist Studies* 18, 1, Spring 1992; Gail Omvedt, "India's movements for democracy: peasants, 'greens,' women and people's power," *Race and Class*, 31 (2), 1989; Jayanta Bandyopadhyay & Vandana Shiva, "Development, poverty and the growth of the green movement in India," *The Ecologist*, Vol. 19, 3, 1989.

25. Cited in Alan Durning,"Action at the grassroots: fighting poverty and environmental decline," *Worldwatch Paper* 88, January 1989.

26. Vandana Shiva, "Ecology movements in India," op. cit.

27. There are literally thousands of examples of women's grassroots environmental actions. Readers interested in summaries of some of these should turn to: Robbin Lee Zeff, Marsha Love and Karen Stults, eds. *Empowering Ourselves: Women and Toxics Organizing*. Arlington, VA: Citizen's Clearinghouse for Hazardous Wastes, 1989; United Nations Environment Programme & WorldWIDE. *Success Stories of Women and the Environment*. Washington DC: WorldWIDE, 1991; Irene Dankelman and Joan Davidson. *Women and Environment in the Third World*. London: Earthscan Publications, 1988; Paul Ress, "Women's Success in Environmental Management," *Our Planet*, Vol. 4, 1, 1992; Women's Feature Service. *The Power to Change: Women in the Third World Redefine Their Environment*. New Delhi, India: Kali for Women, 1992; Annabel Rodda. *Women and the Environment*. London: Zed Books, 1991; Sally Sontheimer, ed. *Women and the Environment: A Reader*. NY: Monthly Review Press, 1991.

28. Katsi Cook & Lin Nelson, "Mohawk women resist industrial pollution," *Women & Global Corporations*, Vol. 6 3, Summer 1985. "Women & Global Corporations" is a regular feature of *Listen Real Loud*, a newsletter produced by the American Friends Service Committee that carries international news of women's political activism.

29. Adrienne Scott, "Margherita Howe, Environmental Watchdog," *Women and Environments*, Spring 1987.

30. "Arab Women Committee passes resolution on the environment," *WorldWIDE News*, Feb./March 1988.

31. Jean Fazzino, "Jessie DeerInWater: Profile of an activist," *Radioactive Waste News*, Fall 1988.

32. Maggie Jones and Wangari Maathai, "Greening the desert: women of Kenya reclaim land," in Leonie Caldecott and Stephanie Leland, eds. *Reclaim the Earth*, London: Women's Press, 1983; Wangari Maathai, "Foresters Without Diplomas," *MS*, March/April 1991.

33. Alan Durning,"Action at the grassroots."

34. Karel van Wolfenren. *The Enigma of Japanese Power*. NY: Knopf, 1989

35. David Sanger, "Uno finds himself Japan's No. 1 'women's issue'", *New York Times*, July 3, 1989.

36. David Apter & Nagayo Sawa, *Against the State: Politics and Social Protest in Japan*. New Haven: Yale University Press, 1984; Purnendra C. Jain, "Green politics and citizen power in Japan," *Asian Survey*, Vol. xxxi, no 6, June 1991.

37. "We will grow back," *Connexions*, No. 6, Fall, 1982, translated from AGORA, a Japanese feminist journal.

38. Peggy Antrobus, quoted at the Global Assembly of Women for a Healthy Planet, Miami, 1991.

39. Cited in Roberto Suro, "Grass-roots groups show power battling pollution close to home," *New York Times*, July 2, 1989.

40. See the discussion of the Jacksonville environmental crisis in Chapter 1.

41. Sithembiso Nyoni, "Africa's food crisis: price of ignoring village women?" in *Women and the Environmental Crisis*, United Nations Environment Liaison Centre, Nairobi, 1985.

42. Zack Nauth, "How toxic pollution can break down racial barriers," *In These Times*, Dec. 14–20, 1988. There is a burgeoning literature on the race and class matrix of environmental issues. Readers interested in pursuing this further might start with Robert Bullard. *Dumping in Dixie*. Boulder, Co: Westview Press, 1990 and Charles Lee, ed. *Toxic Wastes and Race in the United States*. NY: United Church of Christ, 1987.

43. "Toxic racism," *In These Times*, August 5–18, 1987.

44. Described in William Glaberson, "Coping in the Age of 'Nimby'," *New York Times*, June 19, 1988 and also in "Sign of success?" Citizen's Clearinghouse for Hazardous Waste, *Action Bulletin*, October 1987.

45. "Environmental rights in Emelle," *Greenpeace*, V. 13 2, 1988.

46. Reported in "On the Brink," *New Internationalist*, No. 157, March 1986; also, the *Greenpeace* article, in note 45.

47. These retorts are ubiquitous in women's accounts of their activism: see, for example, the story of Mary Sinclair, "The tongue of angels," in Anne Witte Garland, ed., *Women Activists: Challenging the Abuse of Power*, New York: Feminist Press, 1988; Lou Chapman, "Women who make a difference," *Family Circle*, January 10, 1989; and Lois Gibbs' story in her autobiography.

48. Cited in Lou Chapman, "Women who make a difference," *Family Circle*, January 10, 1989.

49. It is not just grassroots women activists who report being trivialized and challenged. Women with very substantial (conventional) credentials receive the same treatment— women such as Alice Stewart, the British scientist who pioneered studies of the effects of low-level radiation, and Rachel Carson, the American scientist credited with sparking contemporary environmental consciousness, were met with much the same derisive sexism.

50. Lois Gibbs. *Women and Burnout.* Arlington, VA: Citizen's Clearinghouse for Hazardous Wastes, 1988.

51. Chelsea Congdon, "Will Spoiling the Environment Save Our Jobs?" *Equal Means*, Summer 1992.

52. Adrienne Scott, "Margherita Howe, Environmental Watchdog," *Women & Environments,* Spring 1987.

53. An idea prompted by reading Sara Ruddick, "Mothers and Men's Wars," in Adrienne Harris & Ynestra King, eds., *Rocking the Ship of State: Toward a Feminist Peace Politics,* Boulder, Colorado: Westview Press, 1989.

54. A paraphrase of an argument made by Micaela di Leonardo, "Morals, mothers, militarism," *Feminist Studies,* Vol. 11, 3.

55. These examples come from Alan Durning, "Action at the grassroots."

Conclusion

1. These projects are summarized in a document entitled "Success Stories of Women and the Environment," published jointly by the United Nations Environment Programme and the WorldWIDE Network (Washington DC), 1991.

2. Timothy Weiskel of the Harvard University Divinty School first introduced me to this term.

3. Leo Marx, "Post-Modernism and the environmental crisis," *Philosophy and Public Policy,* Volume 10, 3/4, Summer/Fall 1990; another key article in the growing humanities literature is Alexander Roman, "The environmental movement as a transformer of culture," *The Probe Post* (Toronto), September 24, 1989.

4. I borrow this metaphor of the feminist spotlight from Pat Hynes, who first used this turn of phrase in a meeting of environmentalists at MIT in 1992.

Selected Bibliography

Adams, Carol. *The Sexual Politics of Meat*. New York: Continuum, 1990.

Agarwal, Anil. "A Case of Environmental Colonialism," *Earth Island Journal*, Spring 1991.

Agarwal, Anil, Juliet Merrifield, and Rajesh Tandon. *No Place To Run: Local Realities and Global Issues of the Bhopal Disaster*. New Market, Tennessee: The Highlander Center, 1985.

Agarwal, Bina, "The gender and environment debate: lessons from India," *Feminist Studies*, Vol. 18, 1, Spring 1992.

Alston, Dana, ed.*We Speak for Ourselves: Social Justice, Race and the Environment*. Washington DC: Panos Institute, 1990.

Ball, Howard. *Justice Downwind: America's Atomic Testing Program in the 1950s*. NY: Oxford University Press, 1986.

Bandyopadhyay, J. & V. Shiva, "Science and Control: Natural Resources and their Exploitation," in Ziauddin Sardar, ed.. *The Revenge of Athena: Science, Exploitation and the Third World*. London: Mansell, 1988.

Bank Information Center. *Funding Ecological and Social Destruction: The World Bank and the IMF*. Washington, DC: Bank Information Center, 1991.

Bellini, James. *High-Tech Holocaust*. San Francisco: Sierra Club, 1986.

Bertell, Rosalie. *No Immediate Danger*. London: The Women's Press, 1985.

Biehl, Janet. *Rethinking Ecofeminist Politics*. Boston: South End Press, 1991.

Bleier, Ruth, ed. *Feminist Approaches to Science*. NY: Pergamon Press, 1986.

Brown, Wendy. *Manhood and Politics*. NY: Rowman & Littlefield, 1988

Bullard, Robert. *Dumping in Dixie: Race, Class and Environmental Quality*. Boulder, Co: Westview Press, 1990.

Caldecott, Leonie and Stephanie Leland, eds. *Reclaim the Earth: Women Speak Out for Life on Earth*. London: The Women's Press, 1983.

Caldicott, Helen. *If You Love This Planet*. NY: Norton, 1992.

Caldicott, Helen. *Missile Envy: The Arms Race and Nuclear War*. NY: Bantam Books, 1985.

Caldicott, Helen.*Towards a Compassionate Society*. Westfield, NJ: Open Media, 1991.

Chavkin, Wendy, ed. *Double Exposure: Women's Health Hazards on the Job and At Home*. NY: Monthly Review Press, 1984.

Center for the American Woman and Politics, Rutgers University. *The Impact of Women in Public Office*. New Brunswick, NJ, 1991.

Cockburn, Cynthia. *In the Way of Women: Men's Resistance to Sex Equality in Organizations*. Ithaca, NY: ILR Press, 1991.

Cockburn, Cynthia. *Brothers: Male Dominance & Technological Change*. London: Pluto Press, 1983.

Collard, Andree. *The Rape of the Wild*. London: Women's Press, 1988.

Commoner, Barry. "Rapid population growth and environmental stress," *International Journal of Health Services*, Vol. 21, 2, 1991.

Connell, R.W.. *Gender & Power*. Stanford: Stanford University Press, 1987.

Daly, Mary. *Gyn/Ecology: The Metaethics of Radical Feminism*. New York: Harper & Row, 1980.

Danielsson, Bengt & Marie-Therese Danielsson. *Poisoned Reign: French Nuclear Colonialism in the Pacific*. Victoria (Australia): Penguin, rev. ed., 1986.

Dankelman, Irene & Joan Davidson. *Women & Environment in the Third World*. London: Earthscan Publications, 1988.

Davis, John & Dave Foreman, eds. *The Earth First! Reader*. Salt Lake City: Peregrine Smith Books, 1991.

Del Tredici, Robert. *At Work in the Fields of the Bomb*. NY: Harper & Row, 1987.

di Leonardo, Micaela, "Morals, mothers, militarism," *Feminist Studies*, Vol. 11, 3, 1985.

Diamond, Irene & Gloria Orenstein, eds. *Reweaving the World: The Emergence of Ecofeminism*. San Francisco: Sierra Club Books, 1990.

Diamond, Irene & Lisa Kuppler, "Frontiers of the Imagination: Women, History, and Nature," *Journal of Women's History*, Vol. 1, 3, 1990.

Dibblin, Jane. *Day of Two Suns: US Nuclear Testing and the Pacific Islanders*. London: Virago, 1988; NY: New Amsterdam Books, 1990.

Donovan, Josephine,"Animal Rights and Feminist Theory," *Signs: Journal of Women in Culture & Society*, 15, 2 (Winter 1990).

Doubiago, Sharon, "Mama Coyote Talks to the Boys," in Judith Plant, ed. *Healing the Wounds: The Promise of Ecofeminism* Philadelphia PA: New Society Publishers, 1989.

Durning, Alan. *Apartheid's Environmental Toll*. Worldwatch Paper 95, Worldwatch Institute, Washington DC, May 1990.

Durning, Alan. *Action at the grassroots: fighting poverty and environmental decline*. Worldwatch Paper 88, Worldwatch Institute, Washington DC, January 1989.

Easlea, Brian. *Fathering the Unthinkable: Masculinity, Scientists and the Nuclear Arms Race*. London: Pluto Press, 1983.

The Ecologist. Special issue: Feminism, Nature, Development, January/February 1992.

Ehrlich, Anne and John Birks. *Hidden Dangers: Environmental Consequences of Preparing for War.* San Francisco: Sierra Club Books, 1992.

Elshtain, Jean Bethke & Sheila Tobias.*Women, Militarism & War.* Totowa, NJ: Rowman & Littlefield, 1990.

Enloe, Cynthia. *The Morning After: Sexual Politics at the End of the Cold War.* Berkeley: University of California Press, forthcoming 1993.

Enloe, Cynthia. *Bananas, Beaches & Bases: Making Feminist Sense of International Politics.* London: Pandora Press, 1989 and Berkeley: University of California Press, 1989.

Enloe, Cynthia. *Does Khaki Become You? The Militarization of Women's Lives.* London: Pandora Press, 1988.

Environmental Project on Central America (EPOCA). *Militarization: The Environmental Impact.* Green Paper 3, 1988.

Ferguson, Kathy. *The Feminist Case Against Bureaucracy.* Philadelphia: Temple University Press, 1984.

Freudenberg, Nicholas. *Not in Our Backyards: Community Action for Health and the Environment.* NY: Monthly Review Press, 1984.

Garb, Yaakov Jerome, "Perspective or escape? Ecofeminist musings on contemporary earth imagery," in Irene Diamond and Gloria Orenstein, eds. *Reweaving the World: The Emergence of Ecofeminism.* San Francisco: Sierra Club Books, 1990.

Garland, Anne ed. *Women Activists: Challenging the Abuse of Power,* New York: Feminist Press, 1988.

Gibbs, Lois. *Women and Burnout.* Arlington, VA: Citizen's Clearinghouse for Hazardous Wastes, 1988.

Gibbs, Lois. *Love Canal: My Story.* Albany, NY: SUNY, 1982.

Gilligan, Carol. *In A Different Voice.* Cambridge, MA: Harvard University Press, 1982.

Goonatilake, Susantha. *Aborted Discovery: Science and Creativity in the Third World.* London: Zed Books, 1984.

Griffin, Susan. *Woman and Nature: The Roaring Inside Her.* New York: Harper & Row, 1980.

Harding, Sandra. *The Science Question in Feminism.* Ithaca. NY: Cornell Univ Press, 1986.

Hartmann, Betsy. *Reproductive Rights & Wrongs: The Global Politics of Population Control and Contraceptive Choice.* NY: Harper, 1987.

Harris, Adrienne & Ynestra King, eds. *Rocking the Ship of State: Toward a Feminist Peace Politics.* Boulder, Colo: Westview Press, 1989.

Hynes, H. Patricia. *The Recurring Silent Spring.* NY: Pergamon Press, 1989.

Hynes, H. Patricia, "Ellen Swallow, Lois Gibbs and Rachel Carson: Catalysts of the American Environmental Movement," *Women's Studies International Forum,* Vol. 8 4, 1985.

Hynes, H. Patricia, "The pocketbook and the pill: reflections on green consumerism and population control," *Reproductive and Genetic Engineering*, Vol. 4 1, 1991.

Hypatia: A Journal of Feminist Philosophy. Special Issue: Ecological Feminism. Spring 1991.

Isaksson, Eva, ed. *Women and the Military System.* NY: Harvester, 1988.

Jackall, Robert. *Moral Mazes: The World of Corporate Managers.* NY: Oxford University Press, 1988.

Kalechofsky, Roberta, "Metaphors of Nature: Vivisection and Pornography," *On the Issues,* Vol. 9, 1988.

Kamel, Rachel. *The Global Factory.* Philadelphia: American Friends Service Committee, 1990.

Kanter, Rosabeth Moss. *Men & Women of the Corporation. NY: Basic Books,* 1977.

Keller, Evelyn Fox. *Reflections on Gender & Science.* New Haven, CT: Yale University Press, 1984.

Kellert, Stephen & Joyce Berry, "Attitudes, knowledge and behaviors toward wildlife as affected by gender," *Wildlife Society Bulletin,* Vol. 15, 3, Fall 1987.

Kheel, Marti, "Meat and Misogyny: Why Animal Rights is a Feminist Issue," *ISIS: Women's World,* March 1987.

King, Ynestra.*What is Ecofeminism?* New York: Ecofeminist Resources, 1990.

King, Ynestra, "Coming of Age With the Greens," *Zeta Magazine,* February 1988.

King, Ynestra. "Healing the Wounds: Feminism, Ecology, and Nature/Culture Dualism," in Alison Jaggar & Susan Bordo, eds. *Gender/Body/Knowledge.* New Brunswick, NJ: Rutgers University Press, 1989.

Lansbury, Carol. *The Old Brown Dog: Women, Workers, and Vivisection in Edwardian England.* Madison: University of Wisconsin Press, 1985.

Lee, Charles, ed. *Toxic Wastes and Race in the United States: A National Report on the Racial and Socio-Economic Characteristics of Communities Surrounding Hazardous Waste Sites* NY: United Church of Christ, April 1987.

Lester, James ed., *Environmental Politics and Policy: Theories and Evidence.* Durham: Duke University Press, 1989.

Loeb, Paul. *Nuclear Culture: Living and Working in the World's Largest Atomic Complex.* Philadelphia: New Society Press, 1986.

MacCormack, Carol P., "Nature, culture and gender: a critique," in Carol MacCormack & Marilyn Strathern eds., *Nature, Culture and Gender.* NY: Cambridge University Press, 1980.

Mathews, Jessica Tuchman, "Redefining security," *Foreign Affairs,* Vol. 68, 2, Spring 1989.

McCormick, John. *British Politics and the Environment.* London: Earthscan Publications, 1991.

Merchant, Carolyn. *The Death of Nature.* New York: Harper & Row, 1980.

Mische, Patricia, "Ecological security and the need to reconceptualize sovereignty," *Alternatives: Social Transformations & Humane Governance,* Vol. XIV, 4, October 1989.

Momsen, Janet ed. *Women and Development in the Third World.* London: Routledge, 1990.

Moyers, Bill & Center for Investigative Reporting. *Global Dumping Ground.* Washington, DC: Seven Locks Press, 1990.

Murphy, Patrick, "Sex-Typing the Planet," *Environmental Ethics,* Vol. 10, 1988.

National Toxics Campaign Fund. *The US Military's Toxic Legacy.* Boston: National Toxic Campaign Fund, 1991.

National Wildlife Federation. *The Toxic 500.* NY: National Wildlife Federation, 1987.

Norwood, Vera & Janice Monk, eds. *The Desert Is No Lady: Southwestern Landscapes in Women's Writing and Art .* New Haven: Yale University Press, 1987.

Parker, Andrew, Mary Russo, Doris Summer, Patricia Yeager, eds. *Nationalisms and Sexualities*. NY: Routledge, 1992.

Pearce, David ed. *Blueprint 2: Greening the World Economy*. London: Earthscan Publications, 1991.

Perrow, Charles. *Normal Accidents: Living with High-Risk Technologies*. NY: Basic Books, 1984.

Peterson, V. Spike. *Gendered States: Feminist (re)Visions of International Relations Theory*. Boulder, Colorado: Lynne Rienner Pub., 1992.

Plant, Judith, ed. *Healing the Wounds: The Promise of Ecofeminism*. Philadelphia: New Society Publishers, 1989.

Prakash, Padma, "Neglect of Women's Health Issues," *Economic and Political Weekly* (Delhi), December 14, 1985.

Ranney, Sally, "Heroines and Hierarchy: Female leadership in the conservation movement," in Donald Snow, ed *Voices from the Environmental Movement*. Washington DC: Island Press, 1992.

Renner, Michael, "Assessing the military's war on the environment," in Worldwatch Institute, *The State of the World. 1991*. NY: Norton, 1991.

Rhodes, Richard. *The Making of the Atomic Bomb*. NY: Simon & Schuster, 1986

Rodda, Annabel. *Women and the Environment*. London: Zed Books, 1991.

Rogers, Barbara. *The Domestication of Women: Discrimination in Developing Societies*. New York: St. Martin's, 1980.

Rogers, Barbara. *Fifty-Two Percent: Getting Women's Power into Politics*. London: Women's Press, 1983.

Rogers, Barbara. *Men Only: An Investigation Into Men's Organizations*. London: Pandora Press, 1988.

Rosenthal, Debra. *At The Heart of the Bomb: The Dangerous Allure of Weapons Work*. Reading, MA: Addison-Wesley, 1990.

Ruddick, Sara. *Maternal Thinking: Toward a Politics of Peace*. Boston: Beacon Press, 1989.

Russell, Diana, ed. *Exposing Nuclear Phallacies*. NY: Pergamon Press, 1989.

Sagoff, Mark. *The Economy of the Earth: Philosophy, Law and the Environment*. Cambridge: Cambridge University Press, 1988.

Salleh, Ariel Kay, "Deeper than Deep Ecology: The Eco-Feminist Connection," *Environmental Ethics*, Vol. 6, Winter 1984.

Sardar, Ziauddin, ed. *The Revenge of Athena: Science, Exploitation and the Third World*. London: Mansell Publishing Co., 1988.

Scarce, Rik. *Eco-Warriors: Understanding the Radical Environmental Movement*. Chicago: Noble Press, 1990.

Seager, Joni, ed. *The State of the Earth Atlas*. New York: Simon & Schuster, 1990.

Seager, Joni and Ann Olson. *Women in the World: An International Atlas*. NY: Simon & Schuster, 1986.

Seager, Joni, "Operation Desert Disaster: Environmental Costs of the War," in Cynthia Peters, ed. *Collateral Damage: The 'New World Order' At Home and Abroad*. Boston: South End

Press, 1992 and in Haim Bresheeth & Nira Yuval-Davis, *The Gulf War and the New World Order*, London: Zed Bo oks, 1991.

Sen, Gita & Caren Grown. *Development, Crises, and Alternative Visions: Third World Women's Perspectives*. New York: Monthly Review Press, 1987.

Shiva, Vandana. *Staying Alive: Women, Ecology & Development*. London: Zed Books, 1989.

Shiva, Vandana. *The Violence of the Green Revolution*. London: Zed Books, 1991.

Shiva, Vandana, "Ecology movements in India," *Development: Seeds of Change*, 3, 1985.

Shulman, Seth. *The Threat at Home: Confronting the Toxic Legacy of the US Military*. Boston: Beacon Press, 1992.

Sivard, Ruth. *World Military and Social Expenditures*. Washington DC: World Priorities Institute, Annual Reports.

Smith, Hedrick. *The Power Game: How Washington Works*. NY: Random House, 1988.

Snow, Donald, ed. *Inside the Environmental Movement*. Washington DC: Island Press, 1992.

Sontheimer, Sally, ed. *Women and the Environment: A Reader*. NY: Monthly Review Press, 1991.

Stiehm, Judith, ed. *Women and Men's Wars*. Oxford: Pergamon Press, 1983. (Originally published as a special issue of *Women's Studies International Forum*, Volume 5, 3/4, 1982.)

Taylor, Dorceta E. "Blacks and the environment," *Environment & Behavior*, Vol. 21, 2, March 1989.

Tickner, J. Ann. *Gender in International Relations: A Feminist Perspective on Achieving Global Security*. NY: Columbia University Press, 1992.

United Nations Environment Programme.*Women and the Environmental Crisis*. Nairobi, Kenya: Environment Liaison Centre, 1985.

United Nations Environment Programme & WorldWIDE. *Success Stories of Women and the Environment*. Washington DC: WorldWIDE, 1991.

Vickers, Jeanne ed. *Women and the World Economic Crisis*. London: Zed Books, 1991.

Walby, Sophie. *Theorizing Patriarchy*. Oxford: Basil Blackwell, 1990.

Wallsgrove, Ruth, "Greens, Browns, Purples: Motherhood and Party Politics in West Germany," *Off Our Backs*, Vol. xvii, 9, October 1987.

Warren, Karen, "Feminism and Ecology: Making Connections," *Environmental Ethics*, Vol. 9, Spring 1987.

Warren, Karen, "The power and the promise of ecological feminism," *Environmental Ethics*, Vol. 12, Summer 1990.

Weinberg, William. *War on the Land: Ecology and Politics in Central America*. London: Zed Books, 1991.

Weingartner, Eric. *The Pacific: Nuclear Testing and Minorities*. London: Minority Rights Group, 1991.

Weir, David and Mark Schapiro. *Circle of Poison*. SanFrancisco: Food First, 1981.

Westing, Arthur, ed. *Cultural Norms, War and the Environment*. NY: Oxford University Press, 1988. (Originally published as a project of the Stockholm International Peace Research Institute).

Wichterich, Christa, "From the Struggle against 'Overpopulation' to the Industrialization of Human Production," *Reproductive and Genetic Engineering,* Vol. 1, 1, 1988.

Women Working For a Nuclear Free Pacific. *Pacific Women Speak: Why Haven't You Known?* London: Green Line, 1987.

Women's Feature Service. *The Power to Change: Women in the Third World Redefine Their Environment.* New Delhi, India: Kali for Women, 1992.

Women's International Information and Communication Service (ISIS), ed. *Women and Development: A Resource Guide for Organization and Action.* Philadelphia: New Society Publishers, 1984.

World Commission on Environment and Development. *Our Common Future* ("The Brundtland Report"). New York: Oxford University Press, 1987.

World Resources Institute. *1992 Information Please Environmental Almanac.* Boston: Houghton Mifflin, 1992.

Zeff, Robbin Lee, Marsha Love & Karen Stults, eds. *Empowering Ourselves: Women and Toxics Organizing.* Arlington, VA: Citizen's Clearinghouse for Hazardous Wastes, 1989.

Zimmerman, Michael, "Feminism, Deep Ecology, & Environmental Ethics," *Environmental Ethics,* Volume 9, Spring 1987.

Index